Optical Remote Sensing of Land Surfaces

Remote Sensing Observations of Continental Surfaces Set

coordinated by
André Mariotti

Optical Remote Sensing of Land Surfaces

Techniques and Methods

Edited by

Nicolas Baghdadi
Mehrez Zribi

ISTE PRESS

ELSEVIER

First published 2016 in Great Britain and the United States by ISTE Press Ltd and Elsevier Ltd

ISTE Press Ltd
27-37 St George's Road
London SW19 4EU
UK

www.iste.co.uk

Elsevier Ltd
The Boulevard, Langford Lane
Kidlington, Oxford, OX5 1GB
UK

www.elsevier.com

British Library Cataloguing-in-Publication Data
A CIP record for this book is available from the British Library
Library of Congress Cataloging in Publication Data
A catalog record for this book is available from the Library of Congress
ISBN 978-1-78548-102-4

Contents

Foreword

I have been entrusted by ISTE Science Publishing with the responsibility for a multidisciplinary editorial line: Earth System – Environment, and within this framework it gives me great pleasure today to present a set of books dedicated to the topic of remote sensing, compiled and edited by Nicolas Baghdadi and Mehrez Zribi.

Both the content and the organization of this collection have largely been inspired by reflections, analyses and prospective works conducted by almost 200 authors and researchers with a high level of international expertise in this discipline.

This community, which is recognized for its scientific merit, has sought to expand its research activities under the direction the two editing authors founded on a solid effort in the area of acquisition and wider dissemination of knowledge within this field.

This represents a community characterized by the firm commitment to adopting a holistic or even an ecosystem approach within the context of an interdisciplinary science of the Earth system. In this scientific context where the complexity of natural systems is compounded with the complexity of societies, the authors have given careful consideration to depicting a finalizable and "public" type of discipline, open to decision makers, managers and all those beyond the scientific community who are interested in the future of our planet.

Two main tools are necessary in order to satisfy the requirements in terms of understanding and characterizing our environment and its evolution: process modeling and observation.

Remote sensing observations in conjunction with measurements and modeling constitute a discipline that makes it possible to understand the functional properties of the observed system and their dependence on its structural properties. This is one of the key disciplines that allow the analysis and provide access to the understanding of the functioning of our environment: in general, this is dedicated to aspects such as the analysis of climate change, the effects of anthropogenic and demographic pressure, natural disasters, the increasing decline in resources (water, etc.), the degradation of biodiversity across all environments, desertification, the need to nourish the planet (for example, mapping of crops and yield prediction), etc.

Analyzing and understanding these different types of problems is rendered possible:

– by analyzing the detected, structural and functional objects (soils, hydrosystems, vegetation, etc.);

– by understanding the main basic processes, which incorporate these main elements: water flow covering all scales and compartments, erosion, meteorology, crop development, soil pollution, etc;

– by developing indicators in order to evaluate the short-, medium- and long-term evolution of all environmental compartments and variables.

The importance of these scientific questions has led to a general mobilization of the international organizations by means of various international conventions and agreements to protect the environment and meet the specific requirements in terms of observation. Various international networks have been developed over the past few years with the purpose of conducting continuous measurements. However, these punctual measurements could not provide sufficient spatiotemporal monitoring, in particular in difficult-to-access regions. Within this context, spatial observation could be implemented to its full potential, both by means of considerable progress in terms of instrumentation and by means of the development of effective data processing and analysis methods, data whose provision becomes increasingly free of rights.

Under the initiative of numerous space agencies (in particular European, North-American, Japanese, etc.), important space missions were launched for the purpose of conducting Earth observations, among which the following may be mentioned:

– Sentinel, within the framework of the Copernius program (formerly referred to as GMES for Global Monitoring for Environment and Security) implemented in numerous areas such as land and marine environment monitoring, emergency management (for example, natural disasters) and climate change monitoring (radar and optical imaging);

– Landsat;

– ALOS, launched by the Japan Aerospace Exploration Agency, in particular for deforestation monitoring;

– SMOS and SMAP, in particular for the global mapping of soil moisture, etc.

Although remote sensing represents a field in which specialist knowledge is required in order to conduct a better analysis and interpretation of data, this programmatic development is undoubtedly associated with a significant progress with respect to the implementation of space-based Earth observations at the level of an increasing number of laboratories across both developed and emerging nations. This development is likewise associated with new disciplines and thematic backgrounds, among which numerous areas of the humanities and social sciences which enrich and extend the primarily physical foundations of remote sensing may be mentioned.

It would be superfluous to list all remote sensing applications along with the disciplines and scientific questions which adopted this concept, as this would also inevitably result in regrettable omissions: nonetheless, the extensive implementation of spatial observation grants the latter a strong interdisciplinary status.

The launch of new large-scale space missions, the higher degree of convenience, including financial convenience, as well as the access to data will facilitate an intensification and generalization of the use of spatial observation data and products: new scientific subjects, new users (managers, decision makers, etc).

The high demand for educational material containing updated information on the various remote sensing concepts and methods and the main applications thereof, in particular at the level of continental surfaces, are derived therefrom.

It is within this framework that this collection of books is proposed, which aims to provide researchers, students in masters, engineer and PhD programs, as well as decision makers, engineers specialized in management services on a territorial, departmental, regional or national scale and players in the decision-making authorities with a tool which incorporates both the foundations of the physical principles underlying various spatial applications and the implementation methods and exemplification at the level of various applications based on spatial observation.

In these six volumes, Nicolas Baghdadi and Mehrez Zribi have mobilized almost 200 internationally recognized researchers to propose a comprehensive "toolkit", describing the latest scientific methods and actions in terms of the implementation of spatial observation.

The first two volumes describe the physical principles underlying various techniques which cover the frequency spectrum ranging from visible to microwaves. The third volume illustrates the agricultural and forestry applications of spatial observation. The fourth volume presents the applications of spatial observation in the field of continental hydrology. The fifth volume is dedicated to the observation of urban and coastal areas, whereas the final volume presents the implementation of spatial observation within the context of risk assessment and understanding.

Thanks are due to Nicolas Baghdadi and Mehrez Zribi for taking the time to draft, harmonize and partially edit these volumes and committing to this effort in terms of putting this modern and high-quality knowledge across and making it accessible to a diverse and vast scientific audience.

I wish to thank them both for their altruism, perseverance and devotion in service of the success of this endeavor.

André MARIOTTI

June 2016

Acronyms

2D	Two dimensions
3D	Three dimensions
4AOP	Automatized Atmospheric Absorption Atlas Operational
6S	Second Simulation of Satellite Signal in the Solar Spectrum
AATSR	Advanced Along-Track Scanning Radiometer
ACORN	Atmospheric Correction Now
ADC	Analog-to-digital converter
ADCP	Acoustic Doppler Current Profiler
ADEME	French Environment and Energy Management Agency
ADEOS	Advanced Earth Observing Satellite
AERONET	Aerosol Robotic Network
AET	Actual Evapotranspiration
AFRITRON	African Tropical Rainforest Observation Network
AGB	Above-Ground Biomass
AGNES	Agglomerative Nesting
AHS	Airborne Hyperspectral Scanner
AHT	Astronomical High Tide
AIEM	Advanced Integral Equation Model
AirSAR	Airborne Synthetic Aperture Radar
ALB	Airborne LiDAR Bathymeter
ALEXI	Atmosphere-Land Exchange Inverse
ALS	Airborne Laser Scanning

AltBOC	Alternate Binary Offset Carrier
AMARTIS	Advanced Modeling of the Atmospheric Radiative Transfer for Inhomogeneous Surfaces
AMMA	African Monsoon Multidisciplinary Analysis
AMSR	Advanced Microwave Scanning Radiometer
ANA	*Agência Nacionalde Aguas* (Brazilian National Water Agency)
AOL	Airborne Oceanographic LiDAR
APD	Avalanche Photodiode
API	Antecedent Precipitation Index
APOM	Aerosol Plume Optical Model
ARVI	Atmospherically Resistant Vegetation Index
ASAR	Advanced Synthetic Aperture Radar
ASCAT	Advanced Scatterometer
ASDF	Averaged Square Difference Function
ASI	*Agenzia Spaziale Italiana* (Italian Space Agency)
ASI	Agriculture Stress Index
ASTER	Advanced Spaceborne Thermal Emission and Reflection Radiometer
ATCOR	Atmospheric and Topographic Correction
ATLAS	Advanced Topographic Laser Altimeter System
ATM	Airborne Topographic Mapper
ATREM	Atmospheric Removal (atmospheric correction method)
ATSR	Along Track Scanning Radiometers
AUC	Area Under Curve
AVHRR	Advanced Very High Resolution Radiometer
AVIRIS	Airborne Visible/Infrared Imaging Spectrometer
AWiFS	Advanced Wide Field Sensor
BEIDOU/ COMPASS	Chinese Navigation Satellite System
BELA	BepiColombo Laser Altimeter
BGB	Below-Ground Biomass
BIL	Band Interleaved by Line
BIOMOD	Biomolecular Design (Computer platform implemented in R)
BIP	Band Interleaved by Pixel

BLUE	Best Linear Unbiased Estimator
BOC	Binary Offset Carrier
BPSK	Binary Phase Shift Keying
BPT	Binary Partition Tree
BRDF	Bidirectional Reflectance Distribution Function
BRGM	*Bureau de Recherches Géologiques et Minières* (French Geological Survey)
BSA	Back Scatter Alignment
BSQ	Band Sequential
BUI	Build-up Index
BWI	Bassist Wetness Index
C/A	Coarse Acquisition
$CaCO_3$	Calcium Carbonate
CAIC	Consistent Akaike's Information Criterion
CALIOP	Cloud-Aerosol LiDAR with Orthogonal Polarization
CALIPSO	Cloud-Aerosol LiDAR Pathfinder Satellite Observation
CanEx-SM	Canadian Experiment for Soil Moisture
CART	Classification and Regression Trees
CASI	Compact Airborne Spectrographic Imager
CBERS	China–Brazil Earth Resources Satellite
CBOC	Composite Binary Offset Carrier
CCD	Charged Coupled Devices
CCDAS	Carbon Cycle Data Assimilation System
CCI	Climate Change Initiative
CCRS	Canada Center for Remote Sensing
CDF	Cumulative Density Function
CDMA	Code Division Multiple Access
CDOM	Colored Dissolved Organic Matter
CEC	Cation-Exchange Capacity
CEM	Constrained Energy Minimization
CEN	French Snow Studies Center
CEOS	Committee on Earth Observation Satellites
CERES	Crop Environment Resource Synthesis

CESBIO	*Centre d'Études Spatiales de la Biosphère* (Center for the Study of the Biosphere from Space)
CFC	Chlorofluorocarbons
CFD	Constant Fraction Discriminator
CFFDRS	Canadian Forest Fire Danger Rating System
cGNSS	Conventional Global Navigation Satellite Systems
CHAMP	Challenging Minisatellite Payload (German geosciences satellite)
CHL	Chlorophyll Content
CHM	Canopy Height Model
CHRIS	Compact High Resolution Imaging Spectrometer
CIRAD	Agricultural research for development (France)
CLASlite	Carnegie Landsat Analyse System Lite
CLM	Community Land Model
CLMGW	Community Land Model with a Ground Water Parameterization
CLPX	Cold Land Processes Field Experiment
CLS	Collecte Localisation Satellite
CLSM	Catchment Land Surface Model
CM	Code Moderate (for GNSS)
CMC	Canadian Meteorological Centre
CMEM	Community Microwave Emission Model
CNES	*Centre National d'Études Spatiales* (French Space Agency)
CNRM	*Centre National de Recherches Météorologiques* (French National Center for Meteorological Research)
CNRS	*Centre National de Recherche Scientifique* (National French Center for Scientific Research)
COD	Controlled Origin Designation
COG	Center of Gravity
CONUS	Contiguous United States
COST	Cosine Estimation of Atmospheric Transmittance
CR	Continuum Removed
CRI	Carotenoid Reflectance Index
Cryosat	Satellite Radar Altimeter
CSA	Canadian Space Agency
CSR	Centre for Space Research

CTA	Classification Tree Analysis
CTFC	Center for Tropical Forest Science (Cameroon)
CTFS	Center for Tropical Forest Science
CTMF	Cluster Tuned Matched Filter
CW	Continuous Waves
CWFIS	Canadian Wildland Fire Information System
CyGNSS	NASA's Cyclone Global Navigation Satellite System
CZCS	Coastal Zone Color Scanner
CZMIL	Coastal Zone Mapping and Imaging LiDAR
DAM	Dry Aerial Mass
DART	Discrete Anisotropic Radiative Transfer
DATAR	Interministerial Delegation for Territorial Planning and Regional Attractiveness
DBH	Diameter at Breast Height
DC	Drought Code
DCA	Dual Channel Approach
DCM	Digital Canopy Model
DDM	Delay Doppler Map
DEGRAD	Forest Degradation program
DEIMOS	Deep Imaging Multi-Object Spectrograph
DEM	Digital Elevation Model
DEOS	Delft Institute of Earth Observations and Space Systems
DERD	Double bounce Eigenvalue Relative Difference
DETER	*Detecção de Desmatamento em Tempo Real* (Near real-time deforestation detection system)
DGPS	Differential Global Positioning System
DHSVM	Distributed Hydrology Soil-Vegetation Model
DIACT	Inter-ministerial Agency for Spatial Planning and Competitiveness
DIANA	Divise Analysis
DIC	Digital Image Correlation
DIMAP	Digital Image MAP
DInSAR	Differential SAR Interferometry
DisALEXI	Disaggregated Atmospheric Land Exchange Inverse

DLIS	Desert Locust Information Service
DLR	*Deutsches Zentrum für Luft und Raumfahrt* (German Space Agency)
DM	Dry Matter
DMC	Disaster Monitoring Constellation or Duff Moisture Code, depending on the application
DMRT	Dense Media Radiative Transfer
DMSP	U.S. Air Force Defense Meteorological Satellite Program
DMU	De Monfort University
DOAS	Differential Optical Absorption Spectroscopy
DOS	Dark Object Substraction
DOY	Day-of-Year
DPSS	Diode-Pumped Solid-State
DS	Diffuse Scatterer
DSM	Digital Soil Mapping
DSM	Digital Surface Model
DTC	Dry Troposphere Correction
DTED	Digital Terrain Elevation Data
DTM	Digital Terrain Model
DVI	Difference Vegetation Index
DW	Dry Weight
DWBA	Distorted Wave Born Approximation
EAARL	Experimental Advanced Airborne Research LiDAR
ECDC	European Centre for Disease Prevention and Control
ECMWFMMT	European Centre for Medium-Range Weather Forecasts Mobile Mapping Technology
ECV	Essential Climate Variables
EDF	*Électricité de France* (French electric company)
EEA	European Environment Agency
EFA	Effective Field Approximation
EFFIS	European Forest Fire Information System
EGNOS	European Geostationary Navigation Overlay Service
EID-Méditerranée	Interdepartmental Agreement for Mosquito Control on the Mediterranean Coast

EKF	Extended Kalman Filter
ELBARA	ETH L-Band Radiometer
ELUE	Effective Light Use Efficiency
EM	Electromagnetic
ENEA	*Energia Nucleare ed Energie Alternative* (Italian National Agency for New Technologies, Energy and Sustainable Economic Development)
EnKF	Ensemble Kalman Filter
ENMAP	Environmental Monitoring and Analysis Program
ENSO	El Niño Southern Oscillation
ENVEO	Environmental Earth Observation Information
ENVISAT	Environmental Satellite
EOS	Earth Observing System
EPICA	European Project for Ice Coring in Antarctica
EPS	European Polar System
EQeau	Model developed by INRS-Ete with the objective to extract the soil water content from SAR images
ERM	Exact Repeat Missions
ERS	European Remote-sensing Satellite
ESA	European Space Agency
ESCAT	ERS Scatterometer
ESSA	Environmental Science Services Administration Satellite
ET	Evapotranspiration
ETM	Enhanced Thematic Mapper
EUFAR	European Facility for Airborne Research
EUMETSAT	European Organization for the Exploitation of Meteorological Satellites
EVASPA	Evapotranspiration Assessment from Space
EVI	Enhanced Vegetation Index
EWT	Equivalent Water Thickness
EZW	Embedded Zerotrees of Wavelet transforms
FAA	Federal Aviation Administration (USA)
FAI	Floating Algae Index
FAO	Food and Agriculture Organization of the United Nations

FAPAR	Fraction of Absorbed Photosynthetically Active Radiation
FBD	Fine-Beam Double Polarization
FBP	Fire Behavior Prediction
Fcover	Fraction of Vegetation Cover
FDTD	Finite Difference Time Domain
FFMC	Fine Fuel Moisture Code
FFT	Fast Fourier Transform
FI	Fine Particles Index
FIPAR	Fraction of Intercepted Photosynthetically Active Radiation
FLAASH	Fast Line-of-sight Atmospheric Analysis of Spectral Hypercubes (Atmospheric correction)
FLOE	Fish LiDAR Oceanographic Experimental
FMA	Fore-Mid-Aft
FMCW	Frequency Modulated Continuous Waves
FOTO	Fourier-based Textural Ordination
FOV	Field of View
FSA	Forward Scatter Alignment
FST	Spectro-Transfert functions
FT	Fourier Transform
FW	Full Waveform or Fresh Weight, depending on the application
FWHM	Full Width at Height Maximum
FWI	Fire Weather Index
FWS	Fraction of Water Surface
GAGAN	GPS Aided GEO Augmented Navigation system
GAI	Green Area Index
GALILEO	European Global navigation satellite system
GCOS	Global Climate Observing System
GCP	Ground Control Points
GDAL	Geospatial Data Abstraction Library
GDR	Geophysical Data Record
GEO	Group on Earth Observations
GEOCAPI	Geostationary Ocean Color Advanced Permanent Imager
GEOGLAM	Group on Earth Observations Global Agricultural Monitoring

GEOSS	Global Earth Observing System of Systems
GEOSUD	Geoinformation for Sustainable Development (France)
GEOTIFF	Geostationary Earth Orbit Tagged Image File Format
GEROS-ISS	GNSS Reflectometry, Radio Occultation and Scatterometry Onboard International Space Station
GF	Green Fraction
GFO	Geosat Follow-On (radar altimeter satellite)
GFZ	*GeoForschungs Zentrum* (German research centre for Geosciences)
GHG	Green House Gase
GIAM	Global Irrigated Area Map
GIEWS	Global Information and Early Warning System on Food and Agriculture
GIM	Global Ionospheric Model
GIMMS	Global Inventory Modeling and Mapping Studies
GIS	Geographic Information System
GLAI	Green Leaf Area Index
GLAS	Geoscience Laser Altimeter System
GLCM	Grey Level Co-occurrence Matrix
GLDAS	Global Land Data Assimilation System
GlobCover	Global Land-Cover Map
GLONASS	Global Navigation Satellite System (Russia)
GLP	Global Land Project
GLRT	Generalized Likelihood Ratio Test
GLS	Global Land Survey (Landsat)
GM	Geodetic Mission
GMES	Global Monitoring for Environment and Security
GMFS	Global Monitoring of Food Security
GMM	Gaussian Mixture Model
GNSS	Global Navigation Satellite System
GNSS-R	Global Navigation Satellite System – Reflectometry
Go	Giga-Octet (1,000,000,000 octets)
GOCE	Gravity field and Steady-state Ocean Circulation Explorer (ESA)
GOES	Geostationary Operational Environmental Satellite

GOFC Fire IT	Global Observation of Forest Cover Fire Implementation Team
GOM	Geometrical Optics Model
GORS	GNSS Occultation Reflectometry Scatterometry
GPCC	Global Precipitation Climatology Centre
GPP	Gross Primary Production
GPR	Gaussian Process Regression
GPS	Global Positioning System
GRACE	Gravity Recovery and Climate Experiment (satellite)
GRDC	Global Runoff Data Center
GRGS	Space Geodesy Research Group
GtC	Gigatonnes of Carbon (10^9 tons)
GUS	Ground Uplink Stations
GVI	Difference Vegetation Index
GWIS	Global Wildfire Information System
Ha	Hectare (= 10,000 m^2)
HCFC	Hydro-chlorofluorocarbons
HEC-RAS	Hydrologic Engineering Center River Analysis System (hydraulic model)
HgCdTe	Mercury Cadmium Telluride
HiRI	High Resolution Optical Imager
HITRAN	High-Resolution Transmission Database
HOG	Histogram of Oriented Gradient
HPC	High Performance Computing
HRG	High Resolution Geometry
HSC	Height-Scaled Crown
HSCOI	Height-Scaled Crown Openness Index
HSR	High Spatial Resolution
HYMAP	Hyperspectral Mapper (airborne hyperspectral sensor)
HYPXIM	Hyperspectral-X Imagery
HYSPEX	Hyperspectral Imaging System
HyspIRI	NASA's Hyperspectral Infrared Imager
IASI	Infrared Atmospheric Sounding Interferometer
ICA	Independent Component Analysis

ICARE	Inversion Code for urban Areas Reflectance Extraction
ICESat	Ice, Cloud and Land Elevation Satellite
ICF	Interferometric Complex Field
ICP	Iterative Closest Point technique
IDAN	Intensity Driven Adaptative Neighborhood
IEM	Integral Equation Model
IFN	French National Forest Inventory
IFOV	Instantaneous Field of View
IGN	French National Geographic Institute and Forest Information
iGNSS	Interferometric GNSS
IHS	Intensity, Hue, Saturation
IMU	Inertial Measurement Unit
INERIS	French National Institute for Environmental Protection and Industrial Risks
InGaAs	Indium Gallium Arsenide
INPE	*Instituto Nacional de Pesquisas Espaciais* (Brazilian Institute of Space Research)
INRA	French National Institute for Agricultural Research
INSAR	Interferometric Synthetic Aperture Radar
InSb	Indium antimonide
INSEE	French National Institute for Statistics and Economic Studies
IOD	Indian Ocean Dipole
IPCC	Intergovernmental Panel on Climate Change
IPT	Interference Pattern Technique
IRD	French Research Institute for Development
IRNSS	Indian Regional Navigational Satellite System
IRSTEA	French National Research Institute of Science and Technology for Environment and Agriculture
ISBA	Interactions Soil-Biosphere-Atmosphere (model)
ISDC	Integrated System Data Center
ISI	Initial Spread Index
ISODATA	Iterative Self-Organizing Data Analysis Technique
ISRO	Indian Space Research Organisation
ITC	Individual Tree Crown

ITCZ	Intertropical Convergence Zone
ITG	Institute of Theoretical Geodesy
IWPB	Institute of Water Problem of Bishkek, Kyrgyzstan
Jason	Radar altimeter
JAXA	Japan Aerospace Exploration Agency
JECAM	Joint Experiment for Crop Assessment and Monitoring
JPEG	Joint Photographic Experts Group (image format)
JPL	Jet Propulsion Laboratory
JRC	European Commission's Joint Research Centre
KBR	K-Band Microwave Ranging
KLT	Karhunen–Loeve transform
LaDAR	Laser Detection And Ranging
LADS	Laser Airborne Depth Sounder
LAGEOS	Laser Geodynamics Satellite
LAI	Leaf Area Index
LANDSAT	LAND + Satellite
LAUVA	Airborne Ultraviolet Aerosol LiDAR
LAX	Maximum LAI
LBAS	Local Based Augmentation System
LCCS	Land Cover Classification System
LCLU	Land Cover / Land Use
LDAS	Land Data Assimilation System
LEGOS	Laboratory for Studies in Geophysics and Spatial Oceanography (France)
LEnKS	Local EnKF Smoother
LEO	Low Earth Orbit
LEP	Leading Edge Position
LEWIS	L-band for Estimating Water In Soils
LFMC	Live Fuel Moisture Content (%)
LHCP	Left Hand Circular Polarization
LiDAR	Light Detection and Ranging
LISAH	Laboratory for Soil, Agrosystems and Water Systems (France)
LISFLOOD-FP	Two-dimensional Hydrodynamic Model

LMA	Leaf Mass per Area
LMM	Linear Mixed Model
LOADDT	Spatial planning and territorial development
LOLA	Lunar Orbiter Laser Altimeter
LOV	Villefranche Oceanography Laboratory (France)
LPCA	Laboratory for Physico-Chemistry of the Atmosphere
LPRM	Land Parameter Retrieval Model
LRM	Low Resolution Mode
LSCE	Climate and Environment Sciences Laboratory (France)
LSM	Land Surface Model
LSSM	Least Squares 3D Surface Matching
LST	Land Surface Temperature
LULCC	Land Use and Land Cover Change
LUT	Look-Up Table
LWIR	Long-Wave Infrared
LZW	*Lempel-Ziv-Welch* (compression algorithm)
MACCS	Multisensor Atmospheric Correction and Cloud Screening processor
MARS	Monitoring of Agriculture with Remote Sensing
MATISSE	Advenced Earth Modeling for Imaging and Scene Simulation
MaxEnt	Maximum Entropy Method
MBOC	Multiplexed Binary Offset Carrier
MCT	Mercury Cadmium Telluride
MEB	Microwave Emission of the Biosphere
MERIS	Medium Resolution Imaging Spectrometer
MESA	Monitoring of Environment and Security in Africa
METEOSAT	METEO + Satellite
METOP	Meteorological Operational Satellite Programme/Advanced Scatterometer
METRIC	Mapping Evapotranspiration at High Resolution with Internalized Calibration
MGVI	MERIS Global Vegetation Index
MHz	Mega-Hertz (= 1,000,000 Hz)
MIMR	Multichannel Microwave Imaging Radiometer

MIPERS	Multistatic Interferometric Polarimetric Electromagnetic model for Remote Sensing
MIR	Middle Infrared
MISDc	*Modello Idrologico Semi Distribuito in Continuo* (Continuous rainfall-runoff model)
MISR	Multi-angle Imaging Spectro Radiometer
MISTIGRI	Microsatellite for Thermal Infrared Ground Surface Imaging (CNES, France)
MLR	Multiple Linear Regression
MLS	Mobile LiDAR Scanner
MMD	Minimum–Maximum Difference
MMS	Mobile Mapping Systems
MMU	Minimum Mapping Units
MMV	Mobile Mapping Vehicle
MNDWI	Modified Normalized Difference Water index
MNF	Maximum Noise Fraction
Mo	Mega-octet (1,000,000 octets)
MODCOU	Hydrogeological model
MODIS	Moderate Resolution Imaging Spectroradiometer
MODTRAN	Moderate Resolution Atmospheric Transmission
MOLA	Mars Orbiter Laser Altimeter
MPE	Maximum Permissible Exposure
MSAS	Multi-functional Satellite-based Augmentation System
MSG	Meteosat Second Generation
MSI	Moisture Stress Index
MSI	Multispectral Instrument (Sentinel-2)
MTSAT	Multi-functional Transport Satellites
MVSA	Minimum Volume Simplex Analysis
MWIR	Mid-Wavelength Infrared
NAOMI	New AstroSat Optical Modular Instrument
NASA	National Aeronautics and Space Administration (USA)
NBR	Normalized Burn Ratio
NCC	Normalized Cross-correlation
NCEP	National Centers for Environmental Prediction

Nd:YAG	Neodymium-doped Yttrium Aluminium Garnet
NDSI	Normalized Difference Snow Index
NDVI	Normalized Difference Vegetation Index
NDVITM	Normalized Difference Vegetation Index Threshold Method
NDWI	Normalized Difference Water Index
NEBN	Noise Equivalent Beta Naught
NEDT	Noise Equivalent Delta Temperature
NEE	Net Ecosystem Exchange
NEF	Noise Equivalent Flux
NEM	Normalized Emissivity Method
NEP	Net Ecosystem Productivity
NEP	Noise Equivalent Power
NIR	Near Infrared
NLES	Navigation Land Earth Station
NLRI	Near Laser Ranging Investigation
NMC	National Meteorological Center (USA)
NMF	Non-negative Matrix Factorization
NOAA	National Oceanic and Atmospheric Administration
NOHD	Nominal Ocular Hazard Distance
NORUT	Norut Northern Research Institute (Norway)
NPV	Non-photosynthetic Vegetation
NPW	Numerical Weather Prediction models
NSC	NarynSyrdarya Cascade
NSCAT	NASA Scatterometer
NSIDC	National Snow and Ice Data Center (USA)
NWP	Numerical Weather Prediction
OA	Overall Accuracy
OBIA	Object Based Image Analysis
OLCI	Ocean and Land Colour Instrument
OLI	Operational Land Imager
OLS	Operational Linescan System
OM	Organic Matter
OMI	Ozone Monitoring Instrument

ONERA	French Aerospace Research Agency
ONF	French National Forest Office
OS	Open Service
OSCAT	OceanSat-2 Scatterometer
OTB	OrfeoToolBox
PA	Producer's Accuracy
PACE	Pre-Aerosol, Clouds and Ocean Ecosystems
PAH	Polycyclic Aromatic Hydrocarbons
PAI	Plant Area Index
PARIS-IoD	Passive Reflectometry and Interferometry System In Orbit Demonstrator
PCA	Principal Components Analysis
PERSIANN	Precipitation Estimation from Remotely Sensed Information using Artificial Neural Networks
PET	Potential Evapotranspiration
PFT	Phytoplankton Functional Types
PGN	Permanent GNSS Network
PHR	Panchromatic High Resolution
PIF	Pseudo-Invariant Features
piGNSS-R	Partial Interferometric GNSS-R
PLF	Polarization Loss Factor
PLOF	Local Land Tenure Plan (France)
PLSR	Partial Least Squares Regression
PMT	Photomultiplier Tubesv
POA	Polarization Orientation Angle
PODAAC	Physical Oceanography Distributive Active Data Center
POLDER	Polarization and Directionality of the Earth's Reflectances
PolInSAR	Polarimetry-interferometry Synthetic Aperture Radar
PolSAR	Polarimetry SAR
POLYMER	Polynomial Based Algorithm Applied to Meris
POM	Physical Optics Model
PPCDAM	Plan for Preventing and Controlling Deforestation in Amazônia Legal
PRESS	Prediction Sum of Squares

PRF	Pulse Repeat Frequency
PRISMA	*Precursore Iperspettrale della Missione Applicativa* (Italian hyperspecral mission)
PRN	Pseudo Random Noise
PROBA	Project for On-board Autonomy
PRODES	*Programa de Cálculo do Desflorestamento da Amazônia* (Brazilian Amazon Deforestation Monitoring Program)
PROSPECT	Leaf Optical Properties Spectra (radiative transfer model)
PRS	Public Regulated Service
PS	Persistent Scatterers
PSI	Persistent Scatterers Interferometry
PSR	Penalized-Spline Regression
PV	Photosynthetic Vegetation
PVI	Perpendicular Vegetation Index
PWC	Plant Water Content
QBO	Quasi-Biennal Oscillation
QCA	Quasi-Crystalline Approximation
Qgis	GIS software (open source)
QPSK	Quadrature Phase Shift Keying
QUIKSCAT	Quick Scatterometer (NASA)
QZSS	Quasi-Zenith Satellite System
RADAR	Radiodetection and Ranging
RAF	French Altimetric System
RAINFOR	Amazon Forest Inventory Network
RAMSES	ONERA Airborne Multi-frequency SAR Imaging System (France)
RAN	Royal Australian Navy
RANSAC	Random Sample Consensus
RCA	Radio Corporation of America
RCM	RADARSAT Constellation Mission
RCS	Radiometric Control Sets
REDD	Reduction of Emissions from Deforestation and Forest Degradation

REDDAF	Reducing Emissions from Deforestation and Degradation in Africa (European Project)
RENAG	French National GNSS Permanent Networks
RESIF	French Seismological and Geological Network
RF	Random Forests (classifier)
RFI	Radio Frequency Interference
RG	Relative Greenness
RGB	Red Green Blue
rGNSS-R	Reconstructed GNSS-R
RHCP	Right Hand Circular Polarization
RMSD	Root Mean Square Difference
RMSE	Root Mean Square Error
RPAS	Remotely Piloted Aircraft System
RPC	Rational Polynomial Coefficients
RPD	Ratio of Performance to Deviation
RPIQ	Ratio of Performance to Inter-Quartile range
RTK	Real Time Kinematic
RTM	Radiative Transfer Model
RUE	Rain Use Efficiency
RVI	Radar Vegetation Index
RVoG	Random Volume over Ground
RXD	Reed-Xiaoli Detector
SAFY	Simple Algorithm For Yield estimate
SAM	Spectral Angle Mapper
SAMIR	Satellite Monitoring of Irrigation (model)
SAR	Synthetic Aperture Radar
SAR in	SAR interferometric
Saral/Altika	Radar Altimeter (French–Indian altimetry mission)
SASS	SEASAT Advanced Scatterometer System
SAVI	Soil-Adjusted Vegetation Index
SBAS	Satellite Local-based Augmentation System
SBAS	Small Baselines

SCA	Snow Cover Area
SCARAB	Scanning Radiometer for Radiation Balance
SCF	Snow Cover Fraction
SDC	Snow Depletion Curve
SDS	Science Data System (USA)
SEAS	Survey of the Environment Assisted by Satellite
SeaWiFS	Sea-viewing Wide Field-of-view Sensor (satellite)
SEBAL	Surface Energy Balance Algorithm for Land
SEBS	Surface Energy Balance System
SEC	Standard Error of Calibration
SEKF	Self-extended Kalman Filter
SEP	Standards Error of Prediction
SER	Section Efficace Radar
SERD	Single bounce Eigenvalue Relative Difference
SEVIRI	Spinning Enhanced Visible and Infrared Imager
SFCW	Stepped Frequency Continuous Waves
SfM	Surface-from-Motion
SFT	Strong Fluctuation Theory
SGBM	Semi-Global Block Matching algorithm
SHALOM	Spaceborne Hyperspectral Applicative Land and Ocean Mission
SHI	State Hydrological Institute (St. Petersburg, Russia)
SHOALS	Scanning Hydrographic Operational Airborne LiDAR Survey
SHOM	French Navy's Hydrographic and Oceanographic Service
SID	Spectral Information Divergence
SIERRA	Spectral Reflectance Image Extraction from Radiance with Relief and Atmospheric Correction
SIFT	Scale Invariant Feature Transform
SIGMA	Simulation Innovation for Global Monitoring of Agriculture
SLA	*Scanner LiDAR aérien* (aerial LiDAR scanner)
SLC	Single Look Complex
SLR	Single-Lens Reflex
SM	Soil Moisture
SMA	Spectral Mixture Analysis

SMAC	Simplified Method for Atmospheric Correction
SMAP	Soil Moisture Active and Passive mission (Radiometer)
SMAPVEX	SMAP Validation Experiment
SMEX02	Soil Moisture Experiment 2002
SMF	Spectral Matched Filter
SMLR	Stepwise Multiple Linear Regression
SMMR	Scanning Multichannel Microwave Radiometer
SMOS	Soil Moisture and Ocean Salinity mission (satellite)
SNAS	Chinese Satellite Navigation Augmentation System
SNR	Signal to Noise Ratio
SNSB	Swedish National Space Board
SNV	Standard Normal Variate
SPAD	Single-Photon Avalanche Diode
SPM	Small Perturbation Model
SPM	Suspended Particulate Matter
SPOT	Satellites for Earth Observation
SRM	Snowmelt-Runoff Model
SRTM	Shuttle Radar Topography Mission
SSA	Surface Specific Area
SSC	Soil Surface Characteristics
S-SEBI	Simplified Surface Energy Balance Index
SSM	Soil Surface Moisture
SSMI	Special Sensor Microwave Imager (satellite)
SST	Sea Surface Temperatures
STF	Spectrotransfer Functions
STICA	Socio-technical Information and Communication Arrangements
STICS	Crop model
SUCROS	Simple and Universal Crop Growth Simulator
SUHI	Surface Urban Heat Islands
Suomi-NPP	Suomi National Polar-Orbiting Partnership
SURFEX	Surface model platform (Météo France)
SVAT	Soil–Vegetation–Atmospheric Transfer
SVM	Support Vector Machine

SVMR	Support Vector Machine Regression
SVR	Support Vector Regression
SWAT	Soil and Water Assessment Tool
SWE	Snow Water Equivalent
SWI	Soil Wetness Index
SWIR	Short-wave Infrared
SWOT	Surface Water Ocean Topography (satellite)
SYSIPHE	Airborne hyperspectral imaging system
T/P	Topex/Poséidon (Franco-American altimeter)
TDR	Time Domaine Reflectometry
TEC	Total Electron Content
TES	Emissivity Separation algorithm
THEIA	French Land Data Centre
THIRSTY	Thermal Infrared Spatial System (satellite project by CNES and NASA)
TIFF	Tag Image File Format
TIN	Triangular Irregular Network
TiO_2	Titanium dioxide
TIR	Thermal Infrared
TISI	Temperature Independant Spectral Indices
TLS	Terrestrial Laser Scanning
TM	Thematic Mapper
TMBOC	Time Multiplexed Binary Offset Carrier
TNT2	Topography based Nitrogen Transfer and Transformation
To	Tera-octet (1,000,000,000,000 octets)
TOA	Top of Atmosphere
TOC	Top of Canopy
TomoSAR	Tomography SAR
Topex/ Poseidon	Radar altimeter
TOPLATS	Topographic Land Atmosphere Transfer Scheme
TRIP	Total Runoff Integrating Pathways
TRMM	Tropical Rainfall Measuring Mission (satellite)

TSAVI	Transformed Soil Adjusted Vegetation Index
TSEB	Two-Source Energy Balance (model)
TTL	Transistor-Transistor Logic
TWAP	Transboundary Water Assessment Program
TWS	Terrestrial Water Storage
UA	User's Accuracy
UAA	Utilized Agricultural Area
UAV	Unmanned Aerial Vehicles
UHI	Urban Heat Islands
ULICE	Ultraviolet LiDAR for Canopy Experiment
UNEP	United Nations Environment Program
USDA	United States Department of Agriculture
USGS	United States Geological Survey
USO	Ultra-Stable Oscillator
UTC	Coordinated Universal Time
UTM	Universal Transverse Mercator
UV	Ultra-violet
VARI	Visible Atmospherically Resistant Index
VCA	Vertex Component Analysis
VD	Virtual dimensionality
VHF	Very high frequency
VHI	Vegetation Health Index
VHSR	Very High Spatial Resolution
VIC	Variable Infiltration Capacity
VIIRS	Visible Infrared Imager Radiometer Suite
VIS	Visible
VISAT	Video, Inertial, and Satellite GPS
Vis-NIR	Visible and Near Infrared
VITO	Flemish Institute for Technological Research (Belgium)
VLA	Very Large Array
VLBI	Very Large Baseline Interferometry
VOS	Volatile Organic Compound
VPD	Vapor Pressure Deficit

VSDI	Visible and Shortwave Infrared Drought Index
VWC	Vegetation Water Content
WALID	Water LiDAR Simulation Model
WASS	Wide Area Augmentation System
WDI	Water Deficit Index
WGHM	Water GAP Global Hydrology Model
WGS	World Geodetic System
WMA	Winter Metric Anomaly
WMO	World Meteorological Organization
WMS	Wide-area Master station
WRS	Wide-area Reference Stations
WSI	Water stress index
WTC	Wet Troposphere Correction
XML	Extensible Markup Language
ZSSD	Zero-mean Sum of Squared Difference

Introduction

Remote sensing refers to the acquisition of information on a given target without making contact with the target. It uses the entire electromagnetic spectrum, ranging from short wavelengths (for example, ultraviolet) to long wavelengths (microwaves). Two volumes of this set of books describes the physical concepts of the principal remote-sensing techniques used for the observation of continental surfaces as well as analysis methodologies and the interpretation of measurements and images (Optical Remote Sensing of Land Surfaces: Techniques and Methods and Microwave Remote Sensing of Land Surface: Techniques and Methods). The four other volumes illustrate methods and examples of remote-sensing data application for agriculture and forest (Land Surface Remote Sensing in Agriculture and Forest), continental hydrology (Land Surface Remote Sensing in Continental Hydrology), urban and coastal zones (Land Surface Remote Sensing in Urban and Coastal Areas), the environment and risks (Land Surface Remote Sensing: Environment and Risks).

The potential of satellite imagery remains underexploited in numerous specialized fields by certain members of the scientific community, but especially by public actors in charge of public policies. This underexploitation is mainly due to the cost of images and the lack of expertise in the processing and analysis of remote sensing data. For this set of books, in addition to providing researchers and students with a pedagogical work tool that is as complete as possible, we hope to show the

---

Introduction written by Nicolas BAGHDADI and Mehrez ZRIBI.

potential of remote sensing for the observation of continental surfaces and contribute to breaking down the barriers that hinder the use spatial data.

Bear in mind that the adventure of spatial remote sensing for the observation of the Earth began in the United States in particular with the launch of LANDSAT-1 in 1972 and AVHRR in 1978, followed by France with the launch of SPOT-1. In light of the interest in remote sensing for surveillance and the management of the environment and resources, other nations have shown an interest in spatial independence including Europe, Canada, Japan, Germany, Italy, India, Brazil, China, Taiwan, etc.

The considerable progress in instrumentation and in the development of methods for the processing and analysis of data places remote sensing at the center of various international programs for surveillance and tracking the impact of climatic and anthropogenic changes on the environment. Spatial resolution of data ranges from low resolution to high resolution, as well as the very high temporal resolution, allow the enrichment and improvement of models depicting the functioning of the ecosystem for providing, a better understanding of the problems threatening our planet. In effect, remote sensing makes it possible to take measurements and send raise concerns about alarming changes to our planet by providing precious information for large territories all over the globe, even where conditions are extreme (meterological conditions, limited access, etc.).

For the past 30 years, remote sensing has gone from a use mainly based on the photo-interpretation of several images to combining photo-interpretation with the computer-assisted analysis of a large quantity of images and data. We have also gone from remote sensing based exclusively on aerial photography and terrestrial sensors to using spatial vectors, airborne instruments and drones. In addition, remote sensing has evolved from primarily optical (visible and infrared) to using the entire spectrum of wavelengths, from the visible to microwaves.

Although multispectral visible optical remote sensing is now well known by scientists and actors in charge of managing the environment, other remote sensing techniques, including thermal, hyperspectral, lidar and radar, are only known to certain specialists. However, the methodological challenges remain vast for the use of visible optical data with, for example, the analysis and interpretation of information from very high-resolution spatial sensors, which remains a major issue in the processing of images. In addition, faced with the radical increase in the volume of data (Big Data), the value of large volumes of data (not necessarily limited to remote sensing data) has become a significant challenge, prompting the need to find alternatives to traditional methods (for example, data mining).

The pedagogical support needs of the community of users are significant in order to facilitate a better understanding of recent developments in remote sensing, the methods of processing and analysis, the advantages and the limitations of each technique, the fields of application and the possible complementarities between different techniques. This series aims to present the most recent principal developments in terms of remote sensing instrumentation and the analysis techniques that render the observation of the Earth via remote sensing more effective and more adapted to different fields of application.

This volume describes the different techniques of optical remote sensing, with a description of the acquisition systems, the taken measurement and the corrections to be completed. The first part presents optical imaging (multispectral, hyperspectral and thermal) with Chapter 1 describing the instrumentation and physics of the measurement, followed by Chapter 2 on the processing of optical images with, in particular, the principles of image visualization, geometric and radiometric corrections, calculations of different thematic indexes and methods of information extraction (supervised and unsupervised classifications). Chapter 3 is dedicated to the estimation of digital elevation models based on optical images.

The second part of this volume is dedicated to lidar systems, with a chapter on the instrumentation and principles of lidar operation and a further chapter on the processing methods for this type of data. In the latter, the filtering methods and the classification of lidar waveforms are presented as well as the inversion methods for lidar equations. In addition, this section covers the analysis of clouds of 3D lidar points with, in particular, methods of information extraction (3D geometry, etc.), the recognition of forms and classification. The final chapter of this section dedicated to lidar systems covers the extraction of digital elevation models from airborne lidar data.

This book, produced by scientists who are internationally renowned in their fields, is addressed to students (at engineering schools working towards Masters and Doctorates), engineers and researchers interested in the observation of the Earth and imaging systems, whether they are signal and image processors, geophysicists or other specialists (agronomists, pedologists, hydrologists, meteorologists, climatologists, glaciologists, etc.). We hope that through this pedagogical work, we can contribute to breaking down the barriers that hinder the use of remote sending data.

We thank the people who have contributed to the development of this work. First, there are the scientists, certainly the chapter authors, as well as the experts from the scientific committee charged with rereading the chapters and providing corrections. This project was carried out thanks to the support of Irstea (French National Research Institute of Science and Technology for Environment and Agriculture), CNRS (French National Center for Scientific Research) and CNES (French National Center for Spatial Studies).

In addition, there is the patience of our families who have tolerated the long hours we have spent in front of our computers in order to complete this project. Our thanks also go to Mr. André MARIOTTI (Professor emeritus, University Pierre et Marie Curie) and Mr. Pierrick GIVONE (Scientific Director, Irstea) for their encouragement and support in the realization of this project.

1

Radiometry in the Optical Domain

1.1. Introduction

This chapter aims to introduce the physical definitions and concepts essential to the understanding of a signal measured with an optical instrument. For any reader hoping to delve deeper into these foundations, one could suggest the works by Liou [LIO 80] and Lenoble [LEN 93] for radiative transfer theory or Fourest *et al.* [FOU 12] for aspects of image quality.

Radiometry defines the science of measuring radiation. Radiative transfer describes the interaction of electromagnetic radiation and matter. This radiation evolves in the form of a wave propagating at speed c of $2.998.10^8$ m/s. It is characterized by its direction of propagation, its wavelength, its intensity, its phase and its polarization. Each ray propagates with a certain vibration that can be expressed either in wavelengths (m) in the optical field, in frequency (Hz) in the field of hyperfrequencies, or in the number of waves $\nu = \frac{1}{\lambda}$ (m^{-1}) in spectrometry (pulsation $= \frac{2\pi}{\lambda}$). In optics, a ray defined by a single wavelength is called monochromatic radiation.

In this chapter, we will limit ourselves to the optical domain, covering the spectral range from 0.04 up to 12 μm. Figure 1.1 pinpoints the optical field in the electromagnetic spectrum.

After defining the main radiative quantities (section 1.2.2), the optical properties of terrestrial surfaces are presented (section 1.2.3.1),

Chapter written by Xavier BRIOTTET.

followed by atmospheric properties (section 1.2.3.3) in order to introduce the notions of radiative transfer in the Earth–atmosphere system (section 1.2.4). Section 1.3 then describes the principal optical technologies. Finally, in section 1.4, the main inversion methods necessary to extract the optical properties of a scene's elements are presented.

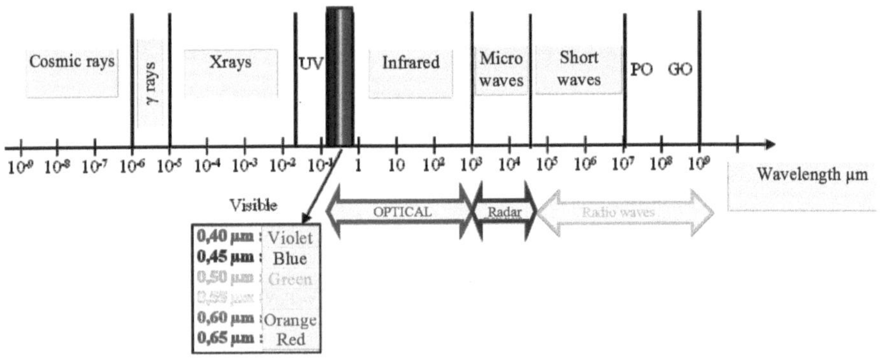

Figure 1.1. *Location of the optical domain in the electromagnetic spectrum. For a color version of this figure, see www.iste.co.uk/baghdadi/1.zip*

1.2. Radiometric quantities

1.2.1. *Preamble*

In order to clearly explain the radiometric quantities involved, it is important to introduce the notion of a solid angle $\Delta\Omega$, centered on a direction Ω and characterized by its zenith angle θ (the angle between the normal to the surface element considered and the direction Ω) and its azimuth angle ϕ (the angle in the horizontal plane between a direction of reference and the projection in this plane of direction Ω). In point O, the solid angle $\Delta\Omega$, delimited by a cone in space, is defined as: $\Delta\Omega = \frac{A}{R^2}$ with A being the area cut by the cone over a sphere of radius R, the center of which is at the apex of the cone. The solid angle unit is the steradian (sr). The elementary solid angle $d\omega$ (Figure 1.2) is defined in a sphere of the radius unit (R = 1) by the relationship: $d\omega = \sin\theta . d\theta . d\varphi$.

As an example, the complete spatial vision around a point corresponds to a solid angle of 4π. For an observer on the Earth, the half-space or hemisphere formed by the celestian arch equals 2π.

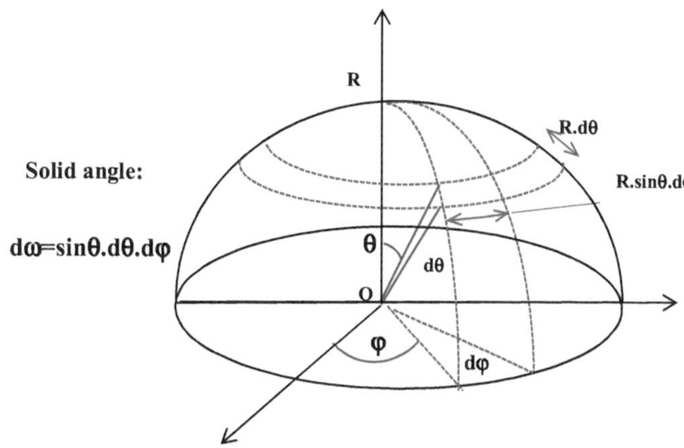

Figure 1.2. *Elementary solid angle $d\omega$ in a sphere of radius R, around the central direction of the azimuth angle φ and the zenith angle θ*

1.2.2. *Radiative quantities*

Radiative quantities can be defined for spectral intervals of varying size or for a given wavelength. In the latter case, these quantities are called spectral and are defined wavelengths.

Radiative energy (in Joules) is transported by photons. A photon, emitted at a wavelength λ, has the energy $E = h.c/\lambda$ where $c = 2.998.10^8$ m.s^{-1} is the speed of light and $h = 6.6261.10^{-34}$ J.s is Planck's constant.

Energy power is defined as the energy transferred per unit of time $dP = dE/dt$ (Watt).

The spectral radiance $L(\lambda)$ of a surface dA is the energy power $dP(\lambda)$ in the elementary spectral interval $[\lambda, \lambda+d\lambda]$, confined in the elementary solid angle $d\omega$, around the direction of propagation (θ, φ) with respect to the normal vector of the surface dA:

$$L(\lambda) = \frac{dP(\lambda)}{\cos\theta.d\omega.dA} \ (W.m^{-2}.sr^{-1}.\mu m^{-1}) \qquad [1.1]$$

A camera records radiance at each pixel in the matrix of detectors.

The density of the spectral energy flux $F(\lambda)$ ($W.m^{-2}.\mu m^{-1}$) is the incident or emerging power of a surface unit per spectral interval unit. It is therefore an integral of the radiance on the solid angle 2π: $F(\lambda) = \int_{2\pi} L(\lambda).\cos\theta.d\omega$. When the flux density is incident on a surface, this quantity is called irradiance, and when it is emitted by the surface, exitance.

As an example, for isotropic radiance in every direction of incidence in the half-sphere, the total irradiance is written as: $F(\lambda) = \pi.L(\lambda)$.

1.2.2.1. *Specific case of the black body*

In 1901, Planck defined the radiation emitted by a cavity (a black body) by assuming that an atomic oscillator can only have a finite number of energy states and that this energy is emitted when the oscillator changes from one energy state to another. Based on this hypothesis, he derived Planck's law, or the law of black body radiation:

$$L_{bb}(\lambda, T) = \frac{2.h.c^2}{\lambda^5(e^{h.c/(K.\lambda.T)}-1)} \text{ (W/m}^2\text{/sr/}\mu\text{m)} \qquad [1.2]$$

where K is Boltzmann's constant = $1.3806.10^{-23}$ $J.K^{-1}$ and T is the temperature of the black body in thermodynamic equilibrium in Kelvin (K). This radiance has the bell shape (Figure 1.3) when represented as a function of the wavelength, with a maximum depending on the temperature T. For a given temperature, the maximum radiance L_{max} is obtained at the wavelength λ_{max} defined by Wien's displacement law:

$$\lambda_{max}.T = 2898\ \mu m.K \text{ and } L_{max} = 4.067.10^{-12}.T^5 \qquad [1.3]$$

The density of energy flux F(T) is a quantity integrated in the entire spectral field and in all directions. It is given in the Stefan–Boltzmann law:

$$F(T) = \sigma.T^4 \text{ (W/m}^2\text{)} \qquad [1.4]$$

where σ is the Stefan–Boltzmann constant ($5.67.10^{-8}$ $W.m^{-2}.K^{-4}$).

As an example, the Sun has a temperature of 5778 K and the wavelength corresponding to the maximum emission is, by applying Wein's displacement law, $\lambda_{max} = 0.502$ μm, situated in the visible field.

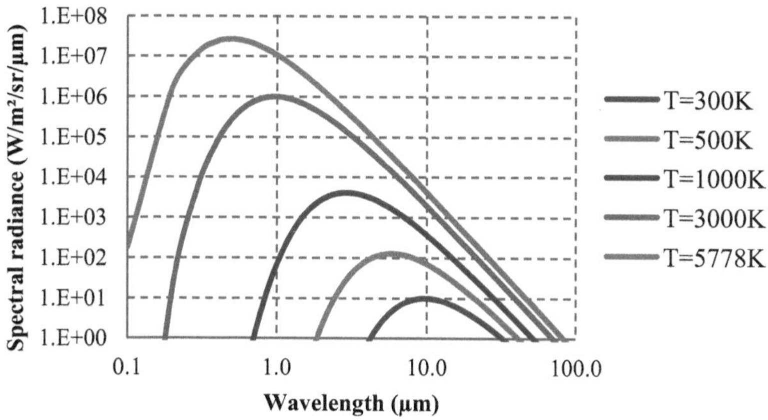

Figure 1.3. *Planck's law as a function of the wavelength for five black body temperatures. For a color version of this figure, see www.iste.co.uk/baghdadi/1.zip*

The average temperature of terrestrial surfaces is 300 K, which corresponds to λ_{max} = 9.60 µm. The spectral range centered around this value is the most adapted domain for the measurement of thermal emission from terrestrial surfaces. It is traditionally called "thermal infrared".

1.2.2.2. *Polarization*

An electromagnetic wave can also be characterized by its polarization, which determines its oscillation behavior during propagation. This behavior follows one or several preferred orientations (linear, elliptical or circular). When the light is partially polarized, as is the case in the Earth–atmosphere system, the Stokes parameters are used to describe its complete state of polarization. This four-dimensional vector has the form $(I, Q, U, V)^T$, where I is the total radiance (sum of non-polarized radiance L_{np} and polarized radiance L_p), $Q = L_p.\cos(2.\Psi)$ with Ψ being the angle between the direction of polarization and an axis of reference, $U = L_p.\sin(2.\Psi)$, and V is the eccentricity that determines the intensity of the circular polarization. In remote sensing, it is recognized that the natural light of the Sun is not polarized, and that its interaction (scattering) with the atmosphere linearly polarizes the wave. Thus, the complete description of this wave is obtained with only the first three components of the Stokes vector (V = 0).

1.2.3. *Radiative properties of matter*

Incident radiation on a surface is subject to three phenomena: absorption, transmission or reflection (for semi-transparent environments, we will use the word diffusion or scattering). The relative importance of these phenomena varies according to the wavelength considered.

Thus, in the visible field, the light that is incident on a sensor is generally not directly issued from its source, but from a diffusion process. This, for example, is what happens when one looks at a scene lit up by the Sun or the sky outside of the solar direction. Thus, diffusion corresponds to a change in direction of incident rays. The term "diffusion" is used when the environment that diffuses the light is composed of particles in a semi-transparent environment, such as the atmosphere or vegetation, and the term "reflection" is used for a surface. Diffusion is often accompanied by an absorption phenomenon. In fact, the traversed medium is not transparent, and can absorb part of the incident energy flux. Thus, droplets of water making up a cloud can greatly diminish the incident solar flux via diffusion and absorption mechanisms. Furthermore, the interaction of the radiation with matter is spectrally dependent. Still in the visible domain, the green color of grass is linked to a greater absorption of photons of wavelengths situated in the blue and red and a higher reflection in the green. It is important to note that absorption is spectrally much more variable than diffusion. Part of the radiation absorbed by this material is reemitted. At an ambient temperature, this radiation finds its maximum emission in the infrared thermal field (Figure 1.3).

Intuitively, one can understand that the signal recorded by a camera will depend upon the radiative properties of the traversed atmosphere, of the observed scene and the wavelength.

When an environment is lit up by a monochromatic source, part of the incident flux will be reflected, part absorbed and part transmitted. By normalizing these last radiative quantities by the incident flux, we obtain the following energy conservation relationship:

$$\rho(\lambda) + t(\lambda) + a(\lambda) = 1 \qquad [1.5]$$

where ρ is the spectral reflectance, t is the transmittance and a is the absorptance.

According to Kirchoff's law, which expresses that an object in thermodynamic equilibrium emits what it absorbs, the absorption coefficient is linked to the emissivity ε by:

$$\varepsilon(\lambda) = a(\lambda) \text{ with } \varepsilon(\lambda) = \frac{L(\lambda,T)}{L_{BB}(\lambda,T)} \qquad [1.6]$$

where $L(\lambda, T)$ is the radiance emitted by the object at temperature T. One can assume that this law is verified for the environments studied in the Earth–atmosphere system at ambient temperatures. Thus, emissivity describes the part of the flux that is emitted after absorption by a body. A distinctive case is the black body, which has an emissivity of 1, while real bodies all have an emissivity below 1.

Some specific cases can occur:

– if the object is a black body: $\varepsilon(\lambda) = a(\lambda) = 1$, $\rho(\lambda) = 0$ and $t(\lambda) = 0$;

– if the object is opaque (in the case of terrestrial surfaces such as bare soil): $t(\lambda) = 0$ and $\rho(\lambda) + \epsilon(\lambda) = 1$. This relationship is the foundation for emissivity measurement methods based on reflectance measurements by reformulating the previous expression: $\epsilon(\lambda) = 1 - \rho(\lambda)$. This expression is also used in the radiative balance equation in the thermal field. A distinctive case is noted in the presence of a perfectly reflective body: $\rho(\lambda) = 1$ and $\varepsilon(\lambda) = a(\lambda) = t(\lambda) = 0$;

– if the object is transparent (in the case of the atmosphere at certain wavelengths): $t(\lambda) = 1$ and $\varepsilon(\lambda) = a(\lambda) = \rho(\lambda) = 0$.

The brightness temperature (or apparent radiative temperature T_b) of a surface is the temperature of a black body delivering the same radiance as the studied surface.

$$L_\lambda(T_b) = \varepsilon(\lambda).\, L_{CN}(\lambda, T) \qquad [1.7]$$

Based on the law of black body radiation (equation [1.2]), the measured flux issued from these processes of reflection, emission and transmission will strongly depend upon the wavelength and temperature. Thus, in the visible, near-infrared field, the flux emitted by surfaces with a near-ambient temperature is negligible with respect to the solar flux reflected by the materials. This is why, in this range (reflective domain: 0.4–2.5 μm), the signal measured will mainly be sensitive to the spectral reflectance of surfaces or to the diffusion of atmospheric particles. However, in the thermal infrared field (commonly defined in the spectral range of 8–12μm), the emitted infrared radiation is preponderant (the contribution of reflected or diffused solar radiation is generally negligible). In addition, the signal acquired by an instrument will depend on the temperature of the measured body and its spectral emissivity.

The notions of reflectance, absorptance, scattering and transmittance will now be detailed, although this introduction is deliberately summarized in order to remain general. For opaque bodies, the common optical properties are reflectance and emissivity. For semi-transparent bodies (such as the atmosphere, water and vegetation), radiative processes such as diffusion, absorption and transmission are considered.

In order to simplify the notations, the spectral dependence of radiometric quantities will be omitted in the remainder of this chapter.

1.2.3.1. *Optical properties of terrestrial surfaces*

1.2.3.1.1. Notions of reflectance

Every surface lit up by a source will reflect light in a more or less selective manner according to the directions of observation Ω_v and incidence Ω_s, and according to the wavelength. The phenomenon of reflection is represented by a spectral optical property called "function of bidirectional reflectance distribution BRDF (Ω_s, Ω_v)", or, more specifically, its integral on the considered interval $\rho(\Omega_s, \Omega_v)$. According to Nicodemus *et al.* [NIC 77], the BRDF defines the relationship between the reflected radiance $L(\lambda,\Omega_v)$ according to the direction Ω_v and the incident irradiance $E(\lambda,\Omega_s)$ according to the incident direction Ω_s on the surface:

$$\mathrm{BRDF}(\lambda, \Omega_S, \Omega_V) = \frac{L(\lambda,\Omega_V)}{E(\lambda,\Omega_S)}\ (\mathrm{sr}^{-1}) \quad [1.8]$$

The BRDF depends on the relative positions of the source, the sensor and the considered surface (Figure 1.4).

As an example, a point source incident on a mirror or the surface of calm water is reflected in a single direction, as defined by the Snell–Descartes law. However, tilled soil will reflect a large part of the incident light in the direction of incidence. A dense lawn that has been cut short reflects light in every direction in an isotropic manner. Such a surface is called Lambertian.

In this latter case, the BRDF of a perfect reflector (like a Spectralon) equals $1/\pi$.

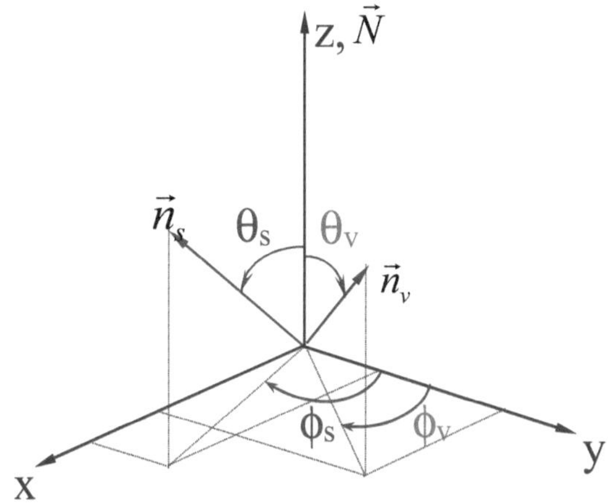

Figure 1.4. *Reference used to define bidirectional reflectance ($\vec{N}$ is the normal to the surface, $\vec{n}_s$ and $\vec{n}_v$ respectively define the direction of incidence and viewing direction)*

Several other types of reflectance can be deduced depending on the considered solid angle of the measured flux and the incident flux. All these quantities can be deduced from the BRDF. Any reader hoping to further investigate this point can refer to the works of Von Schönermark [VON 04] or Schaepman-Strub *et al.* [SCH 06]. As an example, two typical quantities are indicated below.

The bidirectional reflectance is written as: $\rho^{dd}(\lambda, \Omega_S, \Omega_V) = \pi. \text{BRDF}(\lambda, \Omega_S, \Omega_V)$.

The hemispheric (index h) – hemispheric reflectance (index h) is the ratio of the flux reflected in every direction of the hemisphere by the irradiance incident coming from the entire hemisphere:

$$\rho^{hh}(\lambda, \Omega_S, \Omega_V) = \frac{\int_0^{2\pi}\int_0^{\pi/2} \text{BRDF}(\lambda,\Omega_S,\Omega_V).L_S(\lambda,\theta_S.\phi_S).\cos\theta_S.\sin\theta_S.d\theta_S.d\phi_S}{\frac{1}{\pi}.\int_0^{2\pi}\int_0^{\pi/2} L_S(\lambda,\theta_S.\phi_S).\cos\theta_S.\sin\theta_S.d\theta_S.d\phi_S} \quad [1.9]$$

where L_s is the incident radiance in the direction (θ_s, ϕ_s). This quantity integrated over the spectral range is called albedo.

The spectral behavior of reflectance is closely linked to the chemical composition of the considered material. If a photon of wavelength λ is incident on a molecule with the energy state E_o and if the energy (h.c/λ) of the photon is equal to the difference ($E_e - E_o$) between energy states E_e and E_o of the molecule, then the photon is absorbed and the molecule is excited to the level E_e. In the presence of a polychromatic source (natural light) and complex molecules, the absorption spectrum becomes more complex, but the characteristic position of absorption rays depicts the type of material observed.

A difference E_e–E_0 of a few electron volts (eV) (respectively, of several 10^{-1} eV) corresponds to a change in state between electronic levels (respectively, vibrational). This difference, in accordance with the multiples of the energy quantum h.c/λ, corresponds to a spectrum at an absorption band located in the 0.3–1.5 µm range (respectively, in the 1.0–20 µm range). Above 2.5 µm, the absorption signature is also linked to energy differences (in the region of several 10^{-2} eV) between rotational states. The localization of these different fundamental modes depends on the type of molecule.

1.2.3.1.2. Notion of absorptance

If a radiation of radiance L propagates across a non-diffusive environment with a thickness of dx (Figure 1.5), it undergoes an attenuation of dL described by the absorption coefficient σ_a (m^{-1}): $dL = -\sigma_a . L . dx$.

The spectral behavior of the absorption coefficient depends on the chemical nature of the traversed area and its environment. The energy states during the energetic transition can be modified by external causes (enlargement from collision or thermal agitation) and by a natural enlargement from a ray as a result of the duration of finite life of an excited state and the finite width of the energy levels. These processes are described by the absorption coefficient: $\sigma_a(\lambda) = S . f(\lambda - \lambda_0)$ where λ_0 is the central wavelength of the ray, S is its intensity and f is its shape. These quantities depend on the type of medium. For example, such information can be found, for gas, in the HITRAN database [ROT 05].

The integration of this coefficient along the trajectory delineated by the abscissas x_0 and x_1 defines the absorption optical depth τ_a (dimensionless): $\tau_a = \int_{x_0}^{x_1} \sigma_a . dx$. The transmittance t is thus defined by: $t = e^{-\tau_a}$.

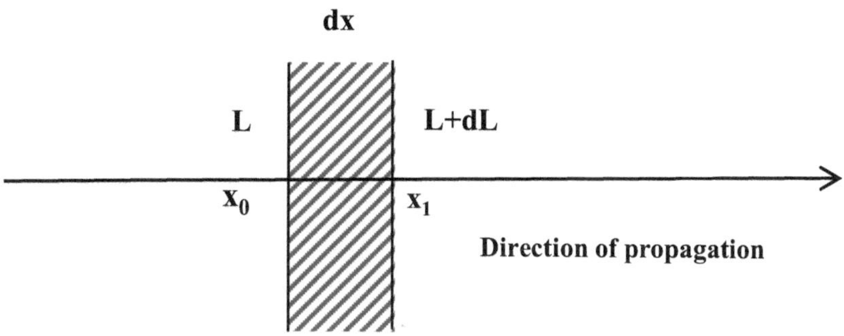

Figure 1.5. *Definition of absorption*

In the presence of a complex medium made up of flat, parallel and homogeneous layers (for example, the atmosphere locally) with an optical depth of $\tau_{a,0}$ along the trajectory perpendicular to these layers, the optical depth along a line of sight of direction θ is equal to $\tau_a(\theta) = \frac{\tau_{a,0}}{\cos(\theta)}$.

1.2.3.1.3. Notion of diffusion

When a radiation traverses a medium composed of diffusive and non-absorbent particles, the incident photons that interact with the environment change direction without changing wavelength (elastic scattering). The attenuation of energy flux crossing such an environment of thickness dx along the direction of propagation is modeled by the diffusion coefficient σ_d (en m^{-1}): $dL = -\sigma_d . L . dx$. As with absorption, integration of this quantity along the trajectory delineated by the abscissas x_0 and x_1 defines the diffuse optical depth τ_d (without dimension): $\tau_d = \int_{x_0}^{x_1} \sigma_d . dx$. The transmittance t_d is thus defined by:

$$t_d = e^{-\tau_d}$$

The energy scattered by these particles is generally not isotropic. Its angular distribution is described by the normalized phase function $\frac{p(\Theta)}{4\pi}$ with the phase angle Θ, defined as the angle between the directions of incidence and diffusion.

1.2.3.1.4. Notion of transmittance

Transmittance expresses the attenuation of the beam along a given trajectory. As we saw earlier, it is linked to two processes: absorption and diffusion. This quantity is defined by the extinction coefficient σ_e:

$$\sigma_e = \sigma_a + \sigma_d \qquad [1.10]$$

Thus, the total spectral transmittance is the product of absorption and diffusion transmittances.

The albedo of simple scattering $\omega_0 = \frac{\sigma_d}{\sigma_e}$ defines the relative importance of absorption and diffusion processes. If ω_0 is close to 1 (respectively, 0), absorption (respectively, diffusion) is negligible.

1.2.3.2. *Examples of optical properties of terrestrial surfaces*

Terrestrial surfaces include the oceans (71% of the surface of the globe) and continental surfaces. The surface of the latter is mainly composed of vegetation (65% of the continental surface), bare soil and water in liquid or solid (cryosphere) status. These different classes can be distinguished in the optical domain by their spectral and directional behaviors.

Water in liquid state and pure ice (Figure 1.6) have very weak reflectance (<0.05) in the reflective range, while snow has a reflectance close to 1 in the short wavelengths and decreases in the 1.0–2.5 μm range. This decrease depends on the size of the snow crystals. With the emissivity of water close to 1 (~0.98) between 8 and 12 μm, its behavior is similar to a black body.

The role of vegetation is fundamental in the carbon cycle, the production of oxygen, the protection of soil, the provision of food for living beings, etc. Figure 1.7 shows the spectra of different types of vegetation (conifers, deciduous trees, dry and green grass).

Green vegetation has a signature presenting a peak in the green and a plateau in the near-infrared range. Its reflectance is reduced globally when the wavelength increases to 2.4 μm. High absorptions around 1.4 and 1.9 μm are due to the presence of water in the plant. In the thermal infrared, green vegetation has a high emissivity around 0.98, which increases in passing from a conifer to a deciduous tree as a result of an increase in roughness. When the plant is dry, its emissivity notably decreases (Figure 1.7).

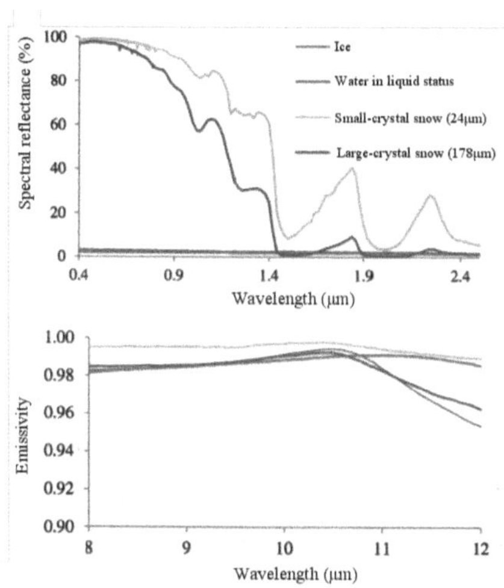

Figure 1.6. *Spectral reflectance of water in liquid and solid status (derived from the ASTER database, http://speclib.jpl.nasa.gov). For a color version of this figure, see www.iste.co.uk/baghdadi/1.zip*

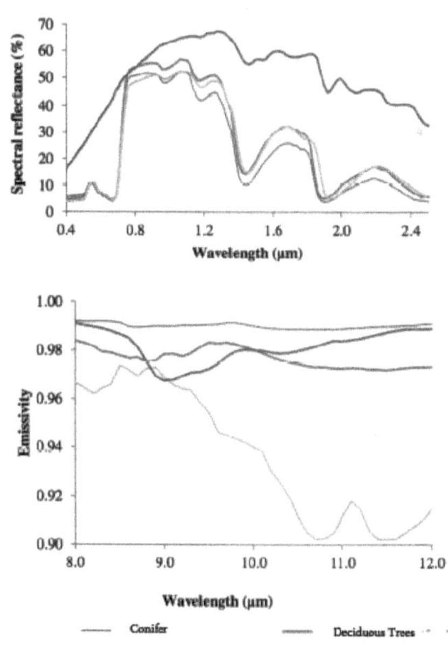

Figure 1.7. *Spectral reflectance of different types of vegetation (derived from the ASTER database, http://speclib.jpl.nasa.gov). For a color version of this figure, see www.iste.co.uk/baghdadi/1.zip*

The spectral analysis of a leaf delivers information on its composition and state. In fact, in its vacuoles, a fresh green leaf contains around 90–95% water and between 5 and 10% dry material (cell walls, cellulose: 15–30%, hemicellulose: 10–30%, proteins: 10–20%, lignin: 5–15%, starch: 0.2–2.7% sugars, etc.), wax (cuticle), chlorophyll a and b (chloroplasts) and other pigments (cytoplasm) such as carotenoids, anthocyanins, flavones, brown pigments, etc. Certain components have specific spectral signatures in the 0.4–2.5 μm range (Figure 1.8).

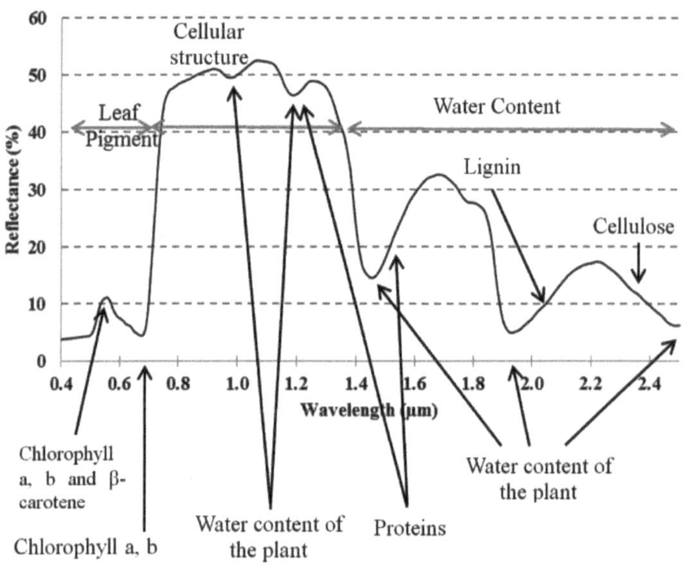

Figure 1.8. *Spectral signature of a leaf and the origin of its spectral behaviors–principal components (derived from the ASTER database, green grass, http://speclib.jpl.nasa.gov)*

Finally, the reflectance of leaves has a specular component, the importance of which depends on its surface roughness. However, the reflectance of the vegetation cover has a backscatter component according to the direction of irradiance. This is due to the fact that a sensor observing a cover according to the antisolar direction does not perceive the shadow of direct solar radiation. This is called the hot spot effect.

Bare soil is composed of minerals and organic materials. Knowledge of the soil composition is necessary in order to understand the phenomena of erosion, soil composition and pollution, or for applications including agriculture or mineral exploration.

Minerals also present a significant spectral signature in the full optical domain. Figure 1.9 indicates the main absorption band characteristic of the presence of more or less complex anions or cations. Thus, ferrous ions (Fe^{2+}) and iron ions (Fe^{3+}) have weak absorption bands located in the short wavelengths (< 1 μm). However, more complex molecules such as carbonates (CO_3^{2-}), silicates (SO_4^{2-}) or the hydroxide ion (OH^-) have absorption bands linked to the vibrational levels for wavelengths > 1 μm. The presence of water in a sand sample strongly influences its signature. The higher the water content, the more the global level of reflectance between 0.4 and 12.0 μm decreases.

Bare soil also has a varying directional behavior. Thus, rough soil has a backscatter component. However, smooth soil, like some manmade materials, has a specular component.

1.2.3.3. *Case of the atmosphere*

The atmosphere is an envelope wrapping around the Earth, largely composed of gases as well as various liquids and solid particles (aerosols, water droplets and ice crystals). It is divided into several layers of varying importance delineated according to the discontinuities in temperature variations as a function of the altitude. Thus, this envelope is divided into four layers separated by isothermal layers (tropopause, stratopause and mesopause) (Figure 1.10). The total thickness between the terrestrial atmosphere and space is generally fixed at 100 km.

1.2.3.3.1. Atmospheric gases

Atmospheric air is composed of a mixture of dry air and water vapor that obeys the ideal gas law: $\mathrm{p.V} = \mathrm{n.R.T}$ with p being pressure, V the volume of gas, T the temperature, n the number of moles of gas of the molar mass M ($kg.mol^{-1}$) and R the ideal gas constant (= 8.3143 $J.K^{-1}.mol^{-1}$). Next, the density defined as n.M/V is: $d = \frac{M.p}{R.T}$.

The atmosphere is composed of five main gases (nitrogen, oxygen, argon, water vapor and carbonic gas) and other gases at trace levels. Each of these responds to the ideal gas law. In addition, by defining the relationship of the density mixture, c_i, for a gas i as $c_i = \frac{V_i}{V}$, we can derive the molar mass of dry air $M = \sum_i c_i . M_i$, where M_i is the molar mass of the gas i, equal to 28.964 $kg.mol^{-1}$. Under the standard atmospheric conditions defined at ground level (T = 273.16 K and p = 1013.25 hPa), the density of dry air is equal to 1.293 $kg.m^{-3}$.

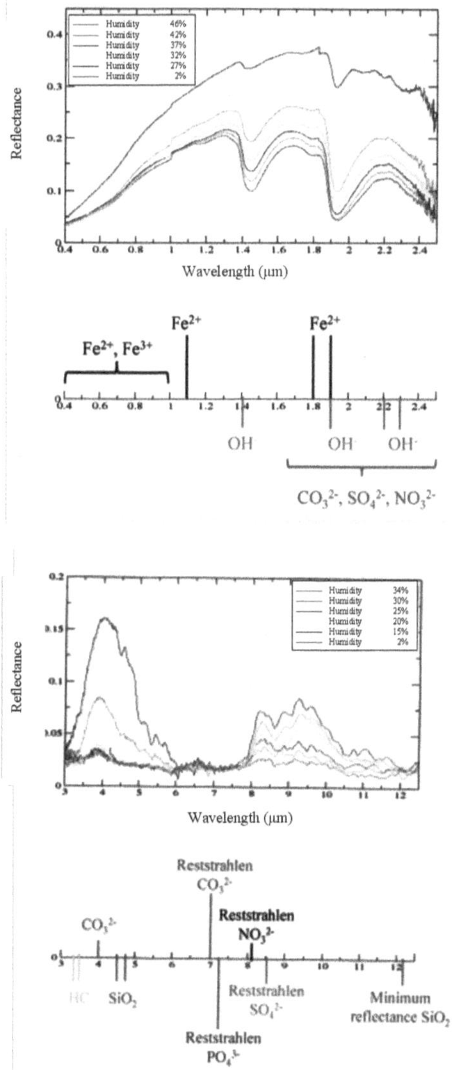

Figure 1.9. *Spectral reflectance of bare soil collected in southwestern France [FAB 15] in the reflective range (high) and the emissive range (low). For a color version of this figure, see www.iste.co.uk/baghdadi/1.zip*

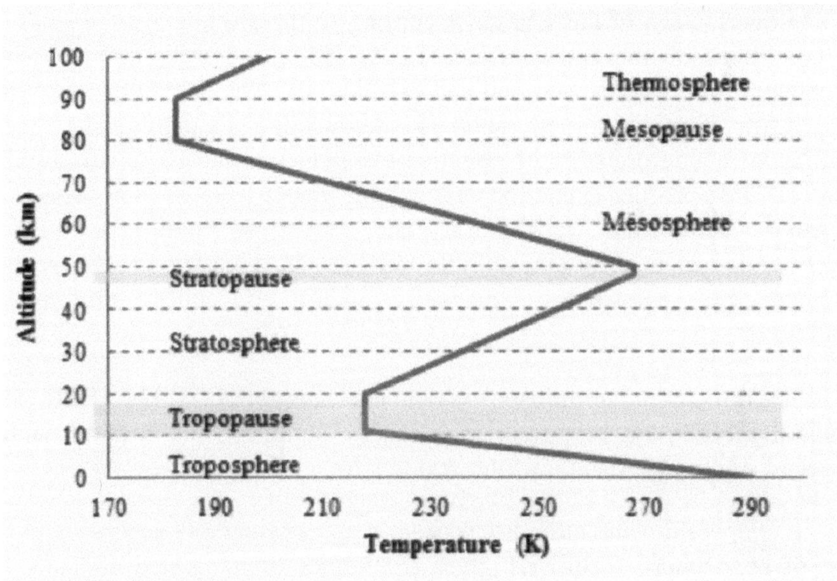

Figure 1.10. *Definition and localization of the atmospheric layers*

Atmospheric gases are generally not evenly distributed throughout the atmosphere. In considering a portion of air between altitudes z and z+dz, with p and p+dp being their respective pressure, the fundamental hydrostatic law is written: $dp = -\rho(z).g.dz$ with g being the normal acceleration of the Earth's gravity. By combining the preceding expression of ρ with this law, and considering that the temperature (fixed at T_0) and the acceleration do not depend on the altitude, we obtain after integration: $p = p_0.e^{-\frac{M.g}{R.T_0}z}$. The pressure thus varies in an exponential function with the altitude. In the lower troposphere, the term $\frac{R.T_0}{M.g}$ is called the "scale factor". It is often close to the value of 8 km. This relationship allows us, for example to estimate the pressure of altitude z, knowing that sea level is generally the level of reference for atmospheric pressure.

Water vapor is a component of the atmosphere that varies significantly geographically and in time. It is mainly located in the lower layers of the atmosphere.

The distribution of these gases according to the altitude also depends on the conditions of pressure and temperature. This is why the atmosphere is completely defined by its vertical profiles of the distribution of gas, temperature, humidity and pressure. This knowledge is a prerequisite for the simulation or interpretation of radiative exchanges in the Earth–atmosphere system. These profiles cannot be deduced by simple laws, as they depend on radiative equilibrium (absorption and emission) and the vertical transportation of heat by atmospheric movements. This knowledge is often obtained from radiosondes. However, standard atmospheric profiles are often used.

In typical radiative transfer codes such as MODTRAN [BER 89], standard profiles are proposed: mid-latitude summer, mid-latitude winter, subarctic summer, subarctic winter and US 76.

The temperature and humidity profiles vary considerably according to the type of atmosphere. Table 1.1 gives the values of integrated contents for standard atmospheric profiles.

Type of atmosphere	mid-latitude Summer	mid-latitude Winter	subarctic Summer	subarctic Winter	Tropical	US 76
Integrated water vapor ($g.cm^{-2}$)	2.92	0.85	2.08	0.42	4.11	1.42

Table 1.1. *Integrated water vapor in the atmosphere according to standard types of atmosphere*

The gases present in the atmosphere will diffuse and/or absorb the incident light and/or emit radiation. As described above, the absorption process is linked to the coefficient of molecular absorption, whose spectral behavior depends on the electronic, vibrational or rotational energy levels characteristic of the gas. Attenuation of the signal along the trajectory through absorption is represented by the atmospheric transmittance.

Figure 1.11 presents a spectral transmittance curve calculated for the entire atmospheric layer where the main absorbent gases are indicated. Water vapor, H_2O, shows very marked absorption bands (for example, around 1.4, 1.9 μm or between 5 and 7.5 μm) as well as an absorption continuum present over the entire optical spectrum.

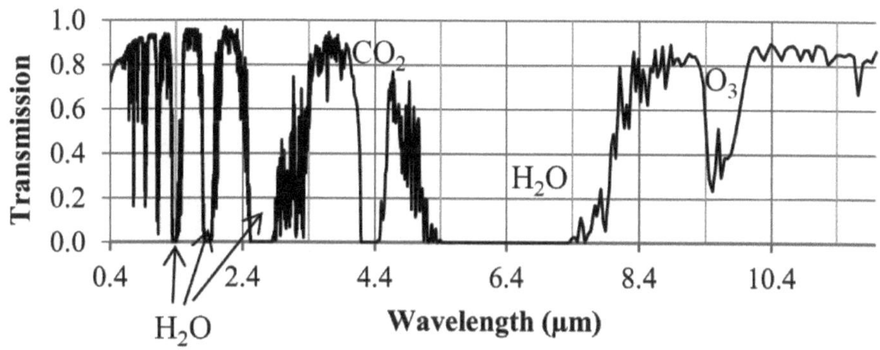

Figure 1.11. *Total spectral atmospheric transmittance (standard model mid-latitude summer) with continental aerosols and an abundance corresponding to a visibility of 23 km*

The scattering due to gaseous particles is characterized by a diffusion coefficient that varies with the wavelength in λ^{-4}. Thus, for short wavelengths (blue), this coefficient is very high and then rapidly decreases when the wavelength increases. Such a behavior explains the blue color of the sky. The optical depth depends on the coefficient and the number of molecules present along the trajectory. For a vertical trajectory, this number is directly linked to the surface pressure. The optical depth of molecular diffusion $\tau_{d,0}$ is often given for a standard atmospheric pressure (p_o = 1013 mbar at sea level). If the user knows the atmospheric pressure p at the moment of acquisition, the optical depth can be corrected by the relationship: $\tau_d = \tau_{d,0}\frac{p}{p_0}$.

The angular diffusion of light by gases is described by the normalized phase function: $\frac{p(\Theta)}{4\pi} = \frac{3}{16\pi}(1 + \cos^2\Theta)$. Its angular dynamic is poor (factor 2 for 4.π sr) compared to aerosols.

1.2.3.3.2. Aerosols

Introduction

Aerosols are defined as a combination of liquid or solid particles suspended in a gaseous or liquid environment. In the atmosphere, these particles are mainly situated in the low layers of the atmosphere (<1.5 km) since aerosol sources are located on the terrestrial surface. However, certain aerosols can still be found in the stratosphere, especially volcanic aerosols ejected into the high altitude layers.

The origin of atmospheric aerosols is either natural or the result of anthropogenic activities. They are usually classified as a function of their composition and origin:

– terrigenous aerosols or dust: these aerosols come from continental surfaces such as deserts or bare soil. They are mainly composed of clay, quartz, feldspar and calcite. They are injected into the atmosphere by the uplifting of particles on the ground due to wind. The smallest particles can travel distances higher than a thousand kilometers. These aerosols are found in the highest quantity in the atmosphere;

– oceanic aerosols and sea spray: as their name indicates, these come from maritime regions. They are formed by waves (sea spray) and by the evaporation of water on the ocean surface;

– soluble aerosols: this term designates a large number of aerosols, especially sulfates, which are very present in urban environments. Soluble aerosols also include nitrates. They mainly come from cities, industrial sites and vegetation;

– combustion aerosols: these aerosols are a result of biomass and domestic fires, industrial emissions and traffic;

– volcanic aerosols: these aerosols are mainly composed of sulfur dioxide emitted by volcanic eruptions and thrown into the stratosphere. As a result of their injection into this part of the atmosphere, they can progressively spread throughout the entire globe.

Microphysical properties of aerosols

The chemical or mineralogical composition, the size and the form of aerosols determine their microphysical properties.

In the optical domain, the parameter characterizing the chemical composition of a particle is its refraction index, which depends on the wavelength. These indices can be found in databases such as the GEISA-aerosols database (http://ara.lmd.polytechnique.fr). The main types of aerosol referenced are soot, sulfates, mineral particles (mixture of clay and quartz), organic materials and marine salts. Each type is associated with a mode corresponding to a specific grain size distribution. Atmospheric aerosols result in a mixture of these different types in variable proportions. In view of the different formation processes of aerosols and their various sources, atmospheric aerosols are characterized by a distribution of trimodal size, corresponding to modes of nucleation (aerosols with a diameter below 0.08 µm), accumulation (aerosols with a diameter between 0.08 and 2.5 µm) and coarseness (aerosols with a diameter higher than or equal to 2.5 µm) with the relative significance proportion depending on the type of aerosol [LEN 93].

For example, marine aerosols have characteristic diameters in the region of 2–4 µm and terrigenous aerosols have diameters in the region of 4–8 µm.

In the radiative transfer codes such as MODTRAN [BER 89], standard aerosol models are proposed: continental, urban, maritime and stratospheric.

Atmospheric aerosols have extremely variable concentrations as a function of the geographic zone and altitude. Their geographic distribution depends on the localization of sources and the meteorological phenomena encountered. Thus, aerosols created in cities are mainly located within the city and their immediate periphery (in the region of several kilometers), while desert or volcanic dust projected higher into the atmosphere can be transported by winds over hundreds or even thousands of kilometers. Climatologies of aerosols are available, such as those proposed by Almeida *et al.* [ALM 91] or the Global Aerosol Data Set (GADS) [KOE 97].

Optical properties of aerosols

Optical properties of aerosols are defined at the level of each particle for an element of volume (simple diffusion albedo, phase function, absorption and diffusion coefficients) or for the entire atmosphere (optical depth and transmittance).

These properties are estimated from their microphysical properties (grain size, composition and shape). When aerosols are approximated to homogeneous spheres, their optical behavior (absorption and diffusion coefficients, and phase function) can be deduced from Mie theory [MIE 08]. For particles with a more complex shape, other tools exist such as T-Matrix methods for particles with an ellipsoid shape. Nevertheless, in the majority of atmospheric aerosols, the spherical approximation is retained.

NOTE.– It is possible to have an estimation of the global diffusion coefficient (molecules and aerosols) from the notion of meteorological visibility. This is defined as the distance between the contrast $(L_{max}-L_{min})/L_{max}$ and 0.02. It is connected to the diffusion coefficient by the relationship: $\sigma_d = \frac{3.912}{V}$. In practice, this diffusion coefficient is defined at 520 or 550 nm. For example, for an average visibility of 23 km, the diffusion coefficient is equal to 0.17 km^{-1}.

The phase function $P(\Theta)$ defines the angular distribution of radiation scattered by aerosols. It describes the probability of the diffusion of a photon by a particle in a given direction. It depends upon the wavelength and the nature of the particle. Among the available models for the phase function, one of the most used is the Henyey–Greenstein function: $\frac{P(\Theta)}{4\pi} = \frac{1}{4\pi} \cdot \frac{1-g^2}{(1+g^2-2.g.\cos(\Theta))^{3/2}}$ where g is the factor of asymmetry. In fact, g describes the scattered direction: $g = \langle \cos(\Theta) \rangle = \frac{\int_{-1}^{+1} \mu.P(\mu).d\mu}{\int_{-1}^{+1} P(\mu).d\mu}$ with μ being the cosine of the phase angle. A negative value (respectively, positive) indicates that the radiation is backscattered (respectively, forwardscattered). When diffusion takes place toward the front and back, this parameter is null. For urban aerosols, g has values ranging from 0.60 to 0.76 at 440 nm, and from 0.45 to 0.64 at 1,020 nm [DUB 02].

From the diffusion and absorption coefficients and the phase function, the albedo of simple diffusion ω_0, the total τ^a, diffused τ_d^a and absorbed τ_a^a optical depth of aerosols and their corresponding transmittances t^a, t_d^a, t_a^a can be deduced.

The simple scattering albedo allows us to quantify the relative importance of the diffusion of aerosols with respect to their absorption. For example, for urban aerosols, Dubovik *et al.* [DUB 02] show that this parameter decreases when the wavelength increases. This is based on measurements made in the city of Greenbelt (USA) and Créteil (France), with a range of variation at 0.44 μm from 0.88 to 1.00 to a range of 0.81–0.97 at 1,020 nm.

The optical depth of aerosols is accessible either directly by measurement or based on radiative transfer code. From the ground, there is for example, the AERONET network (http://aeronet.gsfc.nasa.gov/), which allows for the estimation of optical depths in numerous spectral bands in the visible and near-infrared (VISNIR) from sun and sky viewing. From space, instruments such as MODIS and Sentinel-2 also offer this possibility when looking at specific terrestrial surfaces. Finally, radiative transfer codes estimate an optical depth based on knowledge of the type and abundance of aerosols. An approximated value is possible based on the knowledge of its value at a given wavelength λ_0: $\tau_d^a(\lambda) = \tau_d^a(\lambda_0).\left(\frac{\lambda}{\lambda_0}\right)^{-\alpha}$, with α being the Angström coefficient and λ_0 currently fixed at 550 nm. This parameter provides indications of the grain size and type of aerosol. In fact, when α is close to 0, the optical depth does not significantly vary with the wavelength, as is the case with marine aerosols ($\alpha = 0.4$), which are mainly composed of large particles. However, continental aerosols, which are mainly composed of small particles, have an Angström coefficient between 1 and 2.

The corresponding spectral transmittances can be deduced by calculating the exponential of the optical depth.

Finally, in the simple diffusion regime, diffused radiance is proportional to the product of the albedo of simple diffusion and the phase function.

1.2.4. *Introduction to radiative transfer*

In section 1.2.3, the main radiative quantities describing the radiative interaction between radiation and matter are introduced. They can be summarized as follows: a radiation traversing a medium undergoes attenuation due to absorption and diffusion, and also, two radiative sources can also increase the amplitude modification of the total transmitted radiation associated with the phenomena of diffusion and emission. Thus, the general equation of non-polarized radiative transfer is expressed as:

$$dL = -\sigma_e.L(x).dx + \sigma_e.J(x).dx \qquad [1.11]$$

where J is the source term and σ_e is defined in equation [1.10].

Distinctive simple cases can be considered, in order to understand the radiative phenomena involved as a function of the wavelength.

Along the path from A to B, in the absence of a source (no thermal emission and no scattering), this equation becomes:

$$dL = -\sigma_e . L(x) . dx \text{ leading to the following solution}$$
$$L(B) = L(A) . e^{-\int_A^B \sigma_e dx} \quad [1.12]$$

We recognize the term of optical depth in the exponential. This situation occurs in the spectral range (1.0–2.5 μm) in standard atmospheric conditions where scattering and thermal emission are neglected.

In the absence of scattering, but with a radiance emitted by the medium at temperature T, the equation is written as:

$$dL = -\sigma_e . L(x) . dx + \sigma_e . L_{BB}(T) . dx$$
$$\text{leading to the following solutions } L(B) = L(A) . e^{-\int_A^B \sigma_e dx} +$$
$$\int_A^B L_{BB}(T) . e^{-\int_x^B \sigma_{e(x).dx}} . \sigma_e(x) . dx \quad [1.13]$$

Radiance in B is thus the sum of an attenuation of the incident radiance and a second term representing the flux emitted by the environment at temperature T. This situation occurs in the thermal infrared range (8.0–12.0 μm) where diffusion by particles is negligible but where the flux emitted by the atmosphere not.

If the medium absorbs and scatters, the source term J is thus a diffused radiance L_d and the radiative transfer equation is written as:

$$dL = -\sigma_e . L(x) . dx + \sigma_e . L_d . dx$$
$$\text{leading to the following solutions } L(B) = L(A) . e^{-\int_A^B \sigma_e dx} +$$
$$\int_A^B L_d . e^{-\int_x^B \sigma_{e(x).dx}} . \sigma_e(x) . dx \quad [1.14]$$

The radiance in B is the sum of an attenuation of the incident radiance with a "source" contribution due to the diffusion along the AB path. Such a behavior occurs in the range (0.4–1.0 μm).

These simple expressions allow for the understanding of which flux should be considered according to the spectral range. We retrieved such behaviors in the Earth–atmosphere system, although this environment cannot be considered homogeneous. Indeed, when the path length across the atmospheric layers is low in comparison to the Earth radius, the atmosphere can be accurately described by a set of homogeneous, flat, parallel layers in thermodynamic equilibrium. Otherwise, the resolution of the radiative transfer equation becomes more complex.

Different tools exist delivering an estimation of the radiative contributors directly at ground level and at the level of the acquisition system. These tools require an initial understanding of the state of the atmosphere and the type of landscape (optical properties). Nevertheless, the choice of radiative transfer code depends on the spectral resolution which must be adapted to the problem posed. When a high spectral resolution is required (for example, around 0.1 cm^{-1}), tools such as 4AOP (4aop.noveltis.com/) and FastCode [CLO 86] have to be preferred. Their main drawback is that the simulations are very time-consuming. In addition, when a high spectral resolution is not required, band model codes are preferred, such as MODTRAN (http://www.modtran5.com/) and MATISSE (http://sites.onera.fr/ matisse/), which can be applied in the whole optical domain, or 6S (http://www-loa.univ-lille1.fr), which is limited to the 0.4–3.0 μm range.

Most of these codes estimate the main radiative contributors either at ground level or incident to the instrument. Figure 1.12 details these different terms.

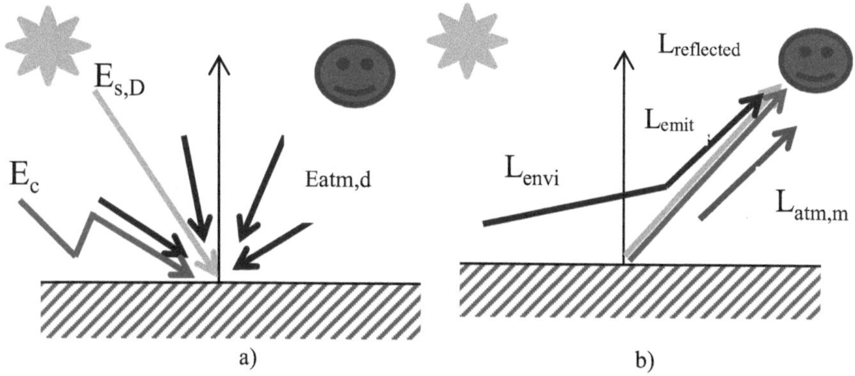

Figure 1.12. *Main radiative contributors incident to the surface a) and at sensor level b)*

Figure 1.12(a) gives the main radiative components incident to the surface:

– the direct solar irradiance (orange line), $E_{s,D}$;

– the downwelling atmospheric irradiance (dark blue line), $E_{atm,d}$, corresponding either to the thermal flux emitted by the atmosphere or with incident photons that have undergone one or more diffusions on the atmospheric particles and are then incident on the point considered; this component results in the integration of the downwelling atmospheric radiance $L_{atm,d}$ over the entire hemisphere;

– the coupling irradiation (red line), E_c, due to multiple reflections between the ground and atmosphere.

Figure 1.12(b) illustrates the incident radiance at the sensor entrance in the instantaneous field of view:

– the reflected radiance (orange line), L_{refl}, which corresponds to the incident irradiance reflected in the sensor direction;

– the radiance emitted (red line), L_{emit}, by the surface at temperature T and emissivity ε;

– the upwelling atmospheric radiance (blue line) $L_{atm,m}$ in the direction of the sensor viewing;

– the upwelling environment radiance (dark blue line), L_{envi}, corresponding to the irradiance incident to the surrounding pixels, then reflected and finally diffused in the viewing direction.

These radiative terms have relative importance depending on the spectral range.

At an ambient temperature in the field (0.4–2.5 µm), assuming a Lambertian surface reflectance ρ and a decoupling between absorption and diffusion phenomena, the signal measured (radiance) by a camera L_T is written as:

$$L_T = T_g(t_m \cdot L_{refl} + L_{atm,m} + L_{envi}) \qquad [1.15]$$

$$\text{with } L_{refl} = t_d \cdot \frac{1}{\pi} \cdot \rho \cdot E_{s,D} \cdot \cos(\theta_s) + \frac{\rho}{\pi}\left[E_{atm,d} + E_c\right] \qquad [1.16]$$

where $E_{s,D}$ is the direct solar irradiance at the top of the atmosphere corrected by the Earth–Sun distance, θ_s is the solar zenith angle and T_g is the total transmission along the downwelling and upwelling paths. The total transmittance on the downwelling trajectory t_d (respectively, upwelling t_m) for a direction of incidence θ_s (respectively, θ_v) is expressed as:

$$t_d(\theta_s) = e^{-\frac{\tau_d}{\mu_s}} + t_{diffus}(\theta_s) \text{ and } t_m(\theta_v) = e^{-\frac{\tau_m(z)}{\mu_v}} + t_{diffus}(z, \theta_v) \quad [1.17]$$

where τ_d (respectively, $\tau_m(z)$) is the total optical depth along the total downwelling (respectively, from the ground to the sensor) path with $\mu_s = \cos\theta_s$ and $\mu_v = \cos\theta_v$.

When the observed scene is flat and homogeneous (same reflectance), the terms of direct, diffuse irradiance and the coupling of equation [1.15] form a geometric series of ρ.s where s is the spherical albedo of the atmosphere. Equation [1.15] becomes:

$$L_T = T_g\left[L_{atm,m} + \frac{E_S.\mu_S}{\pi}\frac{\rho}{1-\rho.s}.t_d(\theta_s).t_m(\theta_v)\right] \quad [1.18]$$

In the presence of a target P of reflectance ρ(P) in a uniform environment of reflectance ρ_{env}, expression [1.15] with [1.17] becomes:

$$L_T = T_g\left[L_{atm,m} + \frac{E_S.\mu_S}{\pi}\frac{t_d}{1-\rho_{env}.s}.\left[\rho(P).e^{-\frac{\tau(z)}{\mu_v}} + \rho_{env}.t_{diffus}(\theta_v)\right]\right] \quad [1.19]$$

Finally, in the case of a target P in a heterogeneous environment, there is no analytical expression. Nevertheless, it is possible to approach the exact solution with a spatial average of reflectances composing the environment, $\langle\rho_{env}\rangle$, weighted by a function that depends on the composition of the atmosphere and the distance from the considered point of the environment to point P. Expression [1.19] is thus written as:

$$L_T = T_g\left[L_{atm,m} + \frac{E_S.\mu_S}{\pi}\frac{t_d}{1-\langle\rho_{env}\rangle.s}.\left[\rho(P).e^{-\frac{\tau(z)}{\mu_v}} + \langle\rho_{env}\rangle.t_{diffus}(\theta_v)\right]\right] \quad [1.20]$$

When the landscape has a three-dimensional shape, the radiative transfer equation is more complex. Nevertheless, there are codes that enable the estimation of the instrument input signal by taking into account such geometry, such as the AMARTIS code [MIE 99].

In these expressions, the effects of absorption and diffusion are dealt with separately. In addition, the total radiance is the product of L_T and the total gaseous transmission on descending and upwelling trajectories.

In the thermal infrared field (8.0–12.0 µm), the general equation is written as:

$$L_T = t_m . L_{emit} + t_m . L_{refl} + L_{atm,m} \qquad [1.21]$$

with $L_{refl} = (1 - \varepsilon) . L_{atm,d}$ the radiance emitted by the atmosphere incident to the ground and then reflected by the surface in the direction of the sensor, $L_{atm,d}$ being the downwelling radiance emitted by the atmosphere, (1-ε) the reflectance of this surface and $L_{emit} = \varepsilon . L_{BB}(T_s)$ the radiance emitted by the surface, of emissivity ε, at temperature T_s.

Figure 1.13 gives an example of the total radiance measured by a sensor situated at the top of the atmosphere observing via the nadir a target composed of green grass under a standard US76 atmosphere with rural aerosols and a visibility of 23 km.

Globally, spectral radiance presents more or less significant discontinuities in part due to the presence of gas in the atmosphere that absorbs radiation. The main absorption gases are water vapor (with absorption bands centered around 0.94, 1.14, 1.38, 1.88, 2.5–3.0 and 5–8 µm as well as a continuum present for the entire spectrum), the oxygen band (0.76 µm), ozone (9.7 µm) and carbon dioxide (2.08 and 4.2 µm). At short wavelengths (<0.6 µm), atmospheric radiance is significant, due to light scattering by molecules and aerosols. In this range, the total radiance is the sum of a reflected radiance and an upwelling atmospheric radiance, with a similar amount. From 1 to 2.5 µm, the main contribution to the total radiance is the reflected radiance. Between 8 and 12 µm, the principal contribution to total radiance is the emitted radiance and, to a lesser degree, the upwelling emitted atmospheric radiance. Between 3 and 5 µm, contributions of the reflected and emitted radiances to the total signal are comparable.

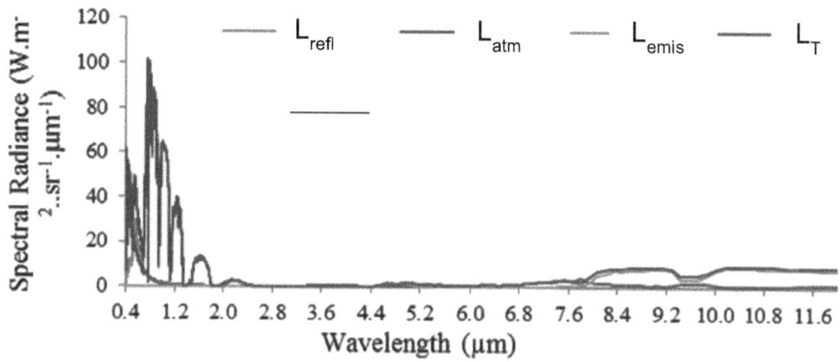

Figure 1.13. *Total radiance L_T, reflected radiance L_{refl}, emitted radiance L_{emit} and upwelling atmospheric radiance $L_{atm,m}$ entering a sensor situated at the top of the atmosphere observing via the nadir a target composed of green grass under a standard US76 atmosphere with rural aerosols with a visibility of 23 km. For a color version of this figure, see www.iste.co.uk/baghdadi/1.zip*

1.3. General presentation of an optical system

In a general manner, an imagery system in the optical field is described by the main elements indicated in Figure 1.14.

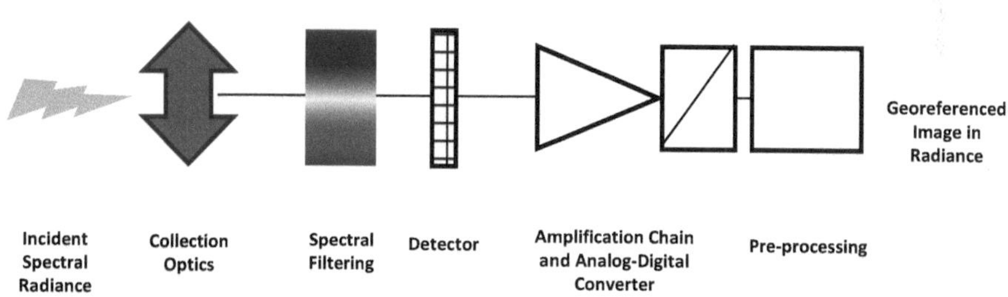

Figure 1.14. *General synoptic of an imagery system*

The signal incident to the system is expressed in a spectral radiance. It is collected by an optical assembly in order to form the image of the landscape in the detector focal plan. Throughout its path in the optical system, the signal is conditioned by spectral filtering in order to select only information in a given spectral range.

The filtered and focused signal recorded by the detector is then converted into an electric quantity. An electronic chain allows for an amplification of the signal and its digitization.

As the instrument is not perfect, the resulting signal is altered by noises and artifacts that must be corrected. The preprocessing step aims to correct the deterministic artifacts linked to the instrument, and then converts the output signal in a georeferenced radiance.

The optic enables us to record the viewed landscape by a detector. Each sensor is characterized by its field of view (FOV) and its instantaneous field of view (IFOV). With this information, we can derive the swath and the spatial resolution of the instrument at ground level once the altitude and the platform are known. Examples of instrumental characteristics are given in Table 1.2.

Instrument	*Spatial resolution at nadir (m)*	*Swath (km)*	*Website*
Pléiades	0.7 (Pan), 2.8 (VISNIR)	20	https://pleiades.cnes.fr/fr/PLEIADES/
Sentinel 2	10 (490, 560, 665, 842 nm), 20 (705, 740, 783, 865, 1,610, 2,190 nm), 60 (443, 945, 1,380 nm)	290	https://sentinel.esa.int/web/sentinel/missions/sentinel-2/
WorldView 3	0.31 (Pan), 1.24 (VISNIR), 0.7 (SWIR)	13.1	https://www.digitalglobe.com/about-us/content-collection/
ASTER	15 (VISNIR), 30 (SWIR), 90 (LWIR)	60	https://asterweb.jpl.nasa.gov/
MODIS	250 (bands 1–2), 500 (bands 3–7), 1,000 (bands 8–36)	2330	http://modis.gsfc.nasa.gov/about/design.php/
Landsat 8	15 (Pan), 30 (VISNIR), 30 (SWIR), 100 (LWIR)	185	http://landsat.usgs.gov/
Spot 6/7	2 (Pan), 6 (VISNIR)	60	https://directory.eoportal.org/web/eoportal/satellite-missions/s/spot-6-7

Table 1.2. *Optical characteristics of some current space remote sensing sensors*

In accordance with the observation system's mission and its sensor characteristics, the observables can be characterized from their physical parameters, such as their geometry, spectral signature or state of polarization.

The use of high spatial resolution instruments (<1 m) gives access to the geometry of the landscape with a high accuracy. Thus, cameras like Pléiades or WorldView 3 have a panchromatic band (wide band) able to sense the geometry of objects with a very high quality.

Multi- or hyperspectral instruments record radiance in several spectral bands, thus providing information on the type of scene being observed and its state. An instrument is called multispectral (respectively, hyperspectral) when the number of bands is reduced, <10 (respectively, >100) with a bandwidth $\Delta\lambda/\lambda$ around 10% (respectively, 1%).

Several optical techniques are used to acquire these spectral images, and several classes can be identified:

– the use of as many cameras as spectral bands: VEGETATION, MISR;

– the use of a filter wheel equipped with several filters: POLDER;

– the separation of flux using a divoli: SPOT;

– the use of matched filters (one per spectral band) configured on the matrix of detectors: Pléiades;

– the use of a grating (Hyperion) or a prism (MERIS);

– the use of a Fourier transformed camera: SYSIPHE.

Examples of instruments are given in Table 1.3. The notion of superspectral system is also discussed. Its characteristics (band number < 20 and spectral resolution around 1–2%) lie between those of multispectral and hyperspectral systems.

Finally, a particular case is that of the POLDER sensor, which, for a given spectral band, has three polarizers of different orientations and can have up to 14 multiangular viewings of the same landscape in order to characterize the aerosols and clouds.

Numerous systems for the observation of the Earth are in operation or will be launched in the coming years. A non-exhaustive summary is given in

Figure 1.15, by mainly focusing on the systems of spatial resolution better than 100 m.

Instrument	Spectral band (μm)
Pléiades	*Panchromatic*: 0.50–0.83 *4 MultiSpectral VISNIR*: 0.45–0.53, 0.51–0.59, 0.62–0.70 and 0.77–0.89
Sentinel 2	***10 MultiSpectral VISNIR***: 0.43–0.45, 0.46–0.52, 0.54–0.58, 0.65–0.68, 0.70–0.71, 0.73–0.75, 0.77–0.79, 0.78–0.90, 0.87–0.88 and 0.93–0.95 **3 SWIR:** 1,370–1,400, 1,570–1,660 and 2,100–2,280
WorldView 3	*Panchromatic*: 0.45–0.80 *8 MultiSpectral VISNIR*: 0.40–0.45, 0.45–0.51, 0.51–0.58, 0.585–0.625, 0.63–0.69, 0.705–0.745, 0.77–0.895 and 0.86–1.04 *8 MultiSpectral SWIR*: 1.195–1.225, 1.55–1.59, 1.64–1.68, 1.71–1.75, 2.145–2.185, 2.185–2.225, 2.235–2.285 and 2.295–2.365
Landsat 8	*9 Multispectral VISNIR -SWIR*: 0.433–0.453, 0.45–0.515, 0.525–0.60, 0.63–0.68, 0.845–0.885, 1.56–1.66, 2.10–2.30, 0.50–0.68 and 1.36–1.39 *2 MultiSpectral LWIR*: 10.6–11.2 and 11.5–12.5
ASTER	*3 MultiSpectral VISNIR*: 0.52–0.6, 0.63–0.69, 0.76–0.86 *6 MultiSpectral SWIR*: 1.6–1.7, 2.145–2.185, 2.185–2.225, 2.235–2.285, 2.295–2.365 and 2.36–2.43 *5 MultiSpectral LWIR*: 8.125–8.475, 8.475–8.825, 8.925–9.275, 10.25–10.95 and 10.95–11.65
MODIS	*20 Multispectral VISNIR -SWIR*: 0.62–0.67, 0.841–0.876, 0.459–0.479, 0.545–0.565, 1.230–1.25, 1.628–1.652, 2.105–2.155, 0.405–0.420, 0.438–0.448, 0.483–0.493, 0.526–0.536, 0.546–0.556, 0.662–0.672, 0.673–0.683, 0.743–0.753, 0.862–0.877, B17: 0.890–0.920, 0.931–0.941, 0.915–0.965 nm and 1.360–1.390 *16 Multispectral MWIR-LWIR*: 3.66–3.84, 3.929–3.989, 3.929–3.989, 4.020–4.080, 4.433–4.498, 4.482–4.549, 6.535–6.895, 7.175–7.475, 8.400–8.700, 9.580–9.880, 10.780–11.280, 11.770–12.270, 13.185–13.485, 13.485–13.785, 13.785–14.085 and 14.085–14.385
EnMap	*Spectral Field*: 0.42–2.45 with a spectral resolution of 6.5 nm in the VNIR, and of 10 nm in the SWIR
Spot 6/7	*Panchromatic*: 0.455–0.745 *4 Multispectral*: 0.455–0.525, 0.530–0.590, 0.625–0.695 and 0.760–0.890

Table 1.3. *Spectral characteristics of space sensors for Earth observation*

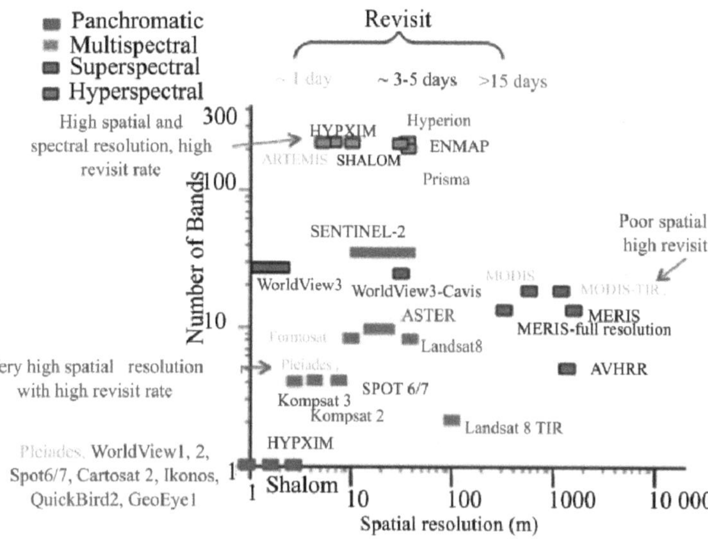

Figure 1.15. *Examples of Earth observation missions that are currently in operation or planned for the future (with the kind authorization of TOTAL-ONERA). For a color version of this figure,* see www.iste.co.uk/baghdadi/1.zip

After spectral filtering, the incident photons are focused on the detector, which converts this radiant energy into an electric signal. There are two main groups of detectors:

– thermal detectors are based on an elevation of temperature induced by the incident photon energy;

– quantum detectors are based on an interaction between light and atoms in the sensor.

Among thermal detectors, we find microbolometers (used, for example, in the IASI mission) and pyroelectric detectors (for example, the SCARAB mission). For quantum detectors or photodetectors, the most widely used technology is based on photovoltaic detectors, such as Charged Coupled Devices in the visible range (for example, the SPOT mission) or HgCdTe (mercury cadmium celluride) and InSb (Indium antimonide) detectors in the infrared range. Each detector technology has a specific spectral sensitivity

(Table 1.4). Photodetectors generally need to be cooled down in order to reduce thermal noise.

Semiconductor	*Range of spectral sensitivity (μm)*
Silicon	0.4–1.0
Mercury cadmium telluride (MCT and HgCdTe)	0.8–25.0
Indium antimonide (InSb)	1.0–5.5

Table 1.4. *Spectral sensitivity range of photodetector technology*

Finally, detectors can also be distinguished by the number and size of photosites available either on a line (for example, the linear detector of 12,000 photosites with the size of 6.5 × 6.5 μm^2 for SPOT 5) or on a matrix (for example, 242 × 548 photosites with a size of 27 × 16 μm^2 for POLDER).

The amplification chain and analog/digital converter aim to format the signal and express it in a quantification step unit.

In order to link the signal X in quantification step units with radiance L_T, an analytical model has to be defined. It depends on the optical configuration of the camera, the spectral sensitivity of the optics, the filters and detectors, the analog/digital conversion, the characteristics of the detector and the amplification stage. In most existing systems, this relationship is considered as linear in the form: X = A.L+b, with A being the absolute calibration coefficient (quantification step/(W.m^{-2}, sr_{-1}, μm^{-1})) and b being a non-centered noise introduced by the instrument. In the visible field, b is the dark current. In the thermal field, this bias is also due to the thermal base emitted by the instrument.

When the detector is composed of several sensitive elements (pixels), coefficient A depends on the position of the pixel, each pixel having its own sensitivity. Thus, in the presence of a spatially uniform landscape in the field of the instrument, the output signal is tainted by a random noise (linked to the electronics) and a fixed spatial noise linked to the different sensitivity of the pixels.

Thus, before using a remote sensing image, some preprocessings are required to overcome the instrument artifact. They are usually organized in several steps which are detailed in the following.

The first level corresponds to the rough image without any processing; this product is not generally distributed by agencies.

The following levels aim to generate an image corrected from radiometric, spectral and geometric defaults introduced by the instrument, and then express this image in a geographic reference. This correction relies on the use of an instrumental model describing these defaults and characterized by their calibration coefficients.

Radiometric calibration aims to convert transmitted signals expressed in the quantification step into a physical quantity, or radiance; this also includes the correction of the signal from instrument artifacts. Conversion into a radiance unit is essential in order to analyze the phenomena present in the Earth–atmosphere system. It is first carried out in a laboratory and is based on a standard source such as an integrated sphere. For instruments in orbit, it is necessary to regularly perform updates for the conversion coefficient as some temporal drift of sensitivity can occur. The methods are based on either the use of an on-board standard (black body, lamp, etc.) or on acquisitions from specific sites (desert, sea, snow, moon, etc.). A presentation of these different methods can be found on the site http://calvalportal.ceos.org/ or in Fourest *et al.* [FOU 12].

The geometric calibration aims to eliminate geometric deformations of the image induced by the optical system. As an illustration, the images acquired by an airborne instrument undergo spatial deformations caused by movements of the plane due to lateral wind or turbulence. The use of stabilizing platforms allows for the reduction of these distortions. Nevertheless, the signature of these distortions on the image depends on the plane's type of movement: pitch, roll and yaw (Figure 1.16). Rolls result in shifted acquisition lines. Pitch motions and yaw result in over- or undersampling portions of the landscape.

Finally, in the presence of a relief, errors in georeferencing appear if the working hypothesis is meant for flat ground (Figure 1.17). Thus, for a viewing angle of 10° (respectively, 20°), and a Δh altitude difference of

10 m (respectively, 50 m) with respect to the average height of the scene, the positioning error dg is 1.8 m (respectively, 18.2 m).

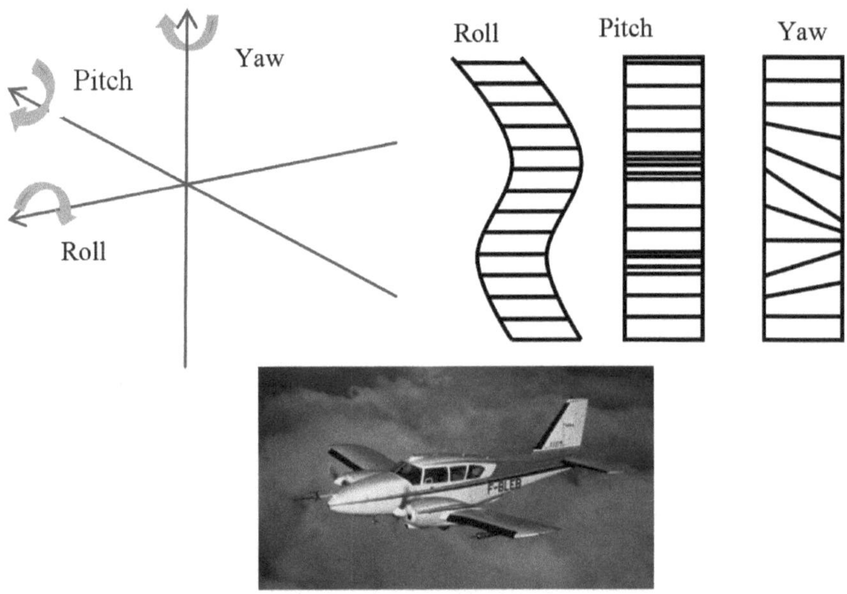

Figure 1.16. *Illustration of the effects of pitch, roll and yaw of a plane (Piper-Aztec, UMS Safire, France) on an image*

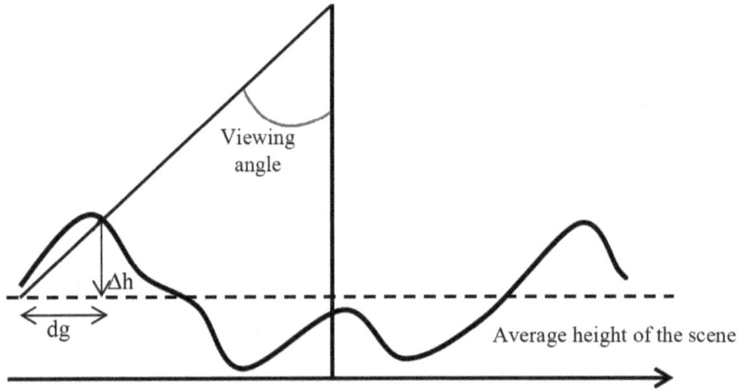

Figure 1.17. *Illustration of a georeferencing error (dg) in the presence of a relief*

These geometric distortions are compensated for in the geometric correction phase by parametric methods, the use of polynomial mathematical models or hybrid models. A detailed description of these methods can be found in Fourest *et al.* [FOR 12].

Spectral calibration is the process that allows us to determine the right wavelength for each spectral band. When this is not the case, significant errors (oscillations) will appear on the Sun's spectrum during the atmospheric correction phase where gaseous absorptions present rapid spectral variations. This type of calibration is applied for hyperspectral instruments. Different methods exist either in the laboratory, based on the standard spectral source, or in flight, using, for example, sharp absorption gas (O_2 absorption hand, for example, at 763 nm [DEL 07]).

1.4. Inversion methods used to retrieve optical properties and the surface temperature

Based on the georeferenced radiance defined at the entrance of the sensor and according to its spectral configuration, it is possible to estimate the optical properties of the surface (reflectance and emissivity) and/or the surface temperature. The estimated quantity depends on the sensor's spectral sensitivity. Thus, there are two main groups of methods depending upon whether one is working in the reflective range or the emissive range. In order to simplify the presentation of these methods, the inversion methods developed in 3.0–5.0 µm where emission and reflection phenomena are comparable will not be considered.

1.4.1. *Inversion methods in the reflective field (0.4–2.5 µm)*

In the 0.4–2.5 µm range, the phenomena to be studied are based on analyzing the reflectance for the Earth's surface or the absorption/scattering for the atmosphere. Also, the estimation of these quantities requires us to eliminate radiative effects due to the observable environment: this step is called atmospheric compensation.

Following the spectral richness of the measurement, different methods have been developed, which are summarized in Table 1.5.

Method type	*Method name*	*Type of instrument*	*Estimated quantity*
Empirical	Normalization of the image with the average image radiance	Several bands <4	Relative reflectance
	Empirical line	Several bands <4	Absolute reflectance
	Dark surface with the presence of clouds	Several bands <4	Absolute reflectance
Based on radiative transfer	Normalization using on particular areas in the image	Several bands <4	Absolute reflectance
	Dark point	Several bands <4	Absolute reflectance
	Inversion of the radiative transfer equation	Multi- to hyperspectral	Absolute reflectance

Table 1.5. *Main methods for estimating reflectance*

1.4.1.1. *Empirical methods*

The first method, developed by Kruse [KRU 88], consists of calculating the average radiance spectrum on the image and then normalizing the radiance of each pixel by this average. The image obtained is equivalent to a spectrum of equivalent reflectance when the scene observed is arid soil. Such a method cannot be applied in the presence of vegetation. An adaptation of this method was proposed by Roberts *et al.* [ROB 86], which assumes that within the image there is a zone that is spectrally uniform. An average radiance is thus calculated over this area and used to normalize the image pixels in order to obtain an image of equivalent reflectance.

These two similar methods do not retrieve real surface reflectance. They do, however, provide information on the global spectral behavior of the materials present. However, certain characteristic absorption bands of the materials present can be eliminated by this normalization. The main advantages of these methods are simplicity and the fact that they do not require ground measurements.

By neglecting the Earth–atmosphere coupling radiative component, the radiance at the sensor level is linearly linked to the surface reflectance: $\rho = a'.L+b'$. Coefficients a′ and b′ can thus be estimated based on ground measurements of reflectance, which are carried out simultaneously with image acquisitions: two targets are used one with a high reflectance level and the other one with a low reflectance level. This empirical line method was developed by Conel *et al.* [CON 87]. While simple in principle, this method is very sensitive to variations in atmospheric conditions between terrain measurements and acquisitions, which will induce artifacts linked to atmospheric absorption bands on the spectrum found.

In the presence of dark surfaces, the difference in radiance between two points composed of the same material, for which one is situated in the shadow of a cloud and the other in the sunlight, allows for the elimination of an additional component diffused in the atmosphere and leading to an estimation of the direct solar irradiance. In normalizing the pixel value by this quantity, an estimation of surface reflectance is deduced [REI 98].

None of these methods require the understanding of absolute instrument calibration, but they do require that atmospheric conditions are stable during the measurements.

These methods are mainly used for multispectral data.

1.4.1.2. *Methods based on radiative transfer*

The first group of methods was developed in the context of atmospheric corrections for images acquired by multispectral instruments, such as Landsat, SPOT, WorldView2, Quickbird, etc.

This first group has the objective of overcoming the atmosphere change between data acquired at different times, for change detection applications. This explains why such methods do not estimate the true reflectance of the surface.

They are based on the normalization defined by a linear law between the reference image acquired at the moment t_0 and the image to be corrected, acquired at the moment t. The PIF (Pseudo-Invariant Features) method selects zones made up of artificial materials (cement, asphalt and bare soil) of invariant reflectance over time and the RCS (Radiometric Control Sets) methods selects dark and bright zones without vegetation.

A second group of methods (DOS, COST and 6S) relies on an estimation of absolute reflectance. These methods assume that the signal to be processed is expressed in a radiance unit.

The DOS (Dark Object Substraction, [CHA 96]) method is an atmospheric correction based on the image. It assumes that there is an area of null reflectance in the image. Thus, the radiance measured by the instrument over these zones of zero reflectance is the upwelling atmospheric radiance $L_{atm,m}$. The reflectance is thus deduced by the following simplified relationship which neglects the other radiative terms: $\rho = \frac{\pi.(L_T - L_{atm,m})}{E_i}$. The incident irradiance, E_i, is the direct solar exoatmospheric irradiance ponderated by the transmittance or knowledge of the atmosphere and a radiative transfer code.

The COST method combines the DOS hypothesis with the assumption that there are some pixels with nearly black zones to estimate the environment component. The corresponding radiance is written as $L_{1\%} = 0.01.\frac{E_{TOA}.\cos\theta_s}{\pi}$ where E_{TOA} is the irradiance at the top of the atmosphere corrected by the Earth–Sun distance, and where θ_s is the solar zenith angle. Thus, the reflectance is written as: $\rho = \frac{\pi.(L_T - L_{atm,m} - L_{1\%})}{E_{s,D}}$. A second method relies on the resolution of the radiative transfer equation issued from the proposed formulation by the radiative transfer code 6S: $\rho = \frac{A.L_T + B}{1 + C(A.L_T + B)}$. Coefficients A, B and C are estimated by 6S by fixing the atmospheric conditions. These methods are used for multispectral images. The RCS, COST and 6S methods produce more stable results than the PIF method.

Finally, the SMAC (Simplified Method for Atmospheric Correction, [RAH 94]) tool is a simplified tool whose radiative parameters which are necessary for atmospheric corrections, have been adapted for a large number of multispectral instruments such as Landsat, RapidEye, Quickbird, Worldview2, Ikonos, Pléiades1A, SEVIRI and DMC-DEIMOS1. Its main advantage is the short processing time, which allows for the processing of a large volume of data. It is based on a semi-empirical model whose coefficients depend on the spectral band.

Vermote *et al.* [VER 02] proposed a new method of atmospheric compensation for the instrument MODIS that would allow for the correction of image contamination from cirrus, from the BRDF and from adjacency effects. The starting equation, considering a Lambertian surface, is equation [1.18], which is modified in order to better take into account the effects of absorption/diffusion coupling. Thus, by dividing this equation by $E_s.\mu_s/\pi$, and separating the diffuse contributions due to molecular diffusion ρ_R and aerosol diffusion ρ_A, the resulting reflectance at the top of the atmosphere becomes:

$$\rho_T = T_g(O_3, O_2, CO_2, M)\left[\rho_R + (\rho_{atm,m} - \rho_R).T_g\left(\frac{U_{H_2O}}{2}, M\right) + \frac{\rho}{1-\rho.s}.t_d(\theta_s).t_m(\theta_v).T_g(U_{H_2O}, M)\right] \quad [1.22]$$

with $M = \frac{1}{\mu_s} + \frac{1}{\mu_v}$.

Here, in order to take into consideration the absorption diffusion coupling, the transmission due to gaseous absorption is processed differently according to the radiative term considered. The different radiative parameters in this equation are estimated based on the transfer code 6S, the input data (surface pressure, ozone and water content) from the National Center for Environmental Prediction (NCEP) or come directly from MODIS, such as the water vapor content or optical depth of aerosols.

Before the inversion of this equation, a test for the presence of cirrus is carried out with the 1.38 μm channel (in this band, the signal issued from the ground is almost completely absorbed, in addition, the measured signal only comes from the potential presence of cirrus). Above a certain threshold, the presence of cirrus is assessed and the following correction is applied in all channels:

$$\rho_T = \rho_T - \rho_{T-1.38}/T_g(U_{H_2O}, 1.38\mu m) \quad [1.23]$$

More recently, the Multisensor Atmospheric Correction and Cloud Screening (MACCS) processor [HAG 15a] was developed in order to generate images in surface reflectance after masking clouds and making atmospheric corrections for multispectral instruments with decametric resolution such as Formosat-2, Landsat 5, 7 and 8, Sentinel-2, Venμs or Spot 4 and 5. The estimation of surface reflectance is based on equation [1.18]. It combines multitemporal and multispectral acquisitions in order to detect clouds and their shadows and estimate the aerosol content and the quantity of water vapor. This method is adapted for missions whose observations are made under near-constant angles. First, the water vapor content is estimated for Sentinel-2 (respectively, Venµs) based on the channel at 940 nm (respectively, 910 nm). For the other satellites, meteorological data are used. Based on this information, the gaseous transmission is computed. Next, based on a temporal acquisition serie a composite image is updated with the last cloud-free pixels acquired. This image serves as a reference for the processing of cloud detection and the estimation of the optical depth of aerosols. Several criteria are used for the detection of clouds [HAG 10]: a channel at 1,380 nm to detect cirrus (Landsat 8 and Sentinel 2), a multitemporal test to detect a sudden augmentation of the reflectance in the blue band, the presence of a cloud and correlation rates for the proximity of a pixel with preceding images in order to reduce all overdetection. Tests are then carried out in order to detect the presence of shadows, water and clouds. The estimation of optical depth combines several criteria [HAG 15b]:

– a multitemporal criterion where the considered geographic zone must have the same reflectance for both the composite image and the image under process;

– a multispectral criterion (the reflectance in blue must be half of the reflectance in red);

– the optical depth must be positive and less than the value obtained by the dark pixel method.

Based on these evaluations, the surface reflectance is estimated from precalculated tables generated by the Successive Orders of Scattering (SOS) transfer code [LEN 07].

The resulting reflectance can be used to update the composite image. In a final step, the composite image is corrected for environmental effects and the effects of slope in the presence of a relief.

With the advent of hyperspectral imagers (such as Hyperion, EnMap and Sysiphe), new methods have been introduced using the spectral richness of acquisitions.

These methods rely on the following radiative transfer equation, assuming that the surface is flat with a Lambertian behavior:

$$\rho = \frac{\pi.(L_T - L_{atm,m} - L_{envi})}{(E_{s,D} + E_{atm,d} + E_c)} \qquad [1.24]$$

Thus, to estimate the surface reflectance based on the radiance measurement L_T, it is necessary to know all the radiative terms of this equation. Different tools have been developed with this objective.

With ATREM code [GAO 93]. This resolution is accomplished in two steps, the first aiming to estimate the radiative atmospheric components and the second estimating reflectance.

The estimation of radiative components depends on the composition of the atmosphere, the effects of which are primarily due to the presence of water vapor (absorption) and aerosols (primarily diffusion). It relies on the use of a radiative transfer code based on a band model. To estimate the integrated water vapor content, the method uses the spectral richness of the acquired image. The principle relies on the relationship of two radiances, one situated at the maximum absorption of water vapor (0.94 and/or 1.14 μm) and the other situated on this absorption band. For a given atmospheric model and altitude sensor, this relationship is proportional to the integrated water content. For aerosols, the user must fix the types of aerosol and their abundance. Such a method is well adapted to flat ground. Adaptations were made in order to improve the estimation accuracy for radiative terms in the absorption bands by using a line by line transfer code [GAO 97] combined with the database HITRAN2000 [ROT 03, ROT 05].

The ACORN code [KRU 04], based on the MODTRAN-4 transfer code [BER 89], provides improvements for the estimation of reflectance of terrestrial surfaces in H_2O absorption bands where water vapor and liquid waterbands overlap in vegetation. In addition, this tool offers several options to remove some artifacts: automatic corrections of eventual spectral shift, removal of bands that are too noisy and filtering of residual errors present in the reflectance after atmospheric correction. Nevertheless, these tools disregard the following radiative components L_{envi} and E_c.

The FLAASH code [ADL 99], similar to the ACORN code, has the main advantage of taking into account radiative components linked to the environment. The starting equation used to estimate the reflectance is drawn from equation [1.19] and is written in the following form:

$$L_T = \left(\frac{A_1.\rho}{1-\rho_{env}.S}\right) + \left(\frac{B_1.\rho_e}{1-\rho_{env}.S}\right) + L_{atm,m}$$

where coefficients A_l and B_l depend on atmospheric and geometric conditions, but not on the surface. A_l, B_l and $L_{atm,m}$ are estimated with the help of the radiative transfer code MODTRAN and a knowledge of the atmosphere. This is defined either from *in situ* measurements of atmospheric characterization or using specific spectral bands in and outside an absorption band of water vapor (here centered at 1.130 µm). The resolution method is then iterative in initially assuming that the scene is uniform ($\rho_{env} = \rho$). The value found is then used as the average environmental reflectance. It can also be noted that the Cochise code [MIE 05] allows us to take into account a non-uniform environment.

In the presence of non-flat ground, the radiative components are no longer directly accessible through classic radiative transfer codes such as MODTRAN. Indeed (Figure 1.18), the radiative terms are modified according to those that we can estimate for flat ground. Direct irradiance has a different solar incidence resulting from the slope of the considered point (the solar zenith angle is defined to the normal vector of the surface). The diffuse irradiance is no longer estimated over $2.\pi$ sr but with a sky solid angle delineated by the relief surrounding the considered point. The coupling irradiance depends on these two irradiances and as a result also changes. Finally, a new irradiance term, E_{refl_envi}, is introduced justified by the reflection of the radiative signal coming from the surrounding pixels and incident to the considered pixel is introduced.

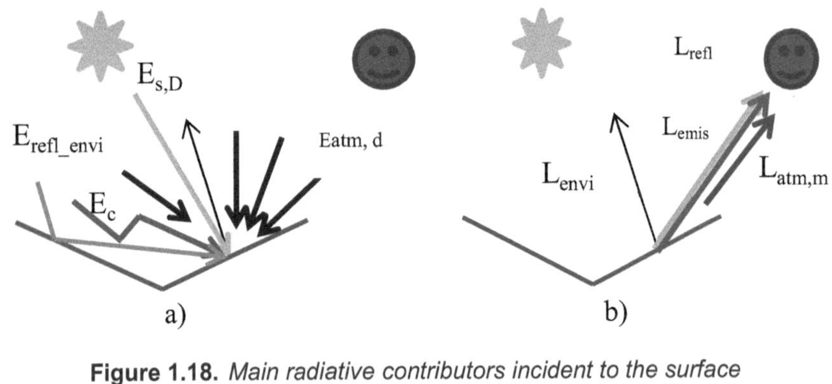

Figure 1.18. *Main radiative contributors incident to the surface in the presence of a relief a) and at the sensor level b)*

New correction codes such as ATCOR [RIC 98, RIC 02] and SIERRA [ACH 09] have been developed in order to take into consideration

Earth–atmosphere coupling effects, as well as to take relief into account through corrections for estimated radiative terms based on MODTRAN. Finally, atmospheric corrections carried out by ATCOR have new features that correct images in the presence of shadows, fog and cirrus [RIC 15].

In conclusion, atmospheric correction tools are now sufficiently mature to be operationally implemented in a ground segment of multispectral imagers (MACCS, [HAG 15a]) or hyperspectral imagers that are either airborne (Eufar: http://www. eufar.net/) or spatial (Enmap, http://www.enmap.org/).

The perspectives are supported on several axes:

– automatic preprocessing of images before atmospheric correction: detection and correction of the presence of fog and clouds;

– consideration of 3D relief for processing acquired images in a mountainous or urban environment;

– operational utilization of meteorological data;

– exploitation of multitemporal data.

1.4.2. *Inversion methods in the emissive range (8.0–12.0 μm)*

In the emissive domain, two quantities are accessible through remote sensing that characterize the state of a surface: the surface temperature T_s and the spectral emissivity of the surface ε. Continuing with equation [1.21], we notice that the total radiance is highly dependent on atmospheric conditions due to the presence of gas (ozone and continuum of water vapor) and the temperature of the atmosphere. Assuming the radiative atmospheric quantities are known, we have at our disposal only one measurement for two unknowns: ε and T_s. The problem is underdetermined.

Several inversion methods have been developed. They depend on the type of instrument carrying out the acquisitions and the type of hypothesis made to solve the problem. Nevertheless, they assume that all acquisitions are carried out in clear-sky conditions. A general view of these methods is described in Table 1.6. This table is non-exhaustive but presents the main classes of method used.

Method Type	*Method Name*
Monoband instrument: estimated quantity T_b	
Instrument with several bands <4: estimated quantity T_s	
Split Window	Split Window
Multi to hyperspectral instrument: estimated quantites ε *and* T_s	
Relative	Spectral relationship
	TISI
	Relative method of α residue
With *a priori* hypothesis on the unknowns	Reference band
	Regression of black body
	Maximum spectral temperature
	Emissivity separation temperature
	Absolute α residual method
Regularization	Gray body emissivity
	Spectral smoothing
	Decomposition of Fourier series

Table 1.6. *Main methods for the estimation of emissivity (ε) and surface temperature (T_S) or brightness temperature (T_b)*

1.4.2.1. *Empirical methods*

Historically, the first instruments had poor spectral resolution and only the surface temperature was researched.

The most popular method is the *split window* method [WAN 96]. This method allows the atmospheric impact to be overcom by using the differential absorption between two channels in the thermal infrared, around 10 and 12 μm. It expresses the temperature at ground level as the weighted sum of brightness temperatures measured in the channels, assuming its emissivity is equal to the unit. In addition, with the exception of scenes such as water, ice or snow, emissivity is not known and the temperature measured is instead a brightness temperature. Improvements have been made in order to take this optical property of the ground into account: spectral ratio method [WAT 92], α-residual method [KEA 93], NDVI method [VAL 96, VAN 93] and estimation based on a classification [SNY 98]. For example, Van de Griend and Owe [VAN 93] proposed an approach that estimates the emissivity based on the vegetation index NDVI: $\varepsilon_0 = a + b.\ln(\text{NDVI})$. The *split window* relationship becomes:

$$T_s = T_i + A_0(T_i + T_j) - A_1 + A_2(1 - \varepsilon) - A_3.\Delta\varepsilon \qquad [1.25]$$

where coefficients A_i depend on atmospheric conditions, ε is the average value of emissivity for channels i and j, Δε is the spectral variation of emissivity and T_i and T_j are brightness temperatures of channels i and j for the same viewing angle.

The following methods assume that the atmospheric correction is carried out previously in order to then estimate the emissivity and surface temperature. To this end, the attenuation terms for the atmosphere on the upwelling path t_m and for upwelling atmospheric radiance $L_{atm,m}$ (see equation [1.21]) are estimated from meteorological data and a radiative transfer code. From these parameters, the Bottom of Atmosphere radiance, L_{BOA}, becomes:

$$L_{BOA} = \varepsilon.L_{BB}(T_s) + (1 - \varepsilon).L_{atm,d} \qquad [1.26]$$

The relative methods do not allow us to retrieve the surface temperature, and the emissivities are defined according to a reference channel.

The spectral ratio method proposed by Watson [WAT 92] relies on the following principle. In writing, the radiance ratio between channels i and j: $\frac{L_i}{L_j} = \frac{\varepsilon_i}{\varepsilon_j}.\frac{L_{CN,i}(T_s)}{L_{CN,j}(T_s)}$, Watson shows that the ratio of black body radiances, $\frac{L_{BB,i}(T_s)}{L_{BB,j}(T_s)}$, only slightly depends on the surface temperature, thus the ratio $\frac{L_i}{L_j}$ only depends on the emissivity ratio. The brightness temperature is estimated in the channel with the highest radiance, and the emissivity ratio can be deduced.

The α-residual relative method by Barducci and Pippi [BAR 96] is as follows. Let α_i be the coefficients defined based on radiance emitted at the ground in band i: $\alpha_i = \lambda_i . \ln L_i - \frac{1}{N}\sum_{j=1}^{N}\lambda_j . \ln L_j + K_i$, where K_i is a wavelength-dependent constant. In expressing the radiance $L_i = \varepsilon_i . L_{BB,i}$ with Wien's approximation of Planck's law : $L_{BB,i}(T) = \frac{C_1}{\lambda_i^5} . e^{-\frac{C_2}{\lambda_i . T}}$ (where C_1 and C_2 are the coefficients of Planck's law), justified in the thermal infrared field for temperatures far less than 1000 K, we obtain: $\alpha_i = \lambda_i . \ln \varepsilon_i - \frac{1}{N}\sum_{i=1}^{N}\lambda_i . \ln \varepsilon_i$.

Thus, the coefficients α_i can be expressed either as a function of the measured radiance or as a function of the emissivities of coefficient Ki, which depends on the characteristics of band i. By using these independent N-1 equations to find the unknown N, we can deduce the ratio of emissivities $\frac{\varepsilon_i^{\lambda_i}}{\varepsilon_j^{\lambda_j}} = e^{[(\alpha_i - \alpha_j) - (K_i - K_j)]}$.

An improvement was proposed by Kealy and Hook [KEA 93] to obtain absolute emissivities based on the empirical function linking emissivities to the variance of coefficients α. Thus, based on the preceding expression of α_i, emissivities are expressed as $\varepsilon_i = e^{\frac{\alpha_i + X - K_i}{\lambda_i}}$ with $X = \frac{1}{N}\sum_{j=1}^{N}\lambda_j . \ln\varepsilon_j$. X is calculated from the empirical function $X = c.(\sigma_a^2)^{1/J}$ where σ_a is the variance of N coefficients α_i and coefficients c and J are obtained through regression along a base of prior measurements.

1.4.2.2. *Methods with a priori hypotheses*

Methods with *a priori* hypotheses on the unknowns are the most commonly used. Indetermination decreases in adding an *a priori* hypothesis on the unknowns. By isolating the emissivity in equation [1.26], we obtain:

$$\varepsilon = \frac{L_{BOA} - L_{atm,d}}{L_{BB}(T) - L_{atm,d}} \text{ where } T = L_{BB}^{-1}\left(\frac{L_{BOA} - L_{atm,d}}{\varepsilon} + L_{atm,d}\right). \quad [1.27]$$

From these equations, several resolution methods are proposed.

To increase indetermination, the reference band method [BAR 96] assumes that the emissivity is *a priori* known in a band. Thus, it is possible to deduce the surface temperature in this band and then estimate the spectral emissivities in the other bands.

The black body regression method assumes that the maximum value ε_{max} of emissivity and the spectral interval in which one finds the corresponding wavelength are known. In addition, two temperature values T_{min} and T_{max} are fixed in order to limit the surface temperature estimation range. Using $T = \frac{T_{min}+T_{max}}{2}$, the spectral emissivity is calculated with equation [1.27] and then compared to ε_{max}. Through successive iterations, the optimal values are those for which the estimated emissivity at the optimal wavelength is equal to the previously defined maximum emissivity.

The method of maximum spectral temperature is similar to the preceding method, but assumes that the emissivity is constant over the spectral interval considered. Thus, the optical surface temperature corresponds to the maximum estimated temperature in the bands. The retained value is then used to estimate spectral emissivity.

For these methods, the emissivity values defined *a priori* come from a database such as the ASTER database.

Gillespie *et al.* [GIL 98] proposed the temperature emissivity separation method (TES method). This method relies on an empirical function linking the minimum value of emissivity with the contrast of the emissivity spectrum. It is carried out in three steps: NEM, spectral ratio and MMD.

The first phase of the algorithm uses the maximum spectral temperature method (or standardized emissivity method NEM) to estimate the surface temperature, T_{NEM}, based on an *a priori* given value of maximum emissivity. Next, the module of spectral ratio calculates $\beta_\lambda = \frac{\varepsilon_\lambda}{\bar{\varepsilon}}$ where $\bar{\varepsilon}$ is the average of emissivities over all the spectral bands. The resulting spectrum thus preserves the form of the spectrum but not its amplitude, an adjustment is also made by the module MMD (Minimum – Maximum Difference) defined by the empirical function: $\varepsilon_{min} = a - 0.687.MMD^b$ with $MMD = \max(\beta) - \min(\beta)$. Parameters a and b are estimated based on a database of spectrums and the instrumental configuration. Once ε_{min} is known, the spectral emissivities in different bands are obtained with the function $\varepsilon_\lambda = \beta_\lambda.\frac{\varepsilon_{min}}{\min(\beta)}$ then the maximum emissivity ε_{max} and lastly the final surface temperature T_{TES} based on module NEM.

1.4.2.3. *Regularization methods*

Regularization methods provide the hypothesis that the spectral behavior of emissivities has slower spectral variations than an atmospheric spectrum. These methods are adapted to data acquired by the imagers with a high number of spectral bands.

The gray body emissivity method [BAR 96] was developed for geosciences in which the objects of study (minerals) present a regularity above 10 µm. From the BOA radiance equation, the cost function (J) aims to minimize the spread between the measurement L_i and the radiance emitted by a body of emissivity ε_i and the surface temperature T_s: $J = \sum_{i=1}^{N}[\langle L_i\rangle - \varepsilon(\lambda_i).L_{BB}(\lambda_i, T)]^2$. This expression disregards the radiative term due to the reflection of incident atmospheric radiance on the surface. In addition, a hypothesis of emissivity regularity is made by limiting M<N to the number of possible values of spectral emissivity.

In the same manner, Borel proposed the *Spectral smoothness* method [BOR 98]. As variations of the atmospheric spectrum are more rapid, they are visible in an incorrect estimation of surface temperature. Thus, the constraint of regularity is fixed on the estimation of spectral emissivity, whose value is optimal for minimal variations of its spectrum. Thus, the optimal temperature is obtained by minimizing the quadratic error $\left\|\mathrm{fit}(\varepsilon(T, \Delta\lambda)) - \varepsilon(T, \Delta\lambda)\right\|$, where the term fit() is the emissivity smoothed on the interval $\Delta\lambda$. From the resulting optimal emissivity, the temperature is deduced. An alternative proposed by Kanani *et al.* [KAN 07] is to find the optimal value from the derivative of the cost function. These methods are adapted to hyperspectral measurements of spectral resolution above 16 cm^{-1}.

1.4.2.4. *Multitemporal methods*

Multitemporal methods exploit acquisitions of the same zone at different moments by assuming that the emissivity remains unchanged.

Becker and Li [BEC 90] proposed the TISI method. In approximating the black body law using Slater's law ($L_{CN,i}(T) = b_i.T^{n_i}$, where coefficients b_i and n_i are estimated based on the black body law as a function of the spectral field and the range of temperature covered, equation [1.26] is written as:

$$L_i = C_i.\varepsilon_i.b_i.T^{n_i} \text{ with } C_i = \left[1 + \frac{1-\varepsilon_i}{\varepsilon_i}.\frac{L_{atm,d,i}}{L_{BB,i}(T)}\right] \qquad [1.28]$$

The index, TISI, is defined as follows:

$$\mathrm{TISI} = \mathrm{M}.\prod_{i=1}^{N} L_i^{a_i} \text{ where } \mathrm{M} = \prod_{i=1}^{N} b_i^{-a_i} \quad [1.29]$$

and in introducing the closure condition, $\sum_i a_i.n_i = 0$, to estimate the coefficients a_i in order to eliminate the system temperature, the TISI becomes:

$$\mathrm{TISI} = \mathrm{C}.\prod_{i=1}^{N} \varepsilon_i^{a_i} \text{ where } \mathrm{C} = \prod_{i=1}^{N} C_i^{a_i} \quad [1.30]$$

Using $\mathrm{TISIE} = \prod_{i=1}^{N} \varepsilon_i^{a_i}$, we obtain a relationship between a term of measurement (TISI) and a term of emissivity:

$$\mathrm{TISI} = \mathrm{C.TISIE} \quad [1.31]$$

If the term$(1 - \varepsilon_i).L_{atm,d,i}$ is negligible in comparison with $L_{CN,i}$, then C is close to 1, and an approximation of C is obtained by choosing for the surface temperature the highest brilliance temperature among the infrared bands to estimate $L_{CN,i}$, and by using a radiative transfer code to estimate the downwelling atmospheric radiance.

1.5. Key points

– In remote sensing, the physics of measurement allow for the estimation of spectral optical properties specific to the materials that make up a scene and its physical state.

– We can distinguish two principal radiative phenomena in the optical range at an ambient temperature: the 0.4–2.5 μm range, which is dominated by reflexion process, and the 8.0–12.0 μm range, dominated by thermal emission processes.

– The current instruments, after radiometric and geometric corrections, deliver radiance at the input of the instrument.

In order to estimate the optical properties of terrestrial surfaces, it is necessary to eliminate the radiative components due to the atmosphere. In order to do so, atmospheric correction methods adapted to the configuration of different types of instruments have been developed.

1.6. Bibliography

[ACH 09] ACHARD V., LENOT X., POUTIER L., "SIERRA: a new approach to atmospheric and topographic corrections for hyperspectral imagery", *Remote Sensing of Environment*, vol. 113, pp. 1664–1677, 2009.

[ADL 99] ADLER–GOLDEN S.M., MATTHEW M.W., BERNSTEIN L.S. *et al.*, "Atmospheric correction for short-wave spectral imagery based on MODTRAN4", in GREEN R.O. (ed.), *Summaries of the Eighth JPL Airborne Earth Science Workshop JPL Publication*, Pasadena, California, Jet Propul. Lab., vol. 99–17, pp. 21–29, 1999.

[ALM 91] ALMEIDA G.A., KOEPKE P., SHETTLE E.P., *Atmospheric Aerosols: Global Climatology and Radiative Characteristics*, A Deepak Pub., p. 561, 1991.

[BAR 96] BARDUCCI A., PIPPI Y., "Temperature and emissivity retrieval from remotely sensed images using the "grey body emissivity" method", *IEEE Transactions on Geoscience and Remote Sensing*, vol. 34, no. 3, pp. 681–695, 1996.

[BEC 90] BECKER F., LI Z.L., "Temperature-independent spectral indices in thermal infrared bands", *Remote Sensing of Environment*, vol. 32, pp. 17–33, 1990.

[BER 89] BERK A., BERNSTEIN L.S., ROBERTSON D.C., MODTRAN: a moderate resolution model for LOWTRAN7, Final report, GL-TR-89-0122, AFGL, Hanscom, p. 42, 1989.

[BOR 98] BOREL C.C., "Surface emissivity and temperature retrieval for a hyperspectral sensor", *Proceedings of IEEE, Conference on Geoscience and Remote Sensing*, IGARSS'98, Seattle, pp. 504–509, 1998.

[CHA 96] CHAVEZ P.S. Jr., "Image-based atmospheric corrections. revisited and improved", *Photogrammetric Engineering & Remote Sensing*, vol. 62, no. 9, pp. 1025–1036, 1996.

[CLO 86] CLOUGH S.A., KNEIZYS F.X., SHETTLE E.P. *et al.*, "Atmospheric radiance and transmittance: FASCOD2", *Extended Abstracts of Sixth Conference on Atmospheric Radiation*, American Meteorological Society, Williamsburg, USA, pp.141–144, 1986.

[CON 87] CONEL J.E., GREEN R.O., VANE G. *et al.*, "AIS-2 radiometry and a comparison of methods for the recovery of ground reflectance", in VANE G. (ed.), *Proceedings of the 3rd Airborne Imaging Spectrometer Data Analysis Workshop JPL Publ.*, Pasadena, CA, Jet Propulsion Laboratory, vol. 87–30, pp. 18–47, 1987.

[DEL 07] DELWART S., PREUSKER R., BOURG L. *et al.*, "MERIS in-flight spectral calibration", *International Journal of Remote Sensing*, Special Issue: MERIS (Medium Resolution Imaging Spectrometer), vol. 28, no. 3–4, pp. 479–496, 2007.

[DUB 02] DUBOVIK O., HOLBEN B., ECK T.F. *et al.*, "Variability of absorption and optical properties of key aerosol types observed in worldwide locations", *Journal of Atmospheric Sciences*, vol. 59, pp. 590–608, 2002.

[FAB 15] FABRE S., BRIOTTET X., LESAIGNOUX A., "Estimation of soil moisture content on spectral reflectance of bare soils in the 0.4 -2.5 µm domain", *Sensors*, vol. 15, pp. 3262–3281, 2015.

[FOU 12] FOUREST S., BRIOTTET X., LIER P. *et al.*, *Satellite Imagery: From Acquisition Principles to Processing of Optical Images for Observing the Earth*, Cepadues Editions, p. 492, 2012.

[GAO 93] GAO B.C., HEIDEBRECHT K.B., GOETZ A.F.H., "Derivation of scaled surface reflectances from AVIRIS data", *Remote Sensing of Environment*, vol. 44, pp. 165−178, 1993.

[GAO 97] GAO B.C., DAVIS C.O., "Development of a line-by-line-based atmosphere removal algorithm for airborne and spaceborne imaging spectrometers", in DESCOUR M.R., SHEN S.S. (eds), *Proc. SPIE, Imaging Spectrometry III*, vol. 3118, pp. 132–141, 1997.

[GIL 98] GILLESPIE A.R., ROKUGAWA S., MATSUNAGA T. *et al.* , "A temperature and emissivity separation algorithm for advanced spaceborne thermal emission and reflection radiometer (aster) images", *IEEE Transactions on Geoscience and Remote Sensing*, vol. 36, no. 4, pp. 1113–1126, 1998.

[HAG 10] HAGOLLE O., HUC M., VILLA PASCUAL D. *et al.*, "A multi-temporal method for cloud detection, applied to FORMOSAT-2, VENµS, LANDSAT and SENTINEL-2 images", *Remote Sensing of Environment,* vol. 114, no. 8, pp. 1747–1755, 2010.

[HAG 15a] HAGOLLE O., HUC M., "Correction Atmosphérique des séries temporelles à haute résolution", *Atelier Tratorria*, Lille, 23–25 March 2015.

[HAG 15b] HAGOLLE O., HUC M., VILLA PASCUAL D. *et al.*, "A multi-temporal and multi-spectral method to estimate aerosol optical thickness over land, for the atmospheric correction of FormoSat-2, LandSat, VENµS and Sentinel-2 Images" *Remote Sensing*, vol. 7, no. 3, pp. 2668–2691, 2015.

[KAN 07] KANANI K., POUTIER L., NERRY F. *et al.*, "Directional effects consideration to improve out-doors emissivity retrieval in the 3–13 μM domain", *Optics Express*, vol. 15, no. 19, pp. 12464–12482, 2007.

[KEA 93] KEALY P.S., HOOK S.J., "Separating temperature and emissivity in thermal infrared multispectral scanner data: implications for recovering land surface temperatures", *IEEE Transactions on Geoscience and Remote Sensing*, vol. 31, no. 6, pp. 1155–1164, 1993.

[KOE 97] KOEPKE P., HESS M., SCHULT I. *et al.*, Global aerosol dataset, Report N 243, Max-Plank-Institut für Meteorologie, Hamburg, p. 44, September 1997.

[KRU 88] KRUSE F.A., "Use of airborne imaging spectrometer data to map minerals associated with hydrothermally altered rocks in the northern Grapevine Mountains, Nevada and California", *Remote Sensing of Environment*, vol. 24, pp. 31–51, 1988.

[KRU 04] KRUSE F.A., "Comparison of ATREM, ACORN, and FLAASH atmospheric corrections using low-altitude AVIRIS data of Boulder, CO", *Summaries of 13th JPL Airborne Geoscience Workshop*, Jet Propulsion Lab, Pasadena, CA, 2004.

[LEN 93] Lenoble J., *Atmospheric Radiative Transfer*, A. Deepak Publishing, Hampton, VA, p. 532, 1993.

[LEN 07] LENOBLE J., HERMAN M., DEUZÉ J.L. *et al.*, "A successive order of scattering code for solving the vector equation of transfer in the Earth's atmosphere with aerosols", *Journal of Quantitative Spectroscopy & Radiative Transfer*, vol. 107, pp. 479–507, 2007.

[MIE 99] MIESCH C., BRIOTTET X., KERR Y. *et al.*, "Monte Carlo approach for solving the radiative transfer equation over mountainous and heterogeneous areas", *Applied Optics,* vol. 38, no. 36, pp. 7419–7430, 1999.

[MIE 05] MIESCH C., POUTIER L., ACHARD V. *et al.*, "Direct and inverse radiative transfer solutions for visible and near-infrared hyperspectral imagery", *IEEE TGARS*, vol. 43, no. 7, pp. 1552–1562, 2005.

[MIE 08] MIE G., "Beiträge zur Optik trüber Medien, speziell kolloidaler Metallösungen", *Annalen der Physik*, Leipzig, vol. 25, pp. 377–445, 1908.

[NIC 77] NICODEMUS F.E., RICHMOND J.C., HSIA J.J. *et al.*, "Geometrical considerations and nomenclature for reflectance", *Library of Congress*, Catalog number: 77-22966, 1977.

[RAH 94] RAHMAN H., DEDIEU G., "SMAC: a simplified method for the atmospheric correction of satellite measurements in the solar spectrum", *International Journal of Remote Sensing*, vol. 15, no. 1, pp. 123–143, 1994.

[REI 98] REINERSMAN P.N., CARDER K.L., CHEN R.F., "Satellite-sensor calibration verification with the cloud-shadow method", *Applied Optics*, vol. 37, pp. 5541–5549, 1998.

[RIC 98] RICHTER R., "Correction of satellite imagery over mountainous terrain", *Applied Optics*, vol. 37, pp. 4004–4015, 1998.

[RIC 02] RICHTER R., SCHLAEPFER D., "Geo-atmospheric processing of airborne imaging spectrometry data, Part 2: atmospheric/topographic correction", *International Journal of Remote Sensing*, vol. 23, no. 13, pp. 2631–2649. 2002.

[RIC 15] RICHTER R., SCHLAPFER D., Atmospheric/topographic correction for Satellite Imagery, ATCOR-2/3 User Guide, Version 9.0.0, 2015.

[ROB 86] ROBERTS D.A., YAMAGUCHI Y., LYON R., "Comparison of various techniques for calibration of AIS data", in VANE G., GOETZ A.F.H. (eds), *Proceedings of the 2nd Airborne Imaging Spectrometer Data Analysis Workshop JPL Publication*, pp. 21–30 (SEE N87-12968 04-43), 1986.

[ROT 03] ROTHMAN L.S., BARBE A., BENNER D.C. *et al.*, "The HITRAN molecular spectroscopic database: edition of 2000 including updates through 2001", *Journal of Quantitative Spectroscopy and Radiative Transfer*, Jet Propulsion Lab, Pasadena, CA, vol. 82, pp. 21–30, 2003.

[ROT 05] ROTHMAN L.S., JACQUEMART D., BARBE A. *et al.*, "The HITRAN 2004 molecular spectroscopic database", *Journal of Quantitative Spectroscopy and Radiative Transfer*, vol. 96, pp. 139–204. 2005.

[SCH 06] SCHAEPMAN-STRUB G., SCHAEPMAN M.E., PAINTER T.H. *et al.*, "Reflectance quantities in optical remote sensing—definitions and case studies", *Remote Sensing of Environment*, vol. 103, pp. 27–42. 2006.

[SNY 98] SNYDER W.C., WAN Z., ZHANG Y. *et al.*, "Classification-based emissivity for land surface temperature measurement from space", *International Journal of Remote Sensing*, vol. 19, pp. 2753–2774, 1998

[VAL 96] VALOR E., CASELLES V., "Mapping land surface emissivity from NDVI: application to European, African and South American areas", *Remote Sensing of Environment*, vol. 57, pp. 167–184, 1996

[VAN 93] VAN DE GRIEND A.A., OWE M. "On the relationship between thermal emissivity and the normalized difference vegetation index for natural surfaces" *International Journal of Remote Sensing,* vol. 14, pp. 1119–1131, 1993.

[VER 02] VERMOTE E.F., EL SALEOUS N.Z., JUSTICE C.O., "Atmospheric correction of MODIS data in the visible to middle infrared: first results", *Remote Sensing of Environment,* vol. 83, pp. 97–111, 2002.

[VON 04] VON SCHÖNERMARK M., GEIGER B., RÖSER H.P. "Reflection properties of vegetation and soil with a BRDF data base", *Wissencchaft und Technik Verl.*, Berlin, 2004.

[WAN 96] WAN Z., DOZIER J., "A generalized split-window algorithm for retrieving land-surface temperature from space", *IEEE Transactions on Geoscience and Remote Sensing*, vol. 34, no. 4, pp. 892–905, 1996.

[WAT 92] WATSON K., "Spectral ratio method for measuring emissivity", *Remote Sensing of Environment*, vol. 42, pp. 113–116, 1992.

2

Multispectral Satellite Image Processing

2.1. Introduction

Having begun in the 1970s with the American Landsat program, the use of optical satellite images for civilian purposes has since experienced significant technological advancements. In the space of 10 years, the images provided by Earth observation satellites have gone from high to very high spatial resolution, from a decametric resolution to a submetric resolution. In addition, the new generation of satellites known as "agile" allow for the reduction of acquisition time above the same area, while providing a synoptic and multispectral view of the territory. Long reserved for scientific use, these data conquer new fields of application and are reaching a much wider audience, competing sometimes with aerial photography.

Optical remote sensing tools are adapting, incorporating rather recent methods developed in other disciplines such as medical imaging, computer vision or statistics. In recent years, constraints due to the increase of the volume of data and the increase in storage volumes involve different areas of Information Technology including data mining or high-performance scientific computing (HPC). Open source software provides users with these powerful algorithms/solutions and therefore participates in the democratization of remote sensing.

However, we must bear in mind that a satellite image is still a complex product. Many factors interact during image acquisition: the atmosphere, the

---

Chapter written by Kenji OSE, Thomas CORPETTI and Laurent DEMAGISTRI.

nature of the surface as well as the satellite attitude and the quality of the sensors. This chapter recalls these principles and presents commonly used methods, from geometric and radiometric correction (section 2.4) up to the extraction of information (section 2.5), passing by the visualization (sections 2.2 and 2.3). While not exhaustive, it gives the reader the knowledge in order to understand optical image processing.

2.2. Concepts and reminders

Many articles and works describe in detail what an image is (in particular [CAL 01]). We will simply recall some of the basics which enable the completion of a remote sensing project: the image seen as a table of data values, the problems of storage and image formats and finally spatial, spectral and temporal resolutions.

In this section, figures will use cropped images from old SPOT 5 HRG products (http://www.geo-airbusds.com/en/23-sample-imagery), consisting of two images of Rome and the surrounding areas, acquired on the 21st of July 2002: a panchromatic image (an image obtained by recording the radiance using one single band for the whole visible spectrum wavelengths) and a multispectral image (an image acquired by recording the radiance using several bands, each one related to a distinctly narrower range of wavelengths).

2.2.1. *Table of values*

An image is a table of pixels ("picture elements"), or a matrix, arranged regularly as L rows of P columns (Figure 2.1). Each pixel contains a numeric value (*Digital Number* or *radiometric value*) which is the mean value of the "energy" returned by the ground surface elements covered by this pixel, after discretization. As we will see later, the definition domain of these values depends on the type of image encoding used. For image display, the numerical value of pixels can be matched with a list of colors.

In Figure 2.2, the pixel values of the image are represented using a scale of gray levels going from black, for a 0 value which means no energy detected at the sensor level, to white, for a 255 value which means saturation of the sensor. Here, 256 values are possible for each pixel of the used images. This can differ for images using other types of encoding (see section 2.2.2.1). A perspective visualization of an image (Figure 2.3) enables us to better assess its informational content, in particular the spatial relations between neighboring pixels according to their Digital Numbers (high or low pixel values). This representation allows us to perceive these image values as a "landform", by analogy to a digital elevation model (which is an image representing the topography of an area), and to gain a better understanding of how core image processing algorithms work: for instance, "thresholding" consists of slicing this landform of radiometric values or peak clippings; "gradient computation" is related to tracking local variations of radiometric values, which can correspond to different types of radiometric relief: for example, plateaus, steep slopes and peaks.

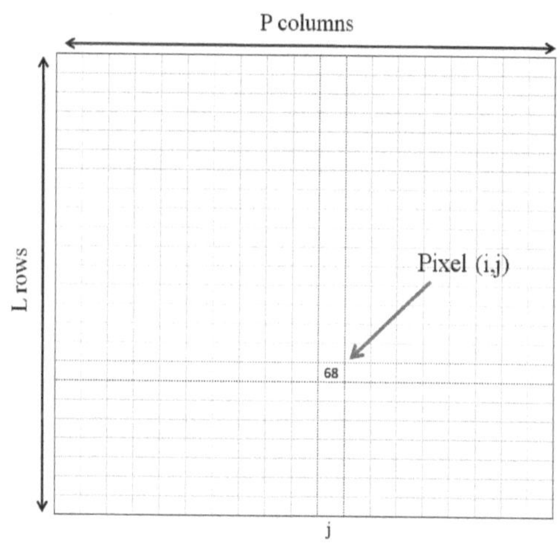

Figure 2.1. *The pixel with the coordinates (i, j) contains the Digital Number 68*

In optical remote sensing, a panchromatic image is a digital record, on one single range of wavelengths overlapping in particular the visual spectrum, of the quantity of electromagnetic radiation measured at pixel level. A panchromatic image is therefore a "single-channel" image.

In the case of a multispectral image, the digital record of the quantity of electromagnetic radiation each pixel receives is performed over several wavelength ranges, often disjointed and narrower in comparison to panchromatic mode. A multispectral image is therefore composed of several channels or bands, each one containing, the amount of radiation measured in very specific wavelength ranges for each pixel (for example, green, red or near infra-red).

Figure 2.2. *SPOT 5 HRG (extract) panchromatic image. The image dynamic has been stretched between the values of 0 and 255 so as to be able to observe the information better. The value of a pixel corresponds to a percentage of gray level between the black (0) and the white (255)*

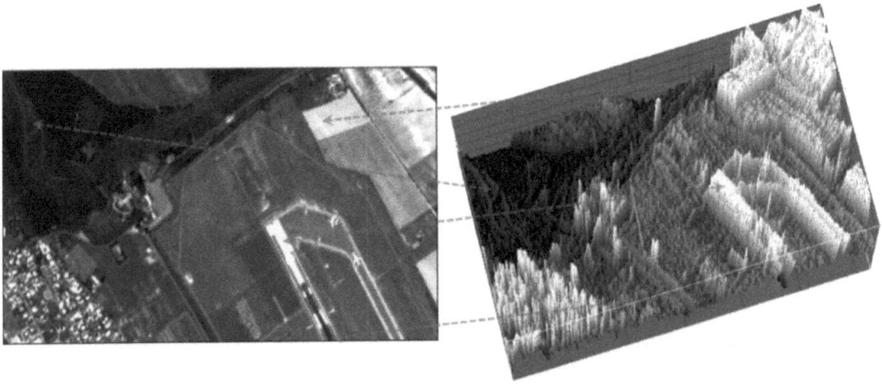

Figure 2.3. *Visualization in perspective of an image gray levels. Intensities are perceived according to the relief: valleys in black, peaks in white*

Image channels or bands are delivered according to the suppliers:

– either in the form of separate “image” files (for example, LANDSAT-TM images) in which case the complete multispectral image is made up of a list of files;

– either stacked into one single file (for example, SPOT-HRG images).

Therefore, a remote sensing image can be seen as a digital three-dimensional matrix: two planimetric dimensions (the rows and columns of the image) and a radiometric dimension (the spectral channels of the image).

In Figure 2.4, the vertical dimension corresponds to the information plans of the different image channels.

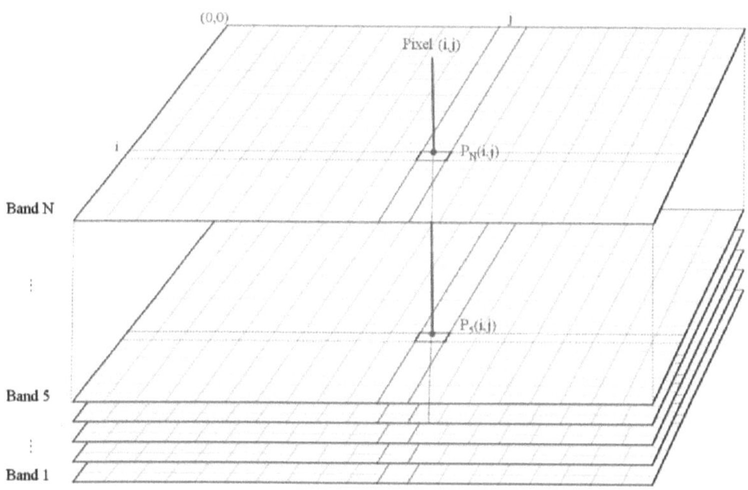

Figure 2.4. *The multispectral image is like a matrix where the information plans that make up the different channels are stacked and superimposed. Pixel P(i,j) is a vector whose components are P1(i,j), P2(i,j), P3(i,j), ... PN(i,j): the pixel values in channels 1 to N*

2.2.2. *Image files storage*

Considerations on the size of remote sensing image files are important in several ways:

– first, when progressing through a project, image files (several acquisitions of the same area, images from various sensors for comparison purposes, etc.) and all intermediary calculation products can occupy a large space in terms of disk usage. Over time, due to the increase in spatial resolution and measurement capacity of new sensors, the size of satellite image files has grown from some tens of megabytes for the first SPOT-HRV or LANDSAT-TM images, to several gigabytes (according to the file format used) for high spatial resolution images (panchromatic band of PLEIADES, or SPOT 6/7 NAOMI images);

– the second point that we consider important to highlight here concerns the software and computer performances with which the images are processed: with files of hundreds of mega-bytes, it can be difficult to open, display or load several images at once. The software or processing algorithms must be adapted, in particular to process images sequentially, block by block, so as not to saturate the computer's RAM. Standard image processing software have evolved and been adapted over recent years. However, these considerations are important, for example, when the use of more specific software is planned, like mathematics or statistics software not directly addressed to image processing; or when processing algorithms need the whole image to be loaded into memory.

This brings us back to the notions of encoding and remote sensing image format.

2.2.2.1. *Encoding of pixels*

The term "encoding" refers to the numeric "type" of variables used to address the pixel values, in the IT sense of the word (bytes, short integers, long, signed, unsigned, float, simple or double precision). For example, it is common to have numerical pixel values between 0 and 255, which correspond to a coding of each pixel value with one byte. During the process of image acquisition, the photographic sensors which measure the information stored afterward in each pixel, first restores an analog signal which must be converted into numerical values. During this analog–digital conversion, there is discretization of the information on a range of numeric values defined during the design of the sensor. The extent of this range of values and the accuracy of measurement desired by the designers set a number of useful values necessary for the storage of these measures. This number of useful values allows us to define the best means of storage: one byte (8 bits) 256 values to be stored (2^8), two bytes (16 bits) 65536 values (2^{16}), etc. To further reduce the weight of image data and disk space used (for example, when data transmission costs are concerned), satellite image providers can decide not to encode pixel values on whole numbers of bytes (1 full byte = 8 bits): 10 bits for NOAA AVHRR images, 11 bits for GEOEYE images, 12 bits for PLEIADES images.

When creating products derived from initial images (creation of masks, calculation of indices, temporary images from processings, etc.), the question of encoding is important as there is the risk of assigning the calculated pixel values to unsuitable computer variable types. For example, vegetation indices calculation produces real numbers ranging from –1 to 1: the simple approach to store these values could be the use of floating point encoding (4 bytes); however, we can also reduce the range of the produced values using linear transforms and then use a lighter encoding type (according to the application context, the loss of precision may be acceptable). Given the performance and storage capacity of current computers, these questions may seem less interesting than they were a few years ago. However, the users of high spatial resolution images available nowadays must bear these questions of encoding in mind when dealing with image files of around 1 Gigabyte.

The greater the number of bytes or bits for pixel value encoding, the bigger the image file will be on the hard drive: the size in bytes, outside of headers is given by (for an uncompressed image):

$$L \times P \times C \times T \qquad [2.1]$$

with:

– L: the number of rows in the image;

– P: the number of columns;

– C: the number of channels;

– T: the number of bytes for encoding.

2.2.2.2. *File formats and metadata*

The size of image files tends to increase as quality improves, which makes their distribution very difficult. Therefore, methods for compressing data are used. The most widespread image formats offer different types of compression, with or without information loss. JPEG2000, which is in itself a compression standard, uses wavelet transforms to reduce the size of images. GeoTIFF, which is often used in an uncompressed way, also encapsulates compression algorithms such as the LZW and JPEG.

For example, a PLEIADES PHR1B panchromatic HiRI image, at 0.50 m ground resolution, composed of 34,000 rows and 29,000 columns, is stored on disk in a file whose size is approximately 1.5 GB in GeoTIFF format (12 bits encoding of the radiometric values and uncompressed option); however, the same image file occupies no more than approximately 500 MB when stored in JPEG2000 format (compression without loss). This compression rate is not constant; it essentially depends on the image and its information contents.

Other image formats can be used, less known but widely spread within specific scientific communities. We can cite as an example the formats used in the field of meteorological data: METEOSAT images, NOAA HRPT file, etc.

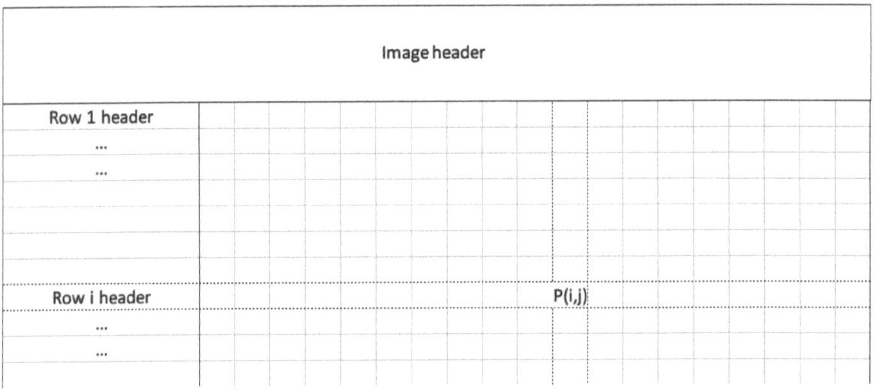

Figure 2.5. *Illustration of the structure of an image*

In a very general way and with a didactic approach, an image can be represented in the following manner (Figure 2.5):

– a header storing aggregate and global information on the whole of the image (date and conditions of acquisition, sensor used, global localization, spatial resolution, etc.);

– optionally a "row header" for each row specific information (location of pixels, calibration parameters by row, etc.);

– the sequence of pixel values composing each row.

For multispectral images, namely files containing multiple channels relating to radiation measurements in different wavelengths, pixel and channel (or band) information can be arranged in three different ways in the file (Figure 2.6):

– BSQ storage ("*Band sequential*"): the channels are stored globally sequentially one after the other.

– BIL storage ("*Band interleaved by line*"): pixel values corresponding to the different channels are stored line by line.

– BIP storage ("*Band interleaved by pixel*"): pixel values of each band are stored sequentially pixel by pixel.

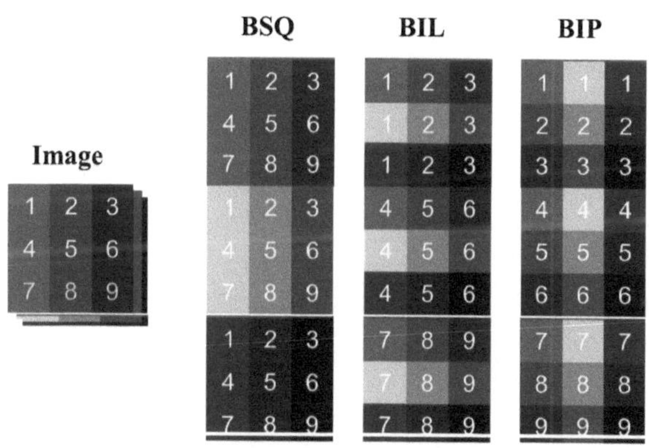

Figure 2.6. *Illustration of the channel layout within the image file*

We notice in the previous descriptions that an image is not only composed of the pixel values. Various descriptive data relating to the acquisition conditions or to the organization of the file are necessary to be able to read and analyze this image. This information on the image data constitutes what is called the metadata of the image.

The amount of metadata differs from one type of image to the other. Some standards have been defined gradually by the various satellite image suppliers but at the moment they are often unique to each of them.

Some metadata can be contained within the image file itself when supported by the format used (for example, storage of information concerning the projection system of the image directly in the "Tags" of the GeoTIFF format): we are talking about "internal" metadata. Other metadata is available through additional files, i.e. "external" metadata: this is the case for DIMAP (XML file) files, for example, for SPOT images or MTL files ("text" file) for LANDSAT images.

Figure 2.7 shows the tree diagram of the information tags contained in a DIMAP SPOT 7 file. As you can see, the XML format involves a hierarchical file structuring with the help of tags relating to different information including radiometric calibration data (Radiometric_DATA) or geometric data (Geometric_DATA), those relating to the encoding or to the processing level, etc.

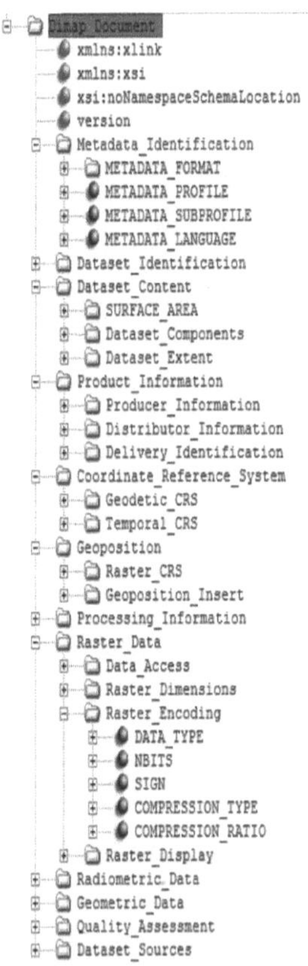

Figure 2.7. *Tree diagram of the tags of a DIMAP SPOT 7 file (metadata file)*

Figure 2.8 presents the output of a command line for accessing the internal metadata contained in a GeoTIFF Landsat 8 file. Here, the image is in the form of separate GeoTIFF channels and the information relates to the channel (panchromatic channel). Among other things here is the image location information (*Corner Coordinates*), the Spatial Reference System (WGS 84 / UTM) as well as information on file encoding.

```
>gdalinfo LC81960302015185LGN00_B8.TIF
Driver: GTiff/GeoTIFF
Files: LC81960302015185LGN00_B8.TIF
Size is 15361, 15621
Coordinate System is:
PROJCS["WGS 84 / UTM zone 31N",
    GEOGCS["WGS 84",
            DATUM["WGS_1984",
            SPHEROID["WGS 84",6378137,298.257223563,
            AUTHORITY["EPSG","7030"]],
            AUTHORITY["EPSG","6326"]],
            PRIMEM["Greenwich",0],
            UNIT["degree",0.0174532925199433],
            AUTHORITY["EPSG","4326"]],
    PROJECTION["Transverse_Mercator"],
    PARAMETER["latitude_of_origin",0],
    PARAMETER["central_meridian",3],
    PARAMETER["scale_factor",0.9996],
    PARAMETER["false_easting",500000],
PARAMETER["false_northing",0],
    UNIT["metre",1,
            AUTHORITY["EPSG","9001"]],
    AUTHORITY["EPSG","32631"]]
Origin = (536992.500000000000000,4899607.500000000000000)
Pixel Size = (15.000000000000000,-15.000000000000000)
Metadata:
    AREA_OR_POINT=Point
Image Structure Metadata:
    INTERLEAVE=BAND
Corner Coordinates:
Upper Left  (  536992.500, 4899607.500) (  3d27'48.03"E, 44d14'55.55"N)
Lower Left  (  536992.500, 4665292.500) (  3d26'51.53"E, 42d 8'19.95"N)
Upper Right (  767407.500, 4899607.500) (  6d20'47.68"E, 44d12' 2.37"N)
Lower Right (  767407.500, 4665292.500) (  6d14' 0.48"E, 42d 5'39.00"N)
Center      (  652200.000, 4782450.000) (  4d52'21.84"E, 43d10'46.18"N)
Band 1 Block=15361x1 Type=UInt16, ColorInterp=Gray
```

Figure 2.8. *Internal metadata of the channel 8 of a Landsat 8 image. Output display of the command line "gdalinfo" from the GDAL Library*

Figure 2.9 shows an example of an external metadata file for a Landsat image 8. Without being XML, the file is also structured with tags that contain information relating to the projection system (SRS), the radiometric calibration and the acquisition processes.

Depending on the processing level in which an image has been produced, all of this metadata (internal or external) allows us to better understand and interpret the content of the image or to go even further in image data processing by integrating, in the processing workflows, parameters delivered by the data provider (for example, radiance conversion of digital numbers, etc.).

```
GROUP = L1_METADATA_FILE
 GROUP = METADATA_FILE_INFO
  ORIGIN = "Image courtesy of the U.S. Geological Survey"
  REQUEST_ID = "0501507048652_00021"
  LANDSAT_SCENE_ID = "LC81960302015185LGN00"
  FILE_DATE = 2015-07-04T17:55:54Z
......
 END_GROUP = METADATA_FILE_INFO
 GROUP = PRODUCT_METADATA
  DATA_TYPE = "L1T"
  ELEVATION_SOURCE = "GLS2000"
  OUTPUT_FORMAT = "GEOTIFF"
  SPACECRAFT_ID = "LANDSAT_8"
  SENSOR_ID = "OLI_TIRS"
  WRS_PATH = 196
  WRS_ROW = 30
......
 END_GROUP = PRODUCT_METADATA
 GROUP = IMAGE_ATTRIBUTES
  CLOUD_COVER = 9.57
  CLOUD_COVER_LAND = 0.27
  IMAGE_QUALITY_OLI = 9
  IMAGE_QUALITY_TIRS = 9
  TIRS_SSM_POSITION_STATUS = "NOMINAL"
  ROLL_ANGLE = -0.001
  SUN_AZIMUTH = 133.78713862
  SUN_ELEVATION = 63.61142041
......
 END_GROUP = IMAGE_ATTRIBUTES
 GROUP = MIN_MAX_PIXEL_VALUE
  QUANTIZE_CAL_MAX_BAND_1 = 65535
  QUANTIZE_CAL_MIN_BAND_1 = 1
......
 END_GROUP = MIN_MAX_PIXEL_VALUE
 GROUP = RADIOMETRIC_RESCALING
  RADIANCE_MULT_BAND_1 = 1.2147E-02
..
  RADIANCE_MULT_BAND_11 = 3.3420E-04
  RADIANCE_ADD_BAND_1 = -60.73679
......
 END_GROUP = RADIOMETRIC_RESCALING
 GROUP = TIRS_THERMAL_CONSTANTS
  K1_CONSTANT_BAND_10 = 774.8853
......
 END_GROUP = TIRS_THERMAL_CONSTANTS
 GROUP = PROJECTION_PARAMETERS
  MAP_PROJECTION = "UTM"
  DATUM = "WGS84"
  ELLIPSOID = "WGS84"
  UTM_ZONE = 31
  GRID_CELL_SIZE_PANCHROMATIC = 15.00
......
 END_GROUP = PROJECTION_PARAMETERS
END_GROUP = L1_METADATA_FILE
END
```

Figure 2.9. *External meta-data file for a Landsat 8 image*

2.2.3. *The four resolutions*

Although the term resolution is used most of the time in reference to the spatial dimension, it is important to remember the four resolutions to consider when working in remote sensing: spatial, spectral, temporal and radiometric resolutions.

2.2.3.1. *Spatial resolution*

Spatial resolution corresponds to the elementary size of the ground surface measured by the different instruments of an embedded sensor. At a particular altitude, an observer (human eye or electronic detector) perceives the ground according to a cone of visibility depicted in Figure 2.10. In this figure, it is clear that the surface area being observed depends strongly on the altitude of the platform: with a similar sensor configuration, at the h^2 altitude (for example, low orbit), the surface observed is smaller than that of altitude h^1 (higher orbit). On a pushbroom sensor (longitudinal along track scanning), the measuring instruments are photographic detectors aligned as a 1 row array perpendicular to the satellite path; at a given moment, t, they measure the information corresponding to the columns at row i. The movement of the satellite along its path allows you to carry out the measurements of the following rows of the image. In the actual sensors, particularly those with very high spatial resolution, the acquisition of any one image row often requires the combination of several detectors arrays so as to obtain the number of pixels required. In other words, during the image acquisition, each sensor has its own visibility cone whose square or rectangular base corresponds to the pixels of the image (Figure 2.11). The minimum dimension (length or width) of the ground projection of a detector via the optical system of the sensor gives the spatial resolution of the system.

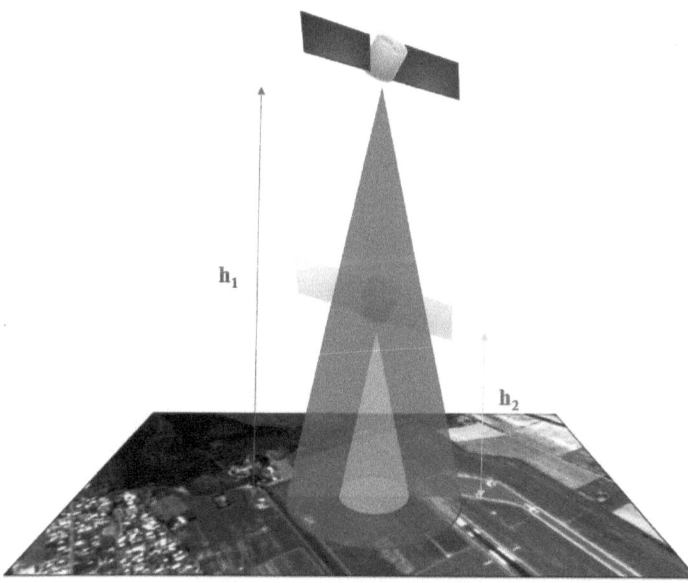

Figure 2.10. *Cone of visibility of a passive sensor embedded at two different altitudes. The height of the platform is important in determining the area of the ground projection*

To be detected, a ground element must have a dimension greater than or equal to the pixel size so that differences from one detector to the next reveal the presence of that element. Note, however, that the measurements obtained by the detectors are the means of energy reflected by all the components of the elementary surface corresponding to the projection of the pixel to the ground. That is why strongly reflective (or very absorbent) elements that are smaller than the image resolution may affect the average radiometry of the pixel that contains them.

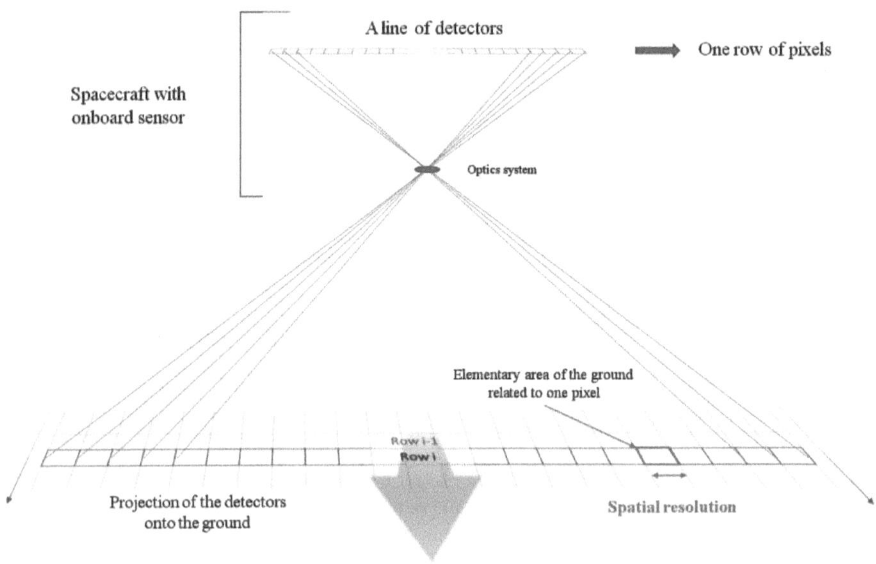

Figure 2.11. *Illustration of the image acquisition process from an array of detectors (pushbroom sensors)*

2.2.3.2. *Spectral resolution*

According to the physical principles of remote sensing, the different elements that constitute our environment (water, snow, bare soils, sand, vegetation, etc.) have specific spectral profiles (reference reflectance curves). This means they reflect or absorb, each in their own way, a certain percentage of light energy that they receive from the sun in the different wavelengths of the electromagnetic spectrum (see Chapter 1). The reference reflectance curves of these different elements are the basis for most of the classification principles; the idea being to try to identify from the images the characteristics of these spectral profiles.

Being able to distinguish different landscape or environments in an image is to be able to pinpoint the radiance differences of these environments to wavelengths for which the spectral profiles differed the most. Therefore, it is not necessarily the number of channels of a multispectral sensor which is important, but rather the spectral bands offered (and the value of their centered wavelength).

Each of the channels of a multispectral sensor contains a measure of the amount of energy which is reflected (or emitted in the case of thermal wavelengths) around a given wavelength: we often talk about the "green" channel, the "red" channel, the "near infrared" channel, etc., it is in fact measures performed in wavelength intervals centered around the "green", the "red" or "near-infrared".

The term "spectral resolution" is often confused with the number of sensor channels since the more channels a sensor has, the more possible it is to draw spectral profiles of specific image components and the more possible it is to identify reference spectral profiles of standard environments or physical elements in an image. However, the primary definition of spectral resolution is more about the width of the "integration" range of the wavelengths around a reference value.

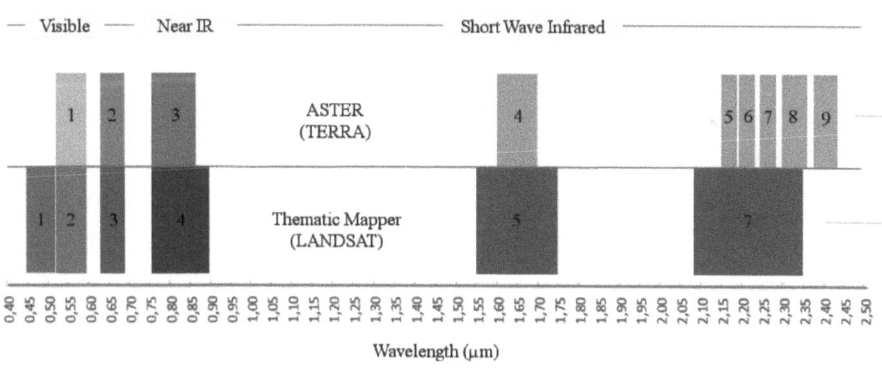

Figure 2.12. *Comparison of LANDSAT TM and TERRA Aster spectral bands. Channels 10 to 14 of Aster (8.125–11.65 μm) and 6 of TM (centered on 11.5 μm) are outside of the range of wavelengths represented*

Figure 2.12 compares the first spectral bands of two sensors: Thematic Mapper from the LANDSAT satellite (channels 1 to 5 and 7) and ASTER (channels 1 to 9) from the TERRA satellite. We note that the near-infrared of ASTER (channel 3) corresponds to a narrower range which is centered on a different value than the near infrared of Thematic Mapper (channel 4). In the short infrared (channel 4 of Aster and 5 of TM), the band width is also narrower. Between 2.05 and 2.35 μm, the two sensors have different designs: TM integrates the wavelengths between 2.08 and 2.35 μm in one single spectral measure (channel 7) where ASTER achieves four (channels 5 to 8).

This simple comparison shows that the relevance of a sensor's spectral resolution can only be assessed from an application point of view: do spectral bands – with their widths and centered values – permit or facilitate (in comparison to other sensors) the distinction of elements or objects of interest in an image?

2.2.3.3. *Temporal resolution*

The temporal resolution is not directly related to a sensor. It corresponds to the frequency of revisits over a site and is therefore dependent on the orbitography of the platform or satellite on which the sensor is hosted. With most of the Earth observation satellites having quasi-polar orbits, this frequency of revisits also depends on the latitude of the area of study. The parameters of an orbit around the Earth (perigee, apogee and tilt), combined with the speed of rotation of the Earth, automatically defines an orbital cycle: the satellite has to pass through a number of orbits (complete revolution around the Earth) before flying over exactly the same area in the same image shooting conditions (16 days for Landsat-8, 26 days for PLEIADES, etc.).

The duration of this orbital cycle or the number of revolutions required prior to returning to the starting point defines this concept of temporal resolution. We speak here of absolute temporal resolution considering the exact same image acquisition conditions (positioning, angle of incidence, etc.). With the advent of "agile" satellites (SPOT 6/7, PLEIADES, GEOEYE, etc.), or with the actual sensors allowing various off-nadir viewing angles, this revisit period can be reduced: it is possible to obtain the image of a same geographical area before the end of the full orbital cycle. The pictured area can be the same but the acquisition conditions are different and the images obtained can present strong disparities. For example, if one tolerates a viewing angle of the sensor of 20°, PLEIADES satellite, it is possible to lower the revisit period to approximately 7 days (5 days for 30° viewing angle).

A strategy used by some data suppliers to further reduce this temporal resolution is to work with a constellation of identical satellites (SPOT 6/7,

PLEIADES 1A/1b, etc.). With the satellite SENTINEL-2A, launched in June 2015, at the equator the revisit period is 10 days, when SENTINEL-2B becomes operational in 2017, this period should go down to 5 days (i.e. 2–3 days for image acquisition at mid-latitudes).

2.2.3.4. *Radiometric resolution*

Radiometric resolution corresponds to the sensitivity of a sensor, i.e. its ability to measure and to enable distinction within the same spectral band of differences in the electromagnetic energy reflected by the elementary ground surfaces.

A sensor with a finer radiometric resolution than another should record more significant intensity levels. Here, we see the link between radiometric resolution and the encoding system of the image that we mentioned above: an encoding with bytes can only store 256 intensity levels, which is often sufficient. Figure 2.13 is extracted from the metadata file delivered with the SPOT 5 data which we used for the previous figures. It is the raw panchromatic channel's histogram of Digital Numbers, before any enhancement of the image dynamics. We notice from these raw values that:

– the encoding extends over 256 levels;

– the values of the Digital Numbers ranged between 20 and 110 and appear centered around the value 45.

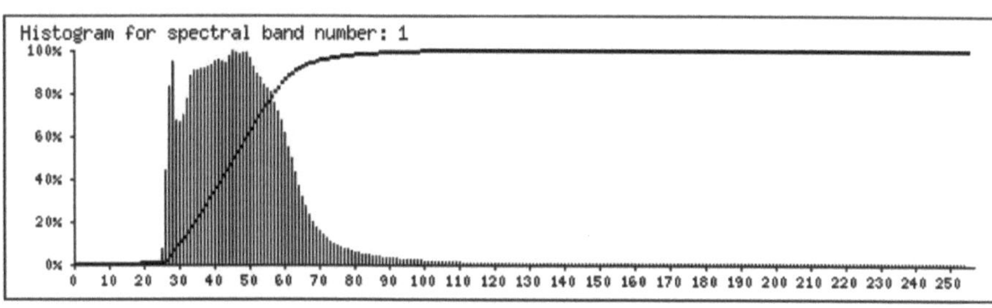

Figure 2.13. *Histogram of the panchromatic channel of a SPOT 5 demonstration image*

For a similar scene, the difference in the radiometric resolution between two sensors is perceived in the width of the original histogram of the spectral band considered; the overall shape of the histogram is similar for the two sensors, but the sensor requiring more radiometric values to record, and therefore which has a finer radiometric resolution, will spread over more values.

2.2.3.5. *Combination of spatial and spectral resolutions: data fusion*

Spatial and spectral resolutions are closely linked; a better spatial resolution automatically leads to a coarser spectral resolution and vice versa. Schematically, while considering the minimum quantity of electromagnetic energy required for a sensor to work, a detector sensitive at only one wavelength (fine spectral resolution) will have to be larger (wider spatial resolution) than a detector sensitive to a wider range of wavelengths (sum of the energy of the whole of the wavelengths of the range). Remote sensing image users would often like to be able to work with images combining the very high spatial resolution of a panchromatic channel with the many bands of a multispectral image. Therefore, data fusion techniques have been developed. The fusion process incorporates spatial information from a panchromatic channel within the different spectral bands of a multispectral image; we are talking here about "*pan-sharpening*" (refining). Among the many techniques that exist, we can cite the *Brovey Transformation* (BT: *Brovey Transform*) which is a commonly used method. The formula for the calculation of the merged channels is given by the following equation [ZHA 02]:

$$Fusion_k = \frac{Multi_k}{Multi_{sum}} \times Pan \qquad [2.2]$$

The spectral bands $Multi_k$ (k is the spectral band index) extends over the spectral wavelength range covered by the panchromatic band Pan, and $Multi_{sum} = \Sigma\ Multi_k$. The $Multi_k$ bands must previously have been resampled to the same spatial resolution as *Pan*.

NOTE.– These methods of fusion are based on the hypothesis that it is possible from multispectral image bands (or from some of them) to simulate a panchromatic channel to the resolution of the multispectral image. The spectral bands of the multispectral channels must therefore cover that of the panchromatic channel and vice versa. This point is a strong assumption which is not always right (for LANDSAT-8 Oli, the panchromatic channel coincides only partially with the blue channel and stops at the red; for LANDSAT-7 ETM+, the panchromatic channel starts from the green and extends beyond the visible until the near infrared, which leads to distortions in the spectral information of the resulting image.

Figure 2.14 shows an extract of a PLEIADES image with the panchromatic band, the multispectral image and the image resulting from the data fusion of the two.

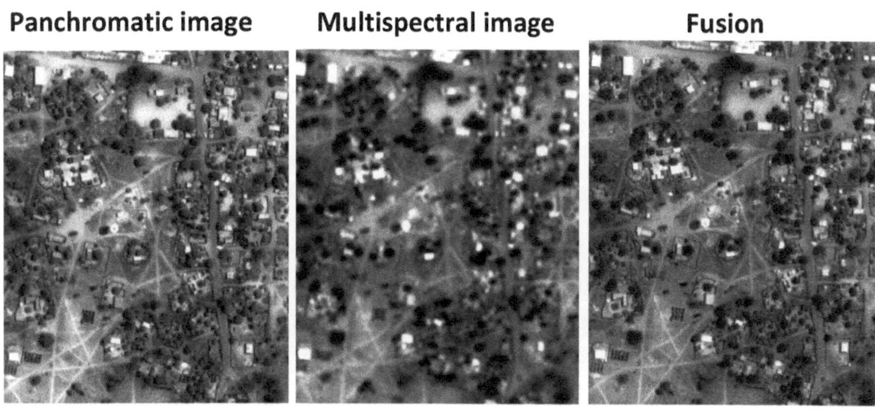

Figure 2.14. *Panchromatic band (0.50 m) of a PLEIADES image on Senegal of 05 June 2014 (left), associated multispectral image (2 m) (center) and data-fusion of the two (right) (© CNES 2014, Airbus distribution D&S, Geosud)*

Other methods exist which operate by substitution, such as where the panchromatic channel replaces the "intensity" channel from the IHS (intensity, hue and saturation) color space. Methods based on wavelets that separate the image information into two low and high frequencies signals are also used to perform the fusion. As the operation is invertible, it is possible to use the high frequency information of a panchromatic channel into the inverse wavelet transform of a multispectral image.

We should note that all of these methods might induce distortions of the fused bands spectral information and thus interfere with thematic analysis such as land use and land cover studies [JAW 13, ZHA 08].

2.3. Viewing of multispectral optical images

An optical satellite image, as well as a photograph, is primarily a visual product subject to interpretation. What do we see on the image? What are the criteria which allow us to identify objects or phenomena? The color? The shape? The texture? The context? All of these questions relating to the photo-interpretation allow us to define *a posteriori* processing and the required algorithms to extract useful information. For example, prior to classifying an image, we design a nomenclature of the landuse/land cover, not only adapted to the data specifications but also to the needs of end users. The visualization is therefore a key step.

At its delivery, the multispectral image is a raw product, often in an unusable state, even for visual interpretation. When an inexperienced user displays the data on the screen, he is often surprised to see nothing. The image is dark and dull, with indistinguishable details. Yet the information is there, in front of us. To make it appear, we simply have to adjust the contrast.

Image-processing software has all the means of visual improvement. In order to understand their principles, we will return briefly to the concept of image in IT. Then, we will discuss the possible improvements in monochromatic and color display.

2.3.1. *Recording of the satellite multispectral image*

A satellite image is composed of multiple channels each corresponding to a spectral band, and in which the pixel value determines the light intensity recorded by the sensor. A spectral band is a matrix of points defined by three dimensions, its coordinates and the intensity relating to the radiance. For practical reasons, radiance, expressed in $W.m^{-2}.sr^{-1}$ (Watt per square meter per steradian), is converted into an integer encoded typically between 8 and 16 bits. The Pleiades satellites sensors, for example, provide four spectral bands encoded in 12 bits, with three bands in the visible domain and one in the near-infrared.

2.3.2. *Monochromatic display and contrast adjustment*

In a spectral band, the range of possible values depends on its encoding. On the screen, the digital values of each pixel are associated by default to gray levels. The minimum corresponds to the black, the maximum to white. Between the two, the gray will go from darkest to lightest. In an 8-bit unsigned encoding, the values extend therefore from 0 to 255, or 256 levels of gray.

To understand the display of the digital values on the screen, we must examine the histogram of the image. This graph associates each digital value (x-axis) to the number of pixels (y-axis) which has this value in the image. The dynamic range of the spectral band is located between the extreme values recorded in the image and only occupies a portion of the available values according to the type of encoding. It then bundles the non-zero elements of the histogram. These are generally found in the lower part, or even the middle, of the x-axis, and this why images appear dark, with less contrast on the display. For example, Figure 2.15 presents the different contrasts applied on a SPOT 6/7 image. The images are delivered in a 16-bit encoding, or 65,536 values. However, the sensors onboard these satellites record the signal on a 12-bit encoding, or 4,096 values, regardless of the spectral band. There are potentially 4-bit untapped. On the histogram, the encoding of a channel is therefore between 0 and 65,535, while the radiometric dynamic will always be between 0 and 4,095.

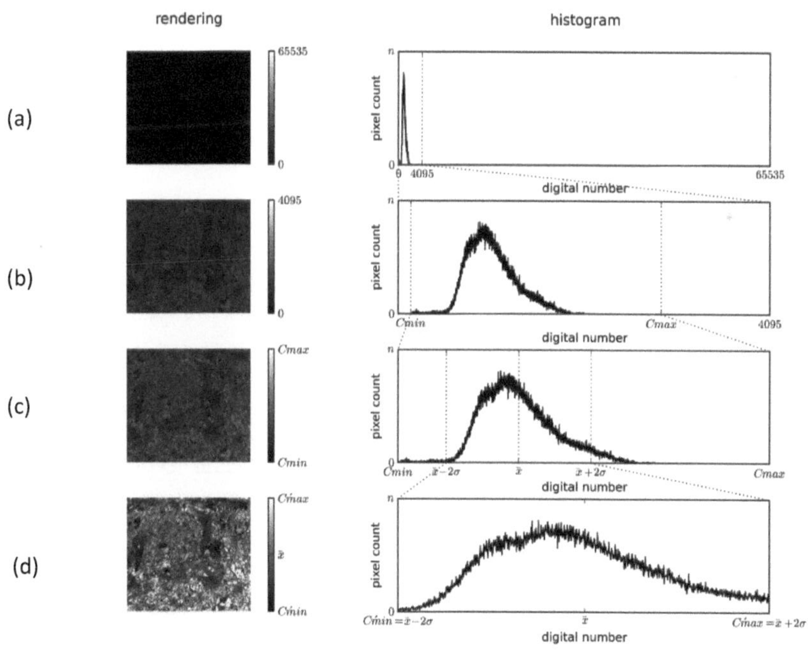

Figure 2.15. *Dynamic range stretching on a SPOT 6 image: a) no stretching of the dynamic, Min and Max bounds correspond to the encoding of the image (16 bit); b) stretching of the dynamic depending on the coding of the sensor (12 bit); c) stretching of the dynamic depending of the min and max values of the image; and d) stretching of the dynamic based on the average and standard deviation of the image. For a color version of this figure, see www.iste.co.uk/baghdadi/1.zip*

As we now understand the difference between the encoding of an image and its dynamic range, we can manipulate the histogram to enhance the contrast of the image.

Contrasting a monochromatic image is consists of spreading the non-zero components of the histogram on the complete authorized dynamic range, that being 256 levels of gray in an 8-bit encoding. We are talking about "stretching of the dynamic". This processing is carried out by means of a table of correspondence or LUT (*Look-Up Table*) in which each non-zero component is associated with a new value for the display. It is carried out "on the fly" and does not change the actual values of the pixels.

Among the techniques of contrast enhancement, the linear transformation, presented in Figure 2.15, is the most commonly used in remote sensing. With D_{min} and D_{max} being the extreme values of the encoding on the screen; C_{min} and C_{max} being extreme values recorded in the channel. The constants of the affine function are calculated as follows (Figure 2.15(b)):

$$a = \frac{D_{max} - D_{min}}{C_{max} - C_{min}} \quad [2.3]$$

$$b = \frac{D_{min}.C_{max} - D_{max}.C_{min}}{C_{max} - C_{min}} \quad [2.4]$$

The correspondence between the digital number DN of the pixel and the DN' value for display is established as follows:

$$DN' = a.DN + b \quad [2.5]$$

As computer screens use 24-bit encoding (RGB, or 3×8-bit), D_{min} and D_{max} correspond, respectively, to the values 0 and 255. The expression can then be simplified:

$$DN' = \frac{255.(DN - C_{min})}{C_{max} - C_{min}} \quad [2.6]$$

This method generally results in better contrast. If, however, the dynamic range of the channel is already at a maximum, often because of a few isolated pixels, the linear transformation does not bring about any great visual improvement. In this case, it is recommended to apply a linear transformation with saturation. The stretching is then applied on a representative part of the dynamic range while the remaining non-zero components are saturated. To do this, it is necessary to determine new bounds C'_{min} and C'_{max}, based on the dominant values ($C_{min} < C'_{min} < C'_{max} < C_{max}$). The latter are assigned to a gray level, while the pixels lower than C'_{min} are saturated in black and those greater than C'_{max} in white.

Bounds C'_{min} and C'_{max} can be entered manually on the histogram or be deduced automatically. Transformation is performed in this way:

$$DN' = \frac{255.(DN - C'_{min})}{C'_{max} - C'_{min}} \text{ where } C'_{min} \leq DN \leq C'_{max}$$

$$DN' = 0 \text{ where } C_{min} \leq DN < C'_{min} \qquad [2.7]$$

$$DN' = 255 \text{ where } C'_{max} < DN \leq C_{max}$$

There are several methods to automatically determine the bounds C'_{min} and C'_{max}. The latter can be calculated, for example, according to the standard deviation σ which measures the dispersion of the pixel values around the average. Closer the bounds C'_{min} and C'_{max} are to the average, the more the image is contrasted and therefore saturated. A 2 σ offset to the average (C'_{min} = average – 2 σ, and C'_{max} = average + 2 σ) generally offers a good compromise.

The use of the standard deviation is effective on uni-modal histograms, similar Gaussian curves. If the image contains contrasting landscapes in terms of radiometry, this method will be less efficient. Indeed, an image with large water surfaces, on the coastline, for example, will have a bimodal histogram in the infrared channel. The first peak corresponds to the water, while the second to the land. The average is likely to be between the two modes and none of the two types of land cover will be properly highlighted. In these conditions, it is preferable to set the min and max bounds manually in order to improve the visualization on certain parts of the image. In case our study focus on agriculture urban or forested areas, bounds will be settled either on either side of the second peak. The water will therefore be saturated in black. However, if we are looking at the turbidity or eutrophication, we favor a contrast on the first peak. The land will therefore be saturated in white.

We have presented here a method of contrast enhancement, linear transformation with or without saturation. There are of course other methods such as the histogram equalization. It is often used when the distribution of pixel values is not uniform. Based on the cumulative histogram, this transformation attempts to give each level of gray the same occurrence of pixels, which levels the histogram. The resulting image is more detailed but can also show noise. Besides, the enhanced parts of the image can sometimes be of poor interest to the user.

Filters can also be applied to images to improve the contrast and reduce/eliminate the noise. The average and median filters are the most used. They replace each pixel value by the average value (or median) of adjacent pixels using a sliding window of Nf × Nf (where Nf is odd: 3, 5, 7 ...). The window is moved along the columns and rows of the image, one pixel at a time, repeating the calculation until the whole image is filtered. By calculating the average around each pixel, a smoothing effect on the image is obtained but also a strong alteration of contours. As for the median filter, contours are preserved by replacing the value of each pixel by the median value of the gray levels of the pixels of the sliding window the sliding window. This last filter allows you to eliminate isolated pixels in fairly homogeneous zones.

2.3.3. *Additive synthesis and color composite images*

We have seen that the visual analysis of a spectral band is already rich in information. Nonetheless, this approach has limitations since only the transformation of the histogram, or quite simply the adjustment of contrast, allows for the differentiation of objects with different reflectance in the same wavelength interval. However, in using the spectral dimension of satellite images, it is possible to make an essential attribute to the photo-interpretation: color. For example, a RapidEye image pixel contains five values of intensity corresponding to certain parts of the electromagnetic spectrum, namely, blue, green, red, red-edge and the near-infrared. The combination of all or some of this information enables us to "color" the displayed image. In remote sensing, we are thus talking about color composite images.

Computer screens are generally based on 32-bit encoding including 24 bits assigned to color restoration. These 24 bits break down into a triplet of 8 bits, each corresponding to the primary colors red, green and blue (RGB) of the additive synthesis. The combination of these three components forms chromatic color. When the three primary colors equal the same value, we obtain gray; black when that value is 0 and white when the value is maximum. If they differ, we obtain all the other colors and shades of the gamut, i.e. subset of colors that a screen allows you to reproduce.

According to this principle, the color composition of satellite images results from the projection of monochromatic spectral bands across three colored filters, red, green and blue. So, for an image with four spectral bands, there are 24 possible arrangements. With the choice of a colored composition being guided by the objects or phenomena studied, the photo-interpreter is used to working with only one or two color composite images. Among the most used, we will present the natural-color composite and the false-color composite (see Figure 2.16).

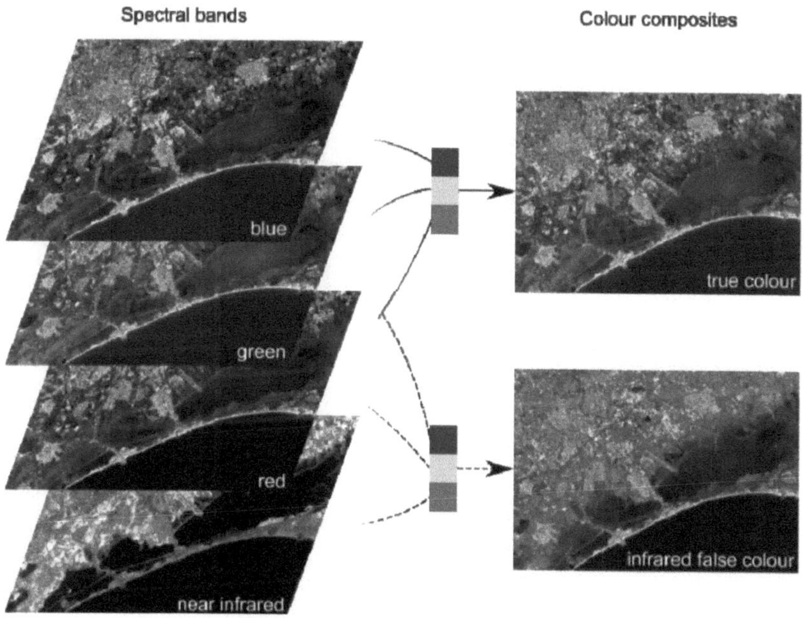

Figure 2.16. *Construction of a color composite images in natural colors and infra-red colors (Spot 6, © Airbus DS 2013, Distribution Geosud). For a color version of this figure, see www.iste.co.uk/baghdadi/1.zip*

The natural-color composite, or "true–color composite", is to assign, respectively, to the filters red, green and blue the spectral bands red, green and blue. In the display, the restitution of colors is very close to what a human could see from space. Vegetation appears in shades of green, water in blue, urban areas in light gray. The creation of such a color composite image thus depends on the characteristics of the satellite sensors. Most of them offer the three primary components of the visible domain. Nevertheless, some, like the SPOT 1 to 5 generation, do not make acquisitions in the blue, therefore rendering it impossible to produce a true-color composite image.

False-color composite images display spectral bands which are different from primary colors. In fact, the color of the objects does not correspond to their "real" color as our eye would perceive it. The false-color is used to highlight certain objects in the image, taking advantage of their spectral signature. Therefore, the composition in infrared false colors is commonly used to study vegetation. It is generally accepted that the infrared, red and green spectral bands are allocated, respectively, to the red, green and blue filters. Vegetation is seen in red since its reflectance is at its maximum in the infrared. The shades vary according to the chlorophyll activity of the species in the environment. To understand this principle, we could change the arrangement of the three spectral bands, for instance by assigning the near infra-red to the green filter. The vegetation would then be colored in green. In a composition in false infrared colors, the water is almost black and does not present any shades because it absorbs most of the electromagnetic radiation. To highlight some of its characteristics, such as turbidity, the visible domain is more appropriate. In conclusion, we must always refer to the spectral profile of studied objects to build an adapted color composition.

2.3.4. *Change detection*

Satellites acquire images regularly and allow the study of the spatial dynamics, and the identification of changes or even anomalies. To compare images acquired in a particular area on different dates, the simplest method is to produce multi-temporal or diachronic color-composite images. In this case, we assign to the filters red, green and blue the same spectral band arising from images acquired on different dates. The resulting image will present colors only in the areas where changes have occurred and gray shades in the invariant areas (Figure 2.17). To do this, it is first necessary to ensure that the images are perfectly super-imposable (with geographical extent and spatial resolution exactly the same) and preferably, radiometrically normalized.

Figure 2.17 presents the extent of the "Rim Fire", a fire which destroyed nearly 100,000 ha of wood in the national forest of Stanislaus (37°51'N 120°05'W, State of California, United States) in 2013. The diachronic colored composition is the synthesis of two near infrared bands acquired before and after the fire by the Landsat 8 sensor Oli. The post-fire image (15 August 2013) is assigned to the red filter, while the pre-fire image (16 September 2013) is assigned to the green and blue filters. The result contains three shades: a) red for the pixels that are passed from a vegetated state to non-vegetated state, b) gray for any invariant pixel between the two dates, and c) cyan for the pixels that have higher values in the green and blue filters than those in the red filter, that is when the reflectance in the near infrared from the post-fire date is higher than that of the pre-fire date.

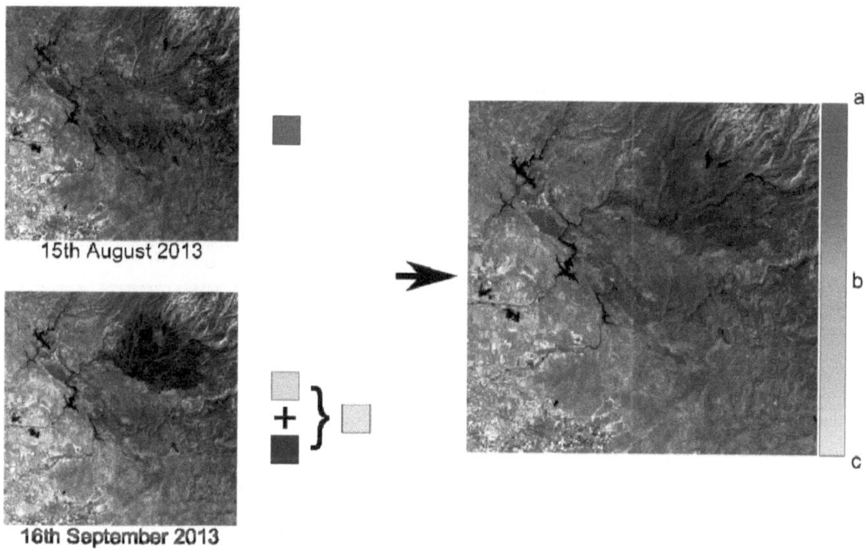

Figure 2.17. *Detection of a burned-out zone on a multi-date colored composition (R = 15 August, V = 16 September, B = 16 September) from Landsat 8 images with 30 m spatial resolution (Distribution U.S. Geological Survey)*

2.4. The preprocessing of multispectral satellite images

Digital image processing often requires prior geometric and radiometric corrections.

2.4.1. *Geometric preprocessing*

2.4.1.1. *Geometric transformations and orthorectification of images*

It is common to work with data from a variety of sources, such as multi-date image data from different sensors and vector data. The analysis of all these data requires that the latter are perfectly superimposable in standardized spatial reference systems (for example, map projection systems). In fact, an image is a projection onto a plane of the observed 3D objects. During the acquisition process, geometric distortions in the image can arise due to different factors, including:

– the attitude of the sensor or the platform (flight path, speed, acceleration, roll, pitch, yaw and viewing angles);

– the lens and dynamic elements activated by the sensor (for example, rotating mirror);

– the Earth (relief, speed of rotation, desired cartographic projection, etc.).

Geometric transformations are in general applied to RAW images. As a result, we obtain an image corrected from distorting effects (or at least from part of them). Two types of models of geometric transformations are used [TOU 04]: on the one hand the physical and rigorous models, and on the other hand, the empirical models.

In the case of physical models, each source of distortion is mathematically modeled and the global geometric transformation needed to correct an image results in a composition of operations (axis and scale transforms, rotations, projections, etc.) corresponding to the various systems involved at each stage of the acquisition process.

Figure 2.11 shows the case of an axial symmetry and Figure 2.18(b) shows a central symmetry detectors of the Polder sensor.

With regard to the empirical models, the most used are based on polynomial functions or rational functions (see sections 2.4.1.2 and 2.4.1.3). In both cases, mathematical equations transform the geographical coordinates of objects into their coordinates within the "image" coordinate system (in that way we speak of indirect models; we speak of direct models for the reverse operation).

Physical models, although more complex and costly in calculation time, have long prevailed. With the arrival of very high spatial resolution data, the empirical models are now widely used [AGU 13].

Figure 2.18 shows the different configurations: the size of the ground surface covered by each detector may vary depending on the viewing angle (or the visibility cone of the detector), altitude (how close the sensor is to the ground), the slope and the aspect of the terrain (directly facing the satellite or away from it). Correcting an image from geometric distortions requires taking into account the combination of all these effects. In particular to take into account the ground elevation, the use of a digital elevation model (DEM) is essential (an image of the topography in which pixel values give the altitude). In this case, it is imperative to know the altimetric accuracy (potential error in altitude) and horizontal accuracy (potential localization error in longitude and latitude) of this DEM; the quality of the geometric corrections depends on how those potential error sources are controlled. Processing an image with a rough digital elevation model whose spatial resolution differs too much from that of the image will give poor results and might generate artifacts in the "corrected" image. For information, the 1 arc second Shuttle Radar Topography Mission (SRTM), a DEM provided by NASA at approximately 30 m of spatial resolution, available on nearly 80% of the land surface, and very often used, is reported to have a planimetric accuracy of 20 m and an absolute altimetric accuracy of 16 m (http://www2.jpl. nasa.gov/srtm/statistics.html).

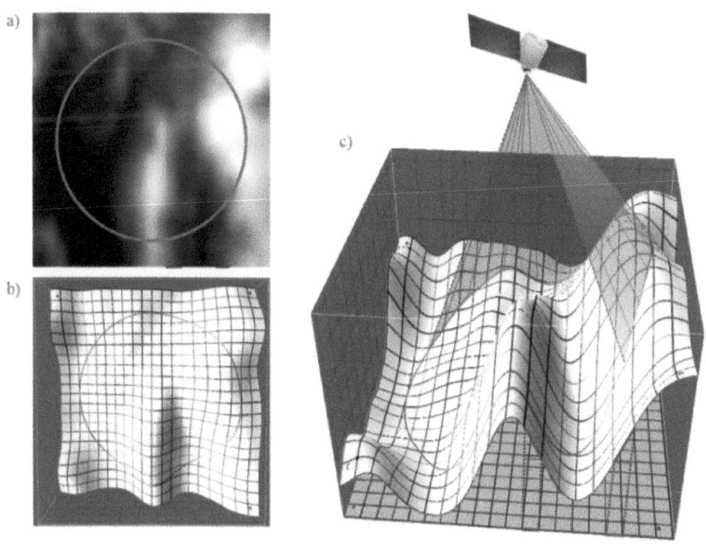

Figure 2.18. *Distortions of a regular grid "draped" on a digital elevation model (DEM). a): DEM, with black and dark gray representing the lower altitudes, and light gray and white, the high altitudes; b) elastic deformations of the grid through a nadir perspective: relief and perspective view distort the squares of the grid; c) three-dimensional representation of the grid deformation: in red, projection of the sides of the grid squares on the terrain. Viewing angle, tilt and aspect of the slope combine and have a direct impact on the area of the ground surface observed by the sensors. For a color version of this figure, see www.iste.co.uk/baghdadi/1.zip*

In terms of the performance of geometric correction models, planimetric accuracies lower than the pixel resolution are often mentioned. However, it is important to bear in mind that the accuracies reported are mean values on a set of control points: a model with a good overall performance does not imply that all the image points are localized with a subpixel precision.

Making an image dataset perfectly superimposable requires the application of geometric corrections, which eliminate viewing angle distortions as much as possible. We talk about orthorectification when an image is transformed so that each image pixel seems to be acquired at nadir (to vertical under the satellite). In Figure 2.18, the orthorectification is the geometric transformation that allows you to display a regular grid after processing (not a distorted one): an image taken by a sensor which would ideally focus each square in the grid perpendicularly to the ground.

Most of the time, geometrical transformations modify the radiometry of images; this is shown in Figure 2.19, where to create an orthoimage which would correspond to the viewing situation illustrated, a choice must be made between the pixel radiometries in the "raw" image to assign radiometric values to the orthorectified pixels: in this figure, four raw pixels correspond to less than three orthorectified pixels. The orthoimage is usually made by using interpolation of the radiometric values of neighboring pixels of the raw image (see section 2.4.1.4).

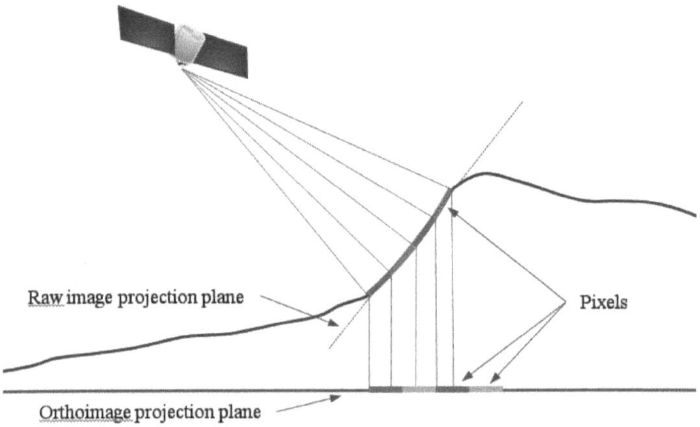

Figure 2.19. *Viewing angle and landforms combined involve modifying radiometric information during orthoimage production. In this case, four pixels of the raw image correspond to less than three pixels of the orthorectified image*

2.4.1.2. *Polynomial transformations*

The polynomial functions are the simplest empirical models of geometric deformations to use. They produce good results on images with low distortions (small viewing angle and smooth landforms):

$$L = \sum_{i=0}^{p} \sum_{j=0}^{q} \sum_{k=0}^{s} l_{ijk} X^i Y^j Z^k \quad C = \sum_{i=0}^{p} \sum_{j=0}^{q} \sum_{k=0}^{s} c_{ijk} X^i Y^j Z^k \qquad [2.8]$$

Equations [2.8] allow you to switch from the geographical or cartographic coordinates (X, Y, Z) to the coordinates in the image reference system (L, C) using l_{ijk} and c_{ijk} coefficients (3D model). When there is no component in altitude Z (s=0), we are talking about 2D models.

The maximum value allowed for the sum of the indices i+j+k is called the "order" of the polynomial function ($l_{ijk}=0$ and $c_{ijk}=0$ for i+j+k > order). Polynomials of a higher order than 3 have no great utility [WHI 04].

For 2D polynomials, order 1 enables correction of translations, rotations, scale changes and perspective distortion; order 2 (Figure 2.20) can also take into account deformations due to twisting and convexity or concavity [TOU 04]. A greater order than 2 is rarely used because it can introduce errors in relation to positioning of pixels [CAL 01]: pixels arranged in a certain order in the original image may change position order in the transformed image.

Polynomial functions of order 1 are written as follows:

$$L = l_{000} + l_{100}X + l_{010}Y + l_{001}Z$$

$$C = c_{000} + c_{100}X + c_{010}Y + c_{001}Z \qquad [2.9]$$

Polynomial functions of order 2 are written as:

$$L = l_{000} + l_{100}X + l_{110}XY + l_{101}XZ + l_{010}Y + l_{011}YZ + l_{001}Z + l_{200}X^2 + l_{020}Y^2 + l_{002}Z^2 \qquad [2.10]$$

$$C = c_{000} + c_{100}X + c_{110}XY + c_{101}XZ + c_{010}Y + c_{011}YZ + c_{001}Z + c_{200}X^2 + c_{020}Y^2 + c_{002}Z^2$$

The coefficients l_{ijk} and c_{ijk} in the distortion model are the unknowns. They are calculated using a list of reference points; remarkable points on the image whose geographic coordinates we know perfectly and unambiguously (e.g. landmarks or Ground Control Points). Each point M of this list generates two equations [2.11]. The system of equations obtained using all of the reference points is then solved by a least squares regression to compute l_{JK} and c_{ijk}:

$$L_M = \sum_{i=0}^{p} \sum_{j=0}^{q} \sum_{k=0}^{s} l_{ijk} X_M^i Y_M^j Z_M^k$$
$$C_M = \sum_{i=0}^{p} \sum_{j=0}^{q} \sum_{k=0}^{s} c_{ijk} X_M^i Y_M^j Z_M^k \quad [2.11]$$

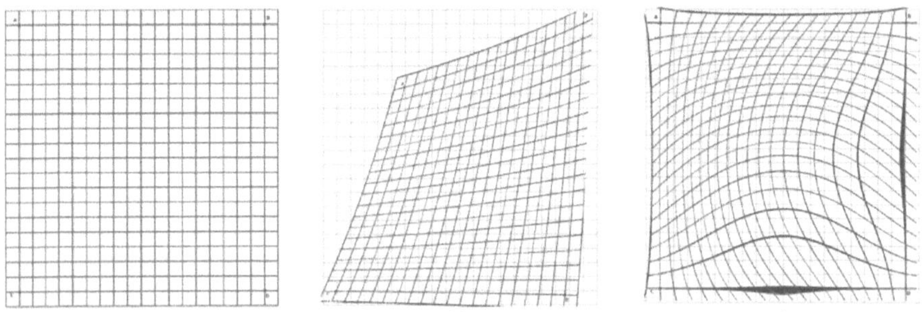

a) b) c)

Figure 2.20. *Examples of deformations (b and c) of a regular grid a) by 2D polynomials of second order. On b) and c), the grid with the dotted line represents the pixels in the new image after correction; their radiometry is calculated based on the distorted pixels of the original image*

These models are very sensitive to the location of the ground control points; they are precise within their close neighborhood, which, depending on the type of deformation to correct, often requires the selection of a large number of points, evenly distributed over the image.

2.4.1.3. *Generalization with rational polynominal functions*

The models based on polynomial functions have been generalized by the use of polynomial ratios (the rational functions):

$$L_n = \frac{\sum_{i=0}^{p} \sum_{j=0}^{q} \sum_{k=0}^{s} p1_{ijk} X_n^i Y_n^j Z_n^k}{\sum_{i=0}^{p} \sum_{j=0}^{q} \sum_{k=0}^{s} p2_{ijk} X_n^i Y_n^j Z_n^k} \quad C_n = \frac{\sum_{i=0}^{p} \sum_{j=0}^{q} \sum_{k=0}^{s} p3_{ijk} X_n^i Y_n^j Z_n^k}{\sum_{i=0}^{p} \sum_{j=0}^{q} \sum_{k=0}^{s} p4_{ijk} X_n^i Y_n^j Z_n^k} \quad [2.12]$$

In contrast to the simple polynomial functions, here the geographic coordinates (X, Y, Z) and the coordinates in the image (L, C) are normalized by using the following equations (where n stands for "normalized"):

$$L_n = \frac{L-L_0}{L_s} \; C_n = \frac{C-C_0}{C_s} \; X_n = \frac{X-X_0}{X_s} \; Y_n = \frac{Y-Y_0}{Y_s} \; Z_n = \frac{Z-Z_0}{Z_s} \qquad [2.13]$$

L_0 .., z_0 and L_s, .., Z_s are constants of the model in order to have, L_n, .., Z_n range between –1 and 1.

These models are very widespread today, particularly with the arrival of very high spatial resolution satellites. Indeed, they have the advantage of being generic and the provision of the set of coefficients of the four polynomials ($p1_{ijk}$, $p2_{ijk}$, $p3_{ijk}$, $p4_{ijk}$) and the parameters of normalization (L_0,.., Z_0 and the L_s, .., Z_s) directly allows their implementation via equations [2.12] and [2.13].

These empirical models have several benefits, in particular:

– an image processing software which implements this geometric correction model can process images from different sensors without knowing the rigorous mathematical model of the deformations. Users do not have to invest in a software or a specialized module every time a new sensor appears;

– the data providers do not have to reveal all the characteristics of their image acquisition system. By approximation of their own mathematical model, they calculate the rational model coefficients and deliver them to users as part of the images files. These are the RPC coefficients (rational polynomial coefficients) that come along with many image products, including GEOEYE, WORLDVIEW, SPOT 6/7 or PLEIADES images.

When necessary, the geometrical accuracy of rational models based on RPC can be improved by adding a second correction stage using a 0 or one-order polynomial function and a small set of Ground Control Points [AGU 13].

2.4.1.4. *Performing a geometric transformation*

If we know the geometric transformation to apply to an image, we can

– browse the pixels one by one;

– calculate their map coordinates using their "image" coordinates (row and column numbers);

– gradually "fill in" the geometrically corrected image.

This approach has the disadvantage of leaving pixels with no values in the corrected image, which are no-data holes, because there is little chance that the map coordinates calculated successively from one pixel to the other follow the regular grid of the corrected image.

For this reason, the geometrical transformations are applied with "indirect" models:

– the geometrically corrected image canvas is browsed pixel by pixel;

– from the map coordinates of these pixels, the coordinates in rows and columns in the initial image are calculated (using the transformation equations);

– as the calculation of these coordinates does not produce whole numbers but coordinates in between pixels, a resampling method has to be selected to decide which radiometric value to assign to each pixel of the corrected image (whether that should be the nearest pixel value, interpolation between neighboring pixels, average value, etc.).

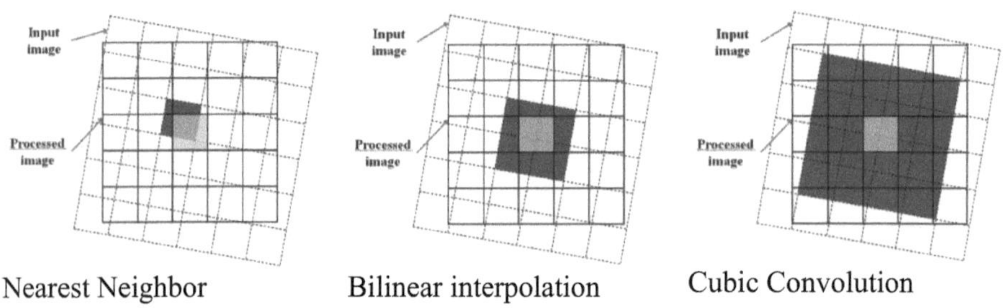

Figure 2.21. *Resampling methods. In each case, dark gray pixels (initial image) are used to calculate the value of the light gray pixel (corrected image)*

There are three main methods of re-sampling (Figure 2.21):

– nearest neighbor: uses for each pixel P of the corrected/edited image, only the value of the pixel that is closest to the location of P within the original image. This is the simplest method and it does not alter the original radiometric pixel value. On account of this, it is preferred if you want to keep the spectral information of the images. However, it can produce some pixel duplication or data loss. This method also tends to produce images with poor visual quality;

– bilinear interpolation: this calculates, using the original image values, a weighted average of the four closest neighboring pixels (window 2×2) of the pixel that is being processed in the corrected image. Averaging original pixel values alters the original signal and creates a completely new value on the final image;

– bicubic interpolation: this calculates, using the original image, a weighted average on a block of 16 pixels (window 4×4), centered on the image coordinates of the pixel that is being processed in the corrected image.

The distances to the neighboring pixels are used to weight the average computations. Unlike the nearest neighbor, the bilinear and bicubic methods produce smoother images in which certain contrasts can be strengthened.

2.4.2. *Radiometric preprocessing*

This section describes the most common methods to correct atmospheric and topographic effects on multispectral satellite images. The objective is to understand all of the settings that come into play. Other mechanisms may alter image radiometry, such as the effects on the ground caused by atmospheric scattering, or even the directional effects related to the solar angle, the angle of observation and to the nature of the surface. Procedures for the correction of these latter effects exist but are not addressed here.

The solar energy reflected by the surface of the Earth and recorded onboard the satellite is the result of many interactions with the atmosphere and the land surface (see section 2.2). The Sun's radiation is attenuated by mechanisms of scattering and absorption during its passage in the atmosphere (see Chapter 1, Volume 1 of this set [BRI 16]). When it reaches the target, the incident energy is absorbed, transmitted and/or reflected by the surface. The reflected energy, which corresponds to the radiance of the surface or Top of canopy (TOC) radiance, crosses the atmosphere again before reaching the Earth observation satellite. The onboard sensor then records an apparent radiance or Top of Atmosphere (TOA) radiance (Figure 2.22). The reflected part of the energy in relation to the incident energy, in other words the reflectance, in a given wavelength, depends on the nature of the target, its environment and the terrain. Also, in remote sensing, it is advisable for certain processing to convert the digital numbers in TOA or TOC reflectance values, or even to correct topographic effects.

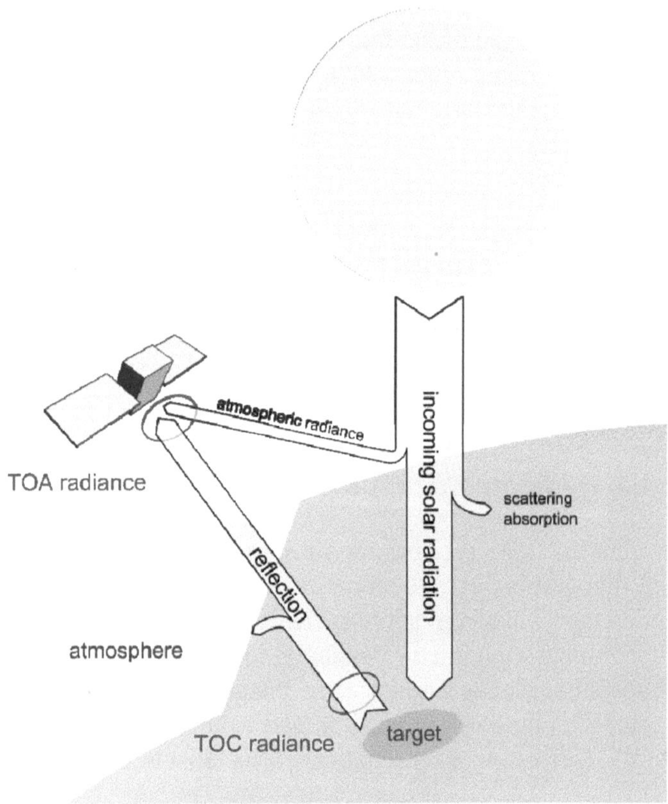

Figure 2.22. *Interactions between solar radiation, the atmosphere and the land surface*

Before engaging in this sometimes long and tedious radiometric preprocessing, it is important to assess their usefulness according to needs and data input. For example, for a study with a set of images acquired above a given zone on different dates, it is better to convert the pixel values in TOC reflectance, to compare the data between them since the radiance lessens when crossing the atmosphere particularly because of aerosol content which constantly varies in time and space.

2.4.2.1. *From digital number to radiance*

As we have seen previously, the encoding of a satellite image is a discrete set of integer values which corresponds to the digital numbers (DN). These values are the numerical transcription of a physical parameter: the radiance recorded by the sensor. The linear relationship, between digital numbers and radiance, is presented as follows:

$$L_k = DN_k . a_k + b_k \qquad [2.14]$$

where L_k represents the apparent radiance, DN_k represents the digital number, a_k represents the gain, b_k represents the offset and k represents the spectral band. These parameters, calculated from the calibration curve of the sensor, evolve with time. They are often provided in the metadata of the image or accessible on the image suppliers' websites.

For SPOT 6/7 images, for example, the Dimap metadata is stored in an XML file. The information relating to the conversion of digital numbers in radiance is available in the <Band_Radiance> section. There are as many <Band_Radiance> sections as spectral bands. Each section contains the spectral band identifier, the update timestamp of the calibration coefficient, the the radiance transfer function, the radiance unit, the gain and the offset (bias):

<Band_Radiance>

<Band_ID>B1</band_ID>

<CALIBRATION_date>09/13/2011 12:00:00</Calibration_date>

<MEASURE_desc>radiometric raw counts (DN) TB TOA Radiance (L).

Formulae The=DN/gain+Bias</MEASURE_desc>

<MEASURE_UNIT>Watt/m2/steradians/micrometers</MEASURE_UNIT>

<MEASURE_uncertainty> 4 </MEASURE_uncertainty>

<Gain> 9.529 </gain>

<bias>0</bias>

For a pixel in the green spectral band (identifier B1), with a digital number 382. The radiance in the L_{B1} is equivalent to:

$$L_{B1} = \frac{382}{9.529} + 0 = 40.08\ \mathrm{W.m^{-2}.sr^{-1}}.\mu m^{-1}$$

The names of the parameters used to convert digital numbers into apparent radiance may differ from a satellite to the other. For Landsat 8, for example, the metadata file is a text file (.txt) in which the gain and the offset correspond, respectively, to radiance_MULT_band_N and radiance_ADD_band_N (where N is the spectral band index).

The radiance calculation is often the first step in the reflectance assessment which is defined as the ratio of the reflected energy (radiance) to solar incoming radiation.

2.4.2.2. *From radiance to apparent reflectance*

The apparent reflectance, or reflectance at the top of the atmosphere (TOA), provides that part of the solar energy reflected by the Earth's surface and the atmosphere at the level of the satellite sensor. Conversion into reflectance enables image normalization in an interval of continuous values between 0 and 1 and thus the comparison of the images. It is performed computing the ratio of apparent radiance (at sensor level) and incident illumination.

The relationship between the apparent radiance L_k and the reflectance R_k at the top of the atmosphere (for a spectral band k) is described as follows:

$$R_k = \frac{L_k . d_{ts}^2 . \pi}{E_{s,k} . \cos \theta_z} \qquad [2.15]$$

d^{Ts} is the distance between the source of light (the Sun) and the target (the Earth), $E_{s,k}$ is the solar irradiance and θ_z is the zenith angle between the radiation main propagation direction and the normal to the surface.

The metadata of a satellite image always provides the solar irradiance for each spectral band, the solar elevation angle (θ_e) and the acquisition date. The zenith angle is obtained by the equation $\theta_z = 90° - \theta_e$. As for the acquisition date, it allows you to deduce an approximate distance (d^{ts}) between the Sun and the Earth in astronomical units (UA):

$$d_{ts} = 1 - 0.01664 \cos\left(0.9856\,(J_{nb} - 4)\right) \qquad [2.16]$$

where J_{nb} is the number of days since 1st January of the acquisition year.

If we take the example of a SPOT 6/7 image with a pixel whose digital number is equal to 382, that is radiance value of 40.08 $W.m^{-2}.sr^{-1}\mu m^{-1}$, it is possible to calculate the TOA reflectance by retrieving the required parameters in the Dimap Metadata:

– acquisition date:

<imaging_date>2014-04-17</Imaging_Date>, here the 107th day of the year 2014

– solar illumination:

<Band_solar_Radiance> <value>1826.087443</value>, here 1826.087 $W.m^{-2}\mu m^{-1}$

– solar elevation angle:

<Solar_Impact>
<sun_ELEVATION>51.1640894962</Sun_ELEVATION>, here 51.164°

The distance between the Sun and the Earth then equals 1.00334222896 UA.

The reflectance at the top of the atmosphere is calculated as follows:

$$R_k = \frac{40.08 \text{ x } 1.00334222896^2 \,.\, \pi}{1826.087443 \cos(90 - 51.1640894962)} \approx 0.089$$

The processing of reflectance images, or radiance, requires encoding in 32-bit to support decimal numbers. Thus, the size of a SPOT 6/7 doubles. To limit the volume of data, 16-bit encoding is recommended using mili-reflectance ranging between 1 and 1000, without decimals.

2.4.2.3. *Corrections of atmospheric effects*

Given that the surface reflectance is dependent on the reflectance and transmittance of the atmosphere, it is difficult to compare a temporal series of images or images acquired on different areas. In fact, the radiance variation for a surface may be due to a change in land cover or even to the atmospheric variations, including the composition and density of aerosols. Besides with the atmosphere not having the same distorting effect according to the wavelength, it therefore amends the relation of reflectance between spectral bands.

The calculation of the surface reflectance (TOC) faces two difficulties; atmospheric conditions are not homogeneous in space and time, and thus the aerosol optical thickness may vary within the same image. On the other hand, the input data to simulate the propagation of electromagnetic radiation in the atmosphere at the time of acquisition are often non-existent or too remote from the target to be sufficiently reliable.

In theory, the most robust approach is to measure the propagation parameters of electromagnetic radiation in the atmosphere at the time of acquisition. The weather data are, however, rarely available. Except if *in situ* measurements are carried out during the satellite acquisition, two approaches are preferable: modeling of the atmospheric effects and empirical adjustment [MOR 92]. In each case, the atmosphere is assumed to be homogeneous over the entire surface of the image, and the result remains an approximation.

2.4.2.3.1. Modeling of the radiative transfer: 6S method

Several physical models exist to simulate the energy received at sensor level in known geometric, atmospheric and spectral conditions. They are therefore intended to resolve a direct problem. However, the correction of the effects of the atmosphere on a satellite image is a problem of reverse type since it must assess the surface reflectance, as the one at the top of the atmosphere is known. The user must input many settings. For example, the 6S model (second simulation of the satellite signal in the solar spectrum, [VER 97]) takes into account the geometric conditions (solar angles and the sensor's viewing angle), atmospheric conditions (pressures, temperature, the proportions of various gases, aerosol compositions, visibility distance and optical thickness at 550 nm), spectral conditions (range of wavelengths) and surface (homogeneous reflectance or not, directional effects and land cover). To simplify parameter entry, the model offers atmospheric profiles depending on the latitude and the season, as well as aerosol profiles according to the predominant land cover [KER 00]. It is also possible to integrate *in situ* data, especially the optical and micro-physical properties of aerosols measured or determined on a daily basis in each point of the Aeronet network (Aerosol Robotic Network). These settings are public and accessible in real time (h+2) on the NASA portal at the following address: http://aeronet.gsfc.nasa.gov/.

The simplified relationship between the apparent radiance (L_k) and the reflectance of surface (R_{sol}), without taking account directional effects, is then estimated for each spectral band (k) in the following way:

$$R_{sol,k} = \frac{A_k.L_k - B_k}{1 + C_k.(A_k.L_k - B_k)} \quad [2.17]$$

where A_k, B_k and C_k are the coefficients for atmospheric correction given in the output of 6S model for each spectral band.

2.4.2.3.2. Empirical adjustments by methods of the invariant targets

The empirical adjustment methods are only based on the information contained in the image to deduce the atmospheric properties. The DOS method (dark object subtraction, [CHA 88, CHA 96]), which is among the best known, assumes that some objects of the land surface are radiometric invariants whose behavior is close to that of a black body. For example, the direct shadows caused by the terrain should have a zero reflectance whatever the wavelength. In actual fact, the value of these pixels is always greater than 0 because of the different radiation scattering mechanisms in the atmosphere. The water masses, clear if possible (without with no suspended matter or macrophytes), are also assimilated to the black body in infrared field. Assuming any environmental effects to be negligible, the apparent radiance of these invariant objects (L_{dos}) corresponds therefore to that of the atmosphere (L_{atm}). The surface reflectance relationship is described as follows:

$$R_{sol,k} = \frac{\pi.(L_k - L_{atm,k})}{T_{v,k}.(E_{t,k}.\cos\theta_z.T_{z,k} + E_{down})} \quad [2.18]$$

where L_{atm} is the atmospheric contribution of the measured radiance at sensor level (path radiance), $T_{v,k}$ is the upward atmospheric transmittance between the target and the sensor, $E_{t,k}$ is the illumination of the target, θz is the zenith solar angle, $T_{z,k}$ is the downward atmospheric transmittance between the sun and the target and, E_{down} is the scattering solar radiation in a downward direction. There are four variants of the DOS based on simplifying assumptions for atmospheric correction variables the L_{atm}, $T_{v,k}$, $T_{z,k}$ and E_{down} (by omitting these four variables, we find the expression of apparent reflectance). For example, for the most basic approach – DOS1 – the upward and downward atmospheric transmittance are assumed to be 100% equal, and E_{down} is ignored.

As the apparent radiance (L_{dos}) and reflectance (R_{inv}) of the invariant targets are not entirely null, the DOS method approximates them. L_{dos} corresponds to the image radiometric minimum for a given spectral band (see equation [2.14]). As for R_{inv}, it amounts to 1% in the visible and 0.1% in the infra-red. The relationship between the atmospheric radiance L_{atm} and the apparent radiance L_{dos} of invariant targets is established as follows:

$$L_{atm,k} = L_{dos,k} - R_{inv,k}.[(E_{t,k}.\cos\theta_z.T_{z,k} + E_{down}).T_{v,k}/\pi] \quad [2.19]$$

Taking into account the constants declared in the DOS1 method, the calculation of the surface reflectance is simplified:

$$L_{atm} = L_{dos} - R_{inv,k}.E_{t,k}.\cos\theta_z/\pi \quad [2.20]$$

$$R_{sol,k} = \frac{\pi.(L_k - L_{atm})}{E_{t,k}.\cos\theta_z} \quad [2.21]$$

However, it is possible if we want to compare two images in the same area but acquired on different dates to stall the radiometry of one compared to the other with the help of linear regressions, without making atmospheric corrections on the images. The reference image must present a minimum cloudiness and a good dynamic range of radiometric values. The supposed invariant targets corresponding to different types of land use/land cover such as buildings, bare soil, roads, etc., can be considered [HAJ 08]. The average reflectance of each target is used to determine the constants of the regression line for each image spectral band:

$$R_{ref,k} = a.R_{img,k} + b \quad [2.22]$$

where $R_{ref,k}$ is the apparent reflectance of the k-band of the reference image and $R_{img,k}$ is the apparent reflectance of the k-band of the image to correct. This method allows us to proceed to reliable multi-temporal analysis while breaking free from the constraints associated with the atmospheric correction.

2.4.2.4. *Corrections of topographic effects*

The terrain causes variations in solar illumination on the surface. Therefore, a type of land cover, according to its position on an illuminated slope or a shaded slope, will not have the same radiance. The methods of topographic effect correction [BAR 10, RIA 03] are trying to bring the reflectance of a steep surface to that of a flat surface. They require as input a digital terrain model to estimate the light intensity of the reflective surface (IL), equal to the cosine of γ_i, which is the angle between the direction of the incident solar radiation and the local configuration of the field (Lambert's cosine law):

$$IL = \cos(\gamma_{i,n}) = \cos(\theta_{p,n}) . cos(\theta_z) + \sin(\theta_{p,n}) . sin(\theta_z) . \cos(\varphi_a - \varphi_{o,n}) \quad [2.23]$$

where θ_p denotes the slope of the ground, θ_z denotes the zenith solar angle, φ_a denotes the azimuthal solar angle, φ_o denotes the angle of exposure and n denotes a pixel from the digital terrain model. The value of IL is between −1 and +1. Negative values indicate shaded areas, which do not receive direct light (see Figure 2.23).

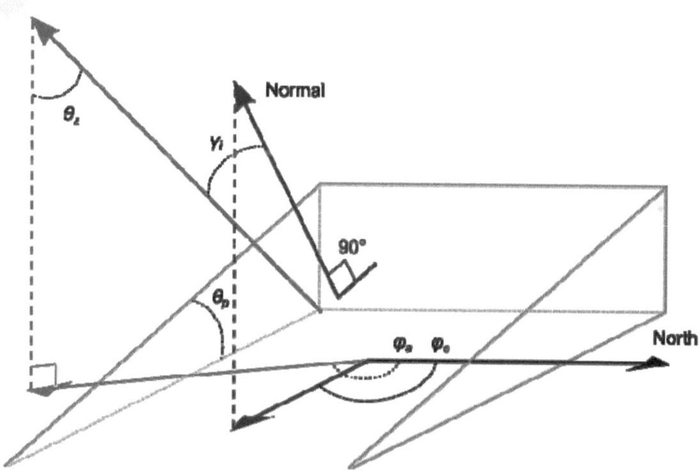

Figure 2.23. *Required angles for the calculation of the light intensity of the reflective surface (IL)*

According to the assumptions on the type of ground surface, the methods are either Lambertian, or non-Lambertian.

2.4.2.4.1. Lambertian correction methods

Lambertian methods assume that the surface reflects incident radiation in all directions equally. The method of cosine correction is the most simple and the best known [TEI 82]. It corrects topographic effects as follows:

$$R_{h,k} = R_k.\frac{cos(\theta_z)}{IL} \quad [2.24]$$

where $R_{h,k}$ is the corrected reflectance for a flat surface, R_k is the reflectance observed on the image and k is the spectral band. Most authors agree on the fact that this method tends to overestimate the corrected reflectance in the areas where the illumination is "low".

2.4.2.4.2. Non-Lambertian correction methods

The best known approach is the result of a semi-empirical equation proposed by Minnaert [MIN 41] to describe the roughness of the lunar surface. The Minnaert correction modifies the cosine correction in the following way:

$$R_{h,k} = R_k.\left(\frac{cos(\theta_z)}{IL}\right)^{K_k} \quad [2.25]$$

where K_k is the Minnaert constant. It represents the roughness of the surface between 0 and 1. If K_k is equal to 1, the surface is a perfect Lambertian reflector and the expression is identical to that of the cosine correction. The Minnaert constant is determined for each spectral band due to linearization of the previous equation [2.25]:

$$\ln(R_K) = \ln(R_{h,k}) + K_k.ln\left(\frac{IL}{cos(\theta_z)}\right) \quad [2.26]$$

In practice, this method needs to collect a number of samples from the image. On this basis, we can build, for each spectral band, a scatter plot on an x-axis showing the light intensity of the reflective surface ($ln\left(\frac{IL}{cos(\theta_z)}\right)$) and on the y-axis the observed reflectance ($\ln(R_K)$). The Minnaert constant then corresponds to the slope of the regression line [2.26]. Similarly, the C-correction method, developed by Teillet *et al.* [TEI 82], introduces a constant C to simulate the scattered illumination and compensate for the overcorrection effects of the cosine correction:

$$R_{h,k} = R_k.\left(\frac{cos(\theta_z)+C_k}{IL+C_k}\right) \quad [2.27]$$

where

$$C_k = \frac{b}{m} \text{ with } R_k = m.IL + b \quad [2.28]$$

These non-Lambertian and semi-empirical approaches apply the constants to the whole image, assuming that the surface of the ground is homogeneous, no matter what the land cover type. That is why numerous improvements are proposed for applying corrections according to the image stratification which takes account of the nature of the ground, topography, etc.

2.5. Extraction of information

In this section, we consider the main techniques for extracting relevant information from multispectral images. The unsupervised and supervised techniques are shown first. With unsupervised techniques, we are dealing with methods of partitioning (more commonly known as clustering) which aim to gather data together (pixels or presegmented objects on the image) in several sets, called clusters.

In a supervised context, the principle is to estimate variables on the whole data set from values known on a subset of this data set. It is then a question of learning, from this subset called a training set, a relationship to bind the data to the variables to be estimated. Once estimated, this relationship is then applied to all other data to perform the full estimation. The variables to estimate may be qualitative (for example, how the soil is being used or the presence of an element in an image) or quantitative (for example, temperature values). In the first case, the associated problem is known as classification (it assigns a class to each input data) while in the second case, we consider regression. Among the families of existing methods, we have made the choice to develop in more detail the principle of support vector machines (SVMs) and support vector regression, which are widely used in practice. This also allows us to discuss the kernel concept used in a number of methods for remote sensing image analysis (and more generally for statistical learning).

All of the techniques presented apply to data that may well be the spectral values from each pixel or quantities calculated on presegmented objects. A discussion pixel/object will then be proposed and we will introduce a few relevant descriptors.

2.5.1. *Unsupervised methods: clustering*

We assume here that we have a set of *N* data $x_i, i = \{1, \dots, N\}$ that can be the *N* pixels of an image or then the *N* vectors describing objects previously segmented. The goal is to partition the data in a number of *k* homogeneous clusters $S_l, l = \{1 \dots, k\}$. Many techniques are available to carry out this task and we include here a few methods among the most used. The readers can refer to Omran *et al.* [OMR 07] for more details.

Among the existing methods, the *k-means* technique is largely used due to its simplicity and performance. This approach seeks to partition the data set in clusters identifiable by their averages (or centers) $\mu_l, l = \{1 \dots, k\}$. The algorithm works in an iterative manner by assigning successively, to the points, the label whose center is the closest, and recalculating the centers after this step of assignment. This principle is described in Figure 2.24:

> – *initialization*: choose k centers $\mu_l, l = \{1 \dots, k\}$ of clusters, randomly
>
> – *allocating stage*: each point $x_i, i = \{1 \dots, N\}$ is assigned to the cluster S_l whose center μ_l is the closest
>
> – *re-calculation of the centers*: with the help of these new assignments, each center is updated: $\mu_l = \sum_{x_j \in S_l} \frac{x_j}{|S_l|}$ where $|S_l|$ is the number of points in the cluster S_l ,
>
> – loop to the first step until convergence.

Figure 2.24. *Principle of the k-means algorithm*

This technique, although simple, performs well in many applications and is therefore widely used. The result depends on the distance criterion taken to compare two vectors. The k-means method is described as *generative* because in this case, a model that represents each cluster (here, its center) is available and can therefore generate representative data from this class (for example, by randomly drawing from around the center). The *generative* term is used as opposed to *discriminative* methods where the goal is to search for the best separation between the clusters (we will return to the principle of discriminative methods when we discuss the SVM technique).

One generative method, which is also commonly used, is the GMM (Gaussian Mixture Model) technique. The underlying theory is to consider that the groups of data follow a Gaussian law whose average and variance must be estimated. The mixture law that follows the sample is therefore:

$$p(x, \alpha, \mu, \Theta) = \sum_{i=1}^{k} \alpha_i g(x, \mu_{i,} \theta_i) \quad [2.29]$$

where α_i are the proportions associated with each class, g is the multi-dimensional Gaussian function taken from the average vector μ_i and the covariance matrix θ_i. The way in which parameters (α, μ, Θ) are estimated is performed using the Expectation-Minimization (EM) algorithm. If the number of clusters is unknown, we can also seek to identify the different peaks of the distribution of points and then to associate a Gaussian function with each peak. In general, this technique works correctly but is limited to the Gaussian models. Other mixtures can be introduced and, according to the complexity of the models, parameter identification problems can arise.

If we wish to have a level of *hierarchy* in the partitioning, ascending and descending methods can be used. The principle of ascending clustering is to associate a different cluster with each piece of data and to regroup these clusters as a hierarchy according to their similarity. The AGNES (AGglomerative NESting) algorithm is often used in this family. Similarly, the descending clustering assigns the same cluster to all data; this cluster is subsequently subdivided or not according to the homogeneity criterion. The DIANA (DIvides ANAlysis) algorithm is based on this principle. Whether we use an ascending or descending algorithm, the question of the distance between clusters is important and can affect the quality of the results. There is no optimum solution for this, but the main approaches used can measure the smallest distance between the points of each cluster, the greatest distance between the points of each cluster, the distance between the mid-point of each cluster or the average distance between all the points. Figure 2.25 illustrates a clustering result using the k-means technique.

2.5.2. *Supervised methods: classification and regression*

Supervised methods enable us to estimate a quantity on an image by relying on a training set. The variables to estimate may be qualitative (such as soil occupation, whether there is an item in the image or not, etc.) or quantitative (temperature, etc.). The first case deals with *classification* while the second case deals with *regression*. As for the training set, it is composed of several samples for which the corresponding output is accessible (measured on the ground, provided by photo-interpretation, etc.). First, we try to establish a link between the input (data images) and output (estimated parameter); and then, this link is applied to all the other input data on which the quantity of interest must be estimated. This principle is illustrated in Figure 2.26.

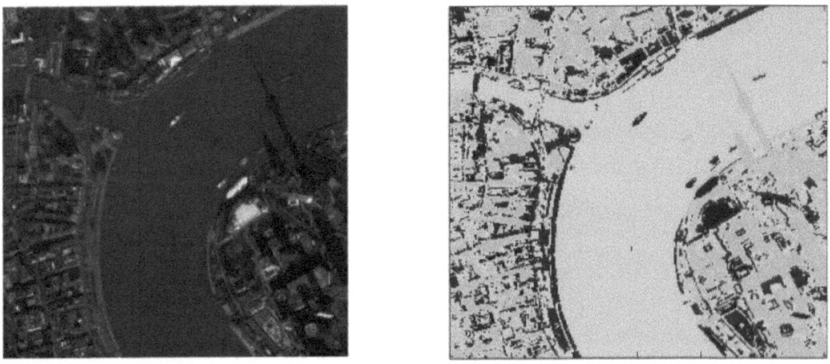

a) Original image   b) Partitioning in 4 clusters

Figure 2.25. *Example of the k-means clustering technique on a Pleiades image representing the City of Shanghai. In the four classes, we distinguish surface water, developed land, roads and highly reflective surfaces*

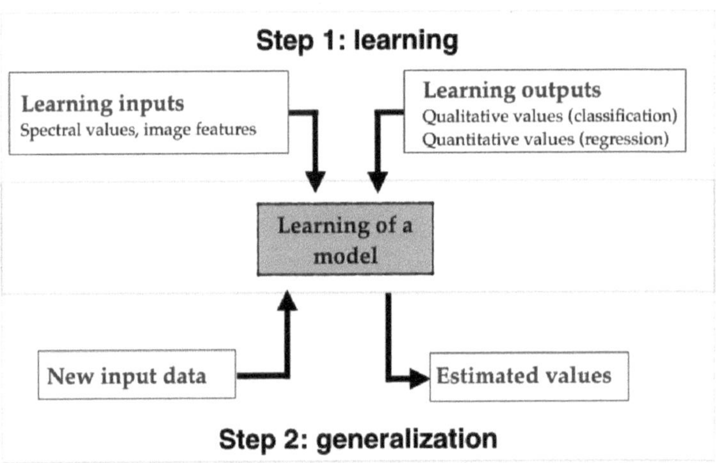

Figure 2.26. *General principle of learning methods*

2.5.2.1. *Classification*

As mentioned above, the methods of classification can be categorized into two major groups: generative and discriminative approaches. A generative method aims to build a joint model associated with the training set and the corresponding labels. When new data occur, we estimate for each label how well it fits in the model and a decision is taken. By contrast, a discriminative method isolates the best discrimination between the different labels. The class assigned to a new point thus depends on its position in relation to the separator. There is no general rule to say that one family is more appropriate than another, but several studies on this subject are available (see, for example, Bishop and Lasserre [BIS 07] and references therein). For classification application alone, only the class is required not the knowledge of the underlying model that led to the class (although it is interesting in some applications). We have therefore made the choice to present here the discriminative methods, which are often used in practice. Among these methods, the *k*-nearest neighbor (KNN) technique is probably the most well known. The principle is simple: to classify a new point, we assign it to the majority class among its neighbors. This technique is effective in many applications, particularly when the boundaries between the classes are clearly distinguished. If this is not the case, the SVM discriminative approaches have proven to be more appropriate and are widely used. These are introduced below.

2.5.2.1.1. Linear classifier: perceptron and support vector machines

The principle of linear classification using the discriminative approach is to find the best linear separator between the training data. If we take the case of two classes (for example, bare soil/covered soil) to extract from multivariate images of dimension P, we must find a hyperplane of dimension P -1 separating at best the training data representing these two classes. This is illustrated in Figure 2.27(a) for a two-dimensional case ($P = 2$ so as to better interpret the problem visually) where each training value is represented in a plan with a color associated with each class. In this figure, we see that there are several solutions (straight black line and dotted lines), but it appears obvious that the straight black line is more suitable, due to its "central" position between the two classes. The dotted lines are derived from the perceptron algorithm (see [FRE 99] for more details), which is one of the most popular algorithms used, which has resulted in isolating a linear separator between the classes. However, it is clear that in spite of good performances, the solutions provided are not optimal, in particular because the separator is very close to one or other of the classes. An approach like this is therefore more sensitive to the noise that is likely to be contained in the data. Here the optimal separator must be as far as possible from each of the classes. In other words, the *margin*, being linked to the distance between the separator and the nearest data, must have a maximum value. This is the principle of the support vector machine (SVM) method. In practice, starting with the training data, we must estimate the equation of the hyperplane (when $P = 2$, we find the equation of a straight line) in such a way that the margin is maximized.

This can be written as the resolution of an optimization problem under constraints: estimate the line which best separates the data set while maximizing the distance between each point and this line (the margin). We can easily demonstrate that the optimization of this problem depends only on some data entries, which in practice are the closest to the best border. These points are called the *support vectors* and are shown in Figure 2.27(b). They are very interesting since they locate the borders that separate the data set. Furthermore, as the solution depends solely on these vector supports, the calculation of the projection of a point on this hyperplane (necessary to calculate the membership of this point to one or the other of the classes) is very fast. It is the theory of the SVM algorithm. It performs well but, as demonstrated, it remains limited to linearly separable data, which is certainly not the case in many applications. In fact, from images with P spectral bands where we wish to differentiate the two types of ground occupation, if we display the points corresponding to each of the classes in a space to P dimensions, there is no reason that the optimum separation should be linear, even if it feels natural that there should indeed be a separation (as we are looking for two different entities). We have two alternatives to overcome this difficulty: find a nonlinear separator or else transform the data into another space where the separation is linear. The first solution remains limited to a parametric form of separation and is not generally applicable. The second idea seems more interesting and can be implemented through the *kernel trick*, presented in the following section.

2.5.2.1.2. Nonlinear classifier, kernel trick

Before going into this in great depth, we will illustrate the principle of kernel approaches on a simple example presented in Figure 2.28(a). There is a set of data in dimension 1 ($x_i \in \mathbb{R}$) on which two labels (blue and red) are represented. If we are looking for a linear separator (a point on the x-axis in this case), we see that it is not possible to find an optimal separator because there will always be a point from each class from one side or the other of the separator. However, if we artificially increase the size of the data by projecting into a space of higher dimension, for example by applying to each point a particular transformation $\phi(x)$: $\mathbb{R} \rightarrow \mathbb{R}^2$, according to the choice of this function, the new data set can become separable. For example, taking $\phi(x) = [x, |x|]$ (we add a dimension corresponding to the absolute value of each point), Figure 2.28(b) reveals that it is possible to find a linear separator (a straight line in $\mathbb{R}^2$) to differentiate the two classes.

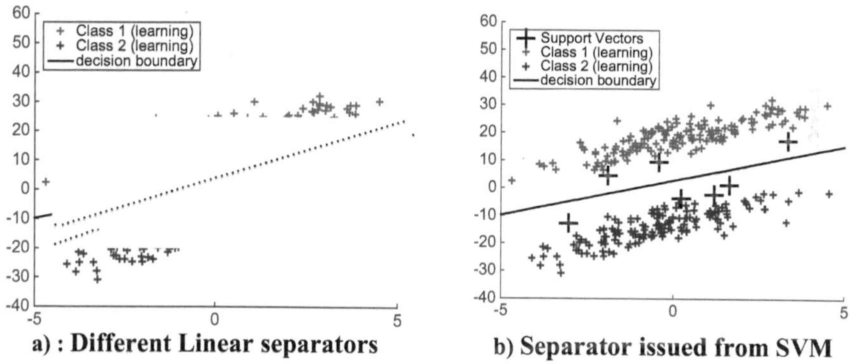

Figure 2.27. *Linear separators. a) Different separators (solid black lines and dotted line) where the best separator is in black and b) support vector machine with the associated support vectors (black crosses)*

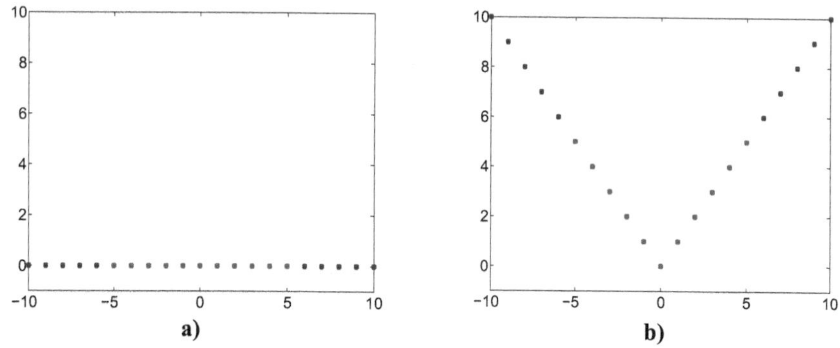

Figure 2.28. *a) Two groups of initial data in $\mathbb{R}$ (blue and red) in dimension 1; non-linearly separable; b) projection in a space of higher dimension (2 here) according to the function $\phi(x) = [x, |x|]$. In this case, data become linearly separable. For a color version of this figure, see www.iste.co.uk/baghdadi/1.zip*

This is the core principle of the kernel trick: artificially increasing the size of the data (by adding redundant information) can improve the separation between the different classes. In practice, given the size and complexity of the multispectral data, more advanced transformations must be applied. Pushed to the extreme, a projection in a space of infinite dimension ($\phi(\boldsymbol{x})$: $\mathbb{R}^P \rightarrow \mathbb{R}^\infty$) would allow optimum separation between the classes. However, such a projection would require too many calculations to be realistic in practice. Therefore, rather than directly calculating such a projection ϕ, we instead use the *kernel* in the space created by this projection. A kernel associated with a mathematical space ($\mathbb{R}, \mathbb{R}^2, \dots, \mathbb{R}^\infty$) defines the way in which the scalar product is calculated in this space (in the usual Euclidean space, the kernel used is $K(\boldsymbol{x}, \boldsymbol{y}) = <\boldsymbol{x}, \boldsymbol{y}> = \boldsymbol{x}^T\boldsymbol{y}$ and is thus linear). Regardless of the size of the space where it is located, the kernel will provide a real (or possibly complex in some cases which we are not discussed here) value and therefore the projection step in a high dimension space can be skipped. Only relying on the kernel trick to work in a potentially very large dimension space is therefore a trick made possible here with the SVM technique because we can easily demonstrate that only scalar products are involved in the search process for the best separator. Thus, the search for the best separator for data projected in a large space (and therefore more easily separable) can be done easily by replacing all the involved scalar products by Kernel functions $K(\boldsymbol{x}, \boldsymbol{y})$. This is what we notice with the kernel trick. It is important to note that in practice, special attention must be given to the function K to ensure that this is actually a kernel (i.e. a scalar product in the space created by the transformation). These properties are related to the positivity, the symmetry and the conditions of nullity. For more information on kernels and the associated theory, we refer readers to Schölkopf and Smola [SCH 02]. In practice, we are only interested in positive definite kernels, the most common of which are written below:

– *linear kernel*: $\boldsymbol{x}, \boldsymbol{y} \in \mathbb{R}^P$; $K(\boldsymbol{x}, \boldsymbol{y}) = \boldsymbol{x}^T\boldsymbol{y}$ (the associated projection function is the identity);

– *polynomial kernel in dimension d*: $\boldsymbol{x}, \boldsymbol{y} \in \mathbb{R}^P$; $K(\boldsymbol{x}, \boldsymbol{y}) = (1 + \boldsymbol{x}^T\boldsymbol{y})^d$ (the associated projection function is more complex to formalize);

– *Gaussian kernel*: $\boldsymbol{x}, \boldsymbol{y} \in \mathbb{R}^P$; $K(\boldsymbol{x}, \boldsymbol{y}) = \exp(-\gamma \parallel \boldsymbol{x} - \boldsymbol{y} \parallel^2)$. The associated projection function cannot be formalized analytically: it is in a space of infinite dimension.

Depending on the applications, one kernel may be more suitable than another. Nevertheless, in practice, a lot of work exists concerning the Gaussian kernel. In particular, since the associated function ϕ projects the data into an infinite space, it usually provides a good class separation. However, it is important to note that in the context of hyperspectral imaging, other kernels have proved to be more efficient, such as the polynomial kernel, spectral kernel or a combination of kernels (see [MER 03]). Figure 2.29 illustrates classifications between two non-linearly separable data sets. It is clear that the use of a non-linear Kernel (here Gaussian) allows us to provide a more realistic separation.

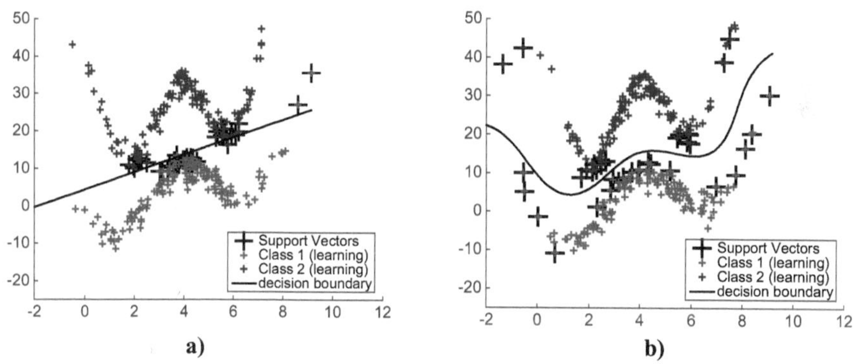

Figure 2.29. *Illustration of the kernel trick. On the same data set (two classes in blue and red), the figure in a) draws the best separation using a linear kernel, while the figure in b) displays the separator obtained with a Gaussian kernel. The points in black are the support vectors. We can see here that the Gaussian kernel allows us to provide a very good separation. For a color version of this figure, see www.iste.co.uk/baghdadi/1.zip*

2.5.2.1.3. SVMs in practice

The SVM technique is used very frequently. For an operational application, it is important to mention the following points:

– up to now, we have analyzed the discrimination between two classes (e.g. bare soil/ covered soil, urban/non-urban, etc.). In order to perform multi-label classification (for example, bare soil/grassland/culture/urban), several strategies are possible. We can refer to an approach such as *one-versus-all* (also called *one-versus-rest*) which means testing each class against all others and retaining the one corresponding to the most represented class as the final label. An alternative approach, the *one-versus-one* strategy, aims to test each pair of classes independently and the class most often chosen will be affected;

– when using a Gaussian kernel, the associated parameter γ can influence the shape of the border. To estimate it automatically, the cross-validation technique is often used. The principle consists of separating the training set randomly into two subsets where the separator is estimated on one set and its accuracy in classification is estimated on the other set. Several combinations of γ are tested. The process is repeated on another random division of the training set. The parameter providing the best result is therefore kept. More details on cross-validation methods can be found in [KOH 95];

– if there are too many samples in the training set or even if the data set is too noisy, there is a strong chance that some training points will not meaningful (located on the wrong side of the optimal border). In this case, the maximization of the margin will lead to too precise a border that will try to separate all data, including noisy ones. For this, the *soft margin* concept was introduced, allowing some points to be inside the margin. The process of minimization remains the same and a tolerance parameter (often denoted by C) is set and can also be estimated by cross-validation;

– if the various components of P dimensions are not in the same space (for example, a vegetation index, a temperature, etc.), then it is important to center-reduce all dimensions. Centering ensures that all the variables have a zero average, while the reduction ensures that they all have a standard deviation of 1. These enable the various components to have a similar role and to be comparable (we are manipulating data without any dimension). By contrast, the variables that would have the greatest value would be above all the others (for example, if in a channel the temperatures ranged from 0 to 30°C and in the other pressures ranged between 800 and 1100 hPa, then the temperature would play a small role even if it is the most discriminative variable). If, on the other hand, we manipulate the same kind of data (for example, only temperature measurements), the question of reduction depends on the application, given that reducing homogenizes the range of variations and can increase the role of a noisy variable. In this case, the Fisher test may be useful. This test allows us to check the equality or inequality of the variance of two variables based on the ratio between the two. If it is around 1, then they are similar and reduction is not necessary; in the opposite case, it is often preferable to reduce. More information on this test is available in [MAD 99];

– there are many easy-to-use libraries for SVMs and associated techniques (soft margin, cross-validation, etc.). Among the existing tools, we can mention in particular the *libsvm*[1] library, which is widely used. Written in C, it interfaces with many languages (Matlab, Scilab, R, Python, CUDA, etc.). We can also call upon the *scikit-learn*[2] package that contains many of the functions of statistical learning for the Python language.

An illustration on satellite images is presented in Figure 2.33. The next section will be dedicated to regression methods.

2.5.2.2. *Regression*

The principle of regression is to estimate a quantity $y \in \mathbb{R}$ from a vector $\boldsymbol{x} \in \mathbb{R}^P$ of dimension P given at input. For this, we have a set N of training samples for which the pairs of values $(\boldsymbol{x}_i, y_i)$ are known ($i = \{1, .., N\}$). The variable y is the *variable to be explained*, while the variable $\boldsymbol{x}$ is the *explanatory variable*. Many studies have been carried out in the field of regression. Linear regression is widely used in the field of remote sensing and by using the kernel trick, a variant of the SVM method applied to regression, called support vector regression (SVR), performs very well with regard to managing complex and nonlinear relationships between explanatory variables and variables to be explained. These principles are introduced in this section. For more details, we refer readers to the many works covering the subject, for example, [FRE 09]. Before presenting the SVR technique, we will briefly present linear regression.

2.5.2.2.1. Linear regression

Linear regression aims to find a linear relationship between the explanatory variables and those to be explained: $\mathrm{y} = \sum_{\mathrm{j}=1}^{\mathrm{P}} \mathrm{w_j}\, \mathrm{x_j} + \mathrm{w_0}$, where x_j is the j-th component of the vector $\boldsymbol{x}$ (of dimension P) and $\boldsymbol{w} = [w_0, w_1, \dots, w_P]^T$ is the vector containing the regression coefficients to be estimated. For this, the least-squares technique, which aims to seek the solution that minimizes the gap between the data and the model, is often used.

1 http://www.csie.ntu.edu.tw/~cjlin/libsvm/.

2 http://scikit-learn.org/stable/.

2.5.2.2.2. Non-linear regression, support vector regression

If the relationship between the explanatory variables and those to be explained is more complex, it is always possible to obtain more sophisticated models. We can, for example, use a polynomial model of order n (written in one dimension: $y = w_n \boldsymbol{x}^n + w_{n-1}\boldsymbol{x}^{n-1} + \cdots + w_1\boldsymbol{x} + w_0$), a logarithmic model (still written in one dimension: $y = w_1 \log(\boldsymbol{x}) + w_0$) or any other model involving an explicit relationship between the explanatory variables and those to be explained. From the moment these models are linearly weighted by coefficients (in each of the models presented, the relationship between y and the coefficients w_i is linear), then the least-squares technique enables these coefficients to be estimate. Such models allow more flexibility in handling complex data, but they remain, however, limited to an explicit form of relationship.

In order to increase the generic nature of the regression methods and to be able to apply them to more complex relationships, it is possible to rely on the SVM principle to extract a regression function. With the same philosophy as for classification, the problem is formulated as the search for a linear hyperplane that adapts to data. The difference with SVM is that now we search the parameters $\boldsymbol{w} = [w_0, w_1, \dots, w_P]^T$ of the minimum margin hyperplane which encompasses all the explanatory variables. In the same way as for the SVM, the solution of this system involves only dot products with *support vectors*. Using the kernel trick, the scalar products involved can be replaced by kernels, to handle nonlinear data. Figure 2.30 illustrates the performance of a regression with a linear model and then with the SVR technique associated with a Gaussian kernel. In practice, the tools allowing the use of SVM techniques (such as *libsvm* or *scikit-learn* presented above) offer the option to apply an SVR regression.

2.5.2.3. *Notes*

The techniques presented above have been used in many applications in multispectral image analysis, either to identify or categorize the type of soil, vegetation, building, etc. We have made the choice to focus primarily on the theory of the kernel methods because the kernel trick is used in many applications for data processing, beyond classification or regression.

Among the classification techniques, we might also see decision trees and random forests, which are also much used. The principle of a decision tree is to create, from all the training data, a succession of tests to bind the explanatory variables (spectral values of the image, object descriptors, etc.) to the variables which must be explained (the different classes). When a new piece of data needs to be classified, it follows different paths depending on the results of these successive tests (these paths are the branches) to arrive at the final class (represented by a leaf). This is shown in Figure 2.31. There are several possibilities for the construction of the tree, the principle in general is to work recursively: starting from a root which contains all the training data, this is subdivided into several nodes if it is impure, i.e. if it contains a significant mixture of several classes. These new nodes are also subdivided if they are not pure and this is repeated up until convergence. Trimming methods can then be applied in order to remove certain subtrees that are not significant. This notion can be extended to regression problems, and this is often included under the term CART (classification and regression trees) [BRE 84].

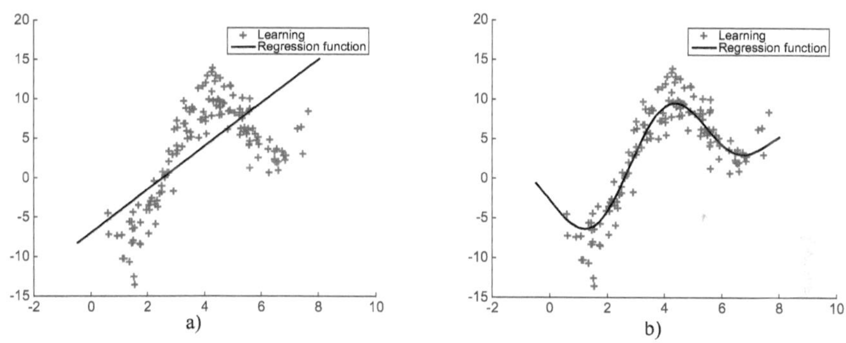

Figure 2.30. *Estimation of a regression function (black curve) from the training (purple) with a linear technique a) and an SVR method with a Gaussian kernel b). For a color version of this figure, see www.iste.co.uk/baghdadi/1.zip*

Each new piece of data to be classified follows a series of tests from the root up to a final class. If, at a given level, there has been an error along the way, the result of the testing is then not significant and this can lead to an erroneous result. Therefore, we can observe an instability of the results provided when we try to classify two similar inputs. To overcome this difficulty, random forests have been introduced. This theory consists of generating a series of decision trees from subsets statistically representative of the training data set. The final class associated with this is the one that has been assigned most often. The random forests enable a significant gain in comparison to decision trees and are now popular techniques of classification (see [PAL 05] for example).

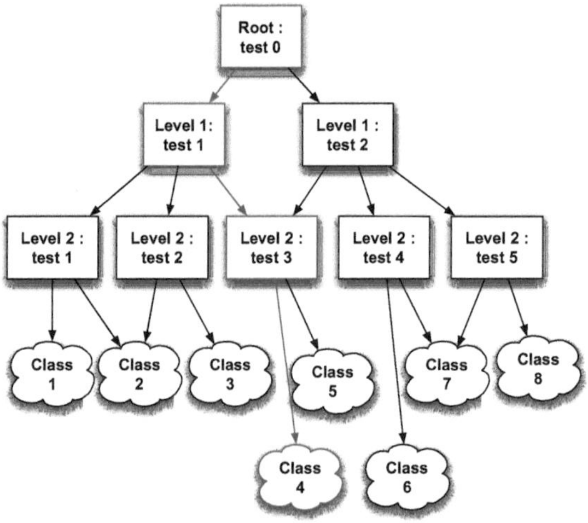

Figure 2.31. *Illustration of the operations in a decision tree*

2.5.2.4. *Spectral values, texture indices, pixel versus subject*

As mentioned previously, depending on the application, the choice of input vector **x** is important. Of course, taking the spectral channels of the images directly is a natural solution. However, it may also be useful to put in the vector data some features derived from spectral values which may be more relevant. Therefore, the vegetation indices (in different contexts of moisture, drought, etc.), snow, dust, etc., can help to better condition the techniques (clustering, regression and classification) of the spectral values of the images. A non-exhaustive list of these indices is presented here.

2.5.2.4.1. Vegetation indices

With infrared (IR), near-infrared (PIR), blue (B), green (V) and red (R)) bands:

– NDVI (Normalized Difference Vegetation Index or Tucker index [ROU 74]):

$$NDVI = \frac{R_{PIR}-R_R}{R_{PIR}+R_R} \quad [2.30]$$

– NDWI (Normalized Difference Water Index [GAO 96]), useful in the context of a dry area:

$$NDWI = \frac{R_V - R_{PIR}}{R_V + R_{PIR}} \quad [2.31]$$

– ARVI (Atmospherically Resistant Vegetation Index [KAU 92]), which reduces atmospheric effects by integrating the blue band:

$$ARVI = \frac{R_{PIR} - R_{RB}}{R_{PIR} + R_{RB}} \text{ avec } R_{RB} = 1 - \gamma(R_{B\_}R_R) \quad [2.32]$$

where γ is a parameter to be set that depends on local conditions (in particular aerosols);

– VSDI (Visible and Short-Wave Infrared Drought Index [ZHA 13]), useful in a drought context:

$$[(R_{IR} - R_B) + (R_R - R_B)] \quad [2.33]$$

2.5.2.4.2. Snow index

With the green band (V):

– NDSI (Normalized Difference Snow Index [SAL 04]):

$$NSDI = \frac{R_V - R_{IR}}{R_V + R_{IR}} \quad [2.34]$$

2.5.2.4.3. Texture indices

With the arrival of very high spatial resolution, the spectral values taken in each pixel are not necessarily optimal to portray the objects contained in the images. As we can see in Figure 2.32, improving the resolution allows us to portray the image content more finely. In this example, we can go beyond the vegetation/urban portrayal to describe the type of vegetation and urban zones in particular in more fine a detail. However, here, the spectral value is no longer the best descriptor and for a good identification of the spatial variability of the images texture indices are often preferred. These indices are calculated in each pixel by a combination of values located in a neighborhood of that pixel. Many indices have been proposed. Here we list the main ones (defined in each pixel i by its neighborhood $\nu(i)$ composed of N points):

– variance (with $\overline{R(\imath)}$ the average reflectance R(i) in the neighborhood v(i)), which measures the overall heterogeneity:

$$V_R(i) = \frac{1}{N}\sum_{l \in v(i)}\left(R(i+l) - \overline{R(\imath)}\right)^2 \qquad [2.35]$$

– skewness, which measures the asymmetry of the ground around the pixel being considered:

$$S_R(i) = \frac{1}{N}\sum_{l \in v(i)}\left(\frac{R(i+l) - \overline{R(\imath)}}{V_R}\right)^3 \qquad [2.36]$$

– kurtosis, which corresponds to a measurement of "flattening":

$$K_R(i) = \frac{1}{N}\sum_{l \in v(i)}\left(\frac{R(i+l) - \overline{R(\imath)}}{V_R}\right)^4 \qquad [2.37]$$

– entropy, which measures the disorder:

$$E_R(i) = -\sum_{l \in v(i)} R(i+l) log(R(i+l) \qquad [2.38]$$

Figure 2.32. *Differences between very high (50 cm, thumbnail image to left) and high spatial resolution (5 m, thumbnail image to the right). We clearly see the detail provided by the very high resolution*

Other indices are also available (compactness, for example), we refer readers to Materka and Strzelecki [MAT 98] for more details. A critical question arises with all these indices: what is the optimal size of the neighborhood around the pixel being considered. This problem is closely linked to the scale analysis where it is positioned. If we have no idea about the different scales involved, we could calculate all of these criteria on several neighborhoods. This increases the size of the data to process (as if the number of spectral bands was artificially increased) and it introduces the idea of a multi-scale concept. In the same spirit, we can also mention the work of Antoine Lefebvre [LEF 11] which puts forward multi-scale indices based on a wavelet decomposition [LEF 08] causing the isolation of the orientation properties and the data scale.

A second possibility to overcome choosing the window size is to calculate these indices on objects (and not on a *raster* image) from a unsupervised a presegmentation of the image. Once these objects are presegmented, the different indices are calculated at the scale of these objects and no longer on a regular neighborhood.

It should be noted, however, that it is sometimes difficult to choose the best index to identify an object. In addition, the technique of multiplying the descriptions by a set of indices is likely to introduce redundancy which could disrupt the proper functioning of the methods discussed earlier. In order to disconnect the information contained in different descriptors, a principal components analysis (PCA) can be performed on all the vector dimensions $\boldsymbol{x}$ at input in order to reduce the redundancy of information. The principle of PCA is presented below.

2.5.2.4.4. Reduction of dimension: principal components analysis

When we manipulate multi-varied data, some information is potentially redundant. Therefore, summarizing the information is extremely relevant not only to better condition the methods to be applied, but it also allows us to reduce the calculations and the volume of data to be processed. Among the many existing methods of reducing the dimension, the most popular is PCA. Let us suppose that N data in dimension P are available. The idea is to make a change of basis of these P dimensions in order to focus the information on only a few components. In practice, when we have P dimensions, we calculate the variance-covariance matrix W (of size $P \times P$) which contains on the diagonal the variance of different dimensions and which contains in

$W(i,j)$ the covariance between the dimensions i and j of the data. We can easily demonstrate that the eigenvector associated with the largest eigenvalue W is the one which explains the most variance of the data, the percentage of variance explained corresponding to the ratio between this value and the sum of all the eigenvalues. In practice, only a few eigenvalues are significant and keeping only a few eigenvectors corresponding to the largest eigenvalues (explaining a given percentage of the total variance) enables us to accurately represent the data. These eigenvectors represent the main components.

To conclude, Figure 2.33 provides three examples of results from these different techniques. In the first case, we have calculated the entropy on a temporal series of radar images (this series being seen as a multi-varied entropy image) and used an SVM classification to extract the plant formations in the wetlands (application in Figure 2.33(a)). The same type of process has been used in Figure 2.33(c) where methods of grasslands management have been identified from LAI (Leaf Area Index, a biophysical parameter derived from images). Finally, Figure 2.33(b) presents a spatialized map of temperature obtained from regressions (SVR type) between satellite data and local sensors.

2.6. Conclusions and prospects

Due to the significant increase in the number of optical satellite images acquired, the improvement of their spatial resolution and the leap forward in computing capabilities, the definition of automated recognition and analysis methods has become an essential issue. Processing is first of all intended to correct (geometry and radiometry) and improve the data quality (noise reduction and color composition) to then derive, from the multispectral values, relevant parameters according to the objects or phenomena studied (indices, classification, etc.). In recent years, the processing has tended to imitate the cognitive process of thematic mappers in no longer considering the image as a series of pixels but as a set of interrelated objects (contextual relationships). This approach is all the more useful as the arrival of the very high spatial resolution has led to a very strong spectral heterogeneity within objects.

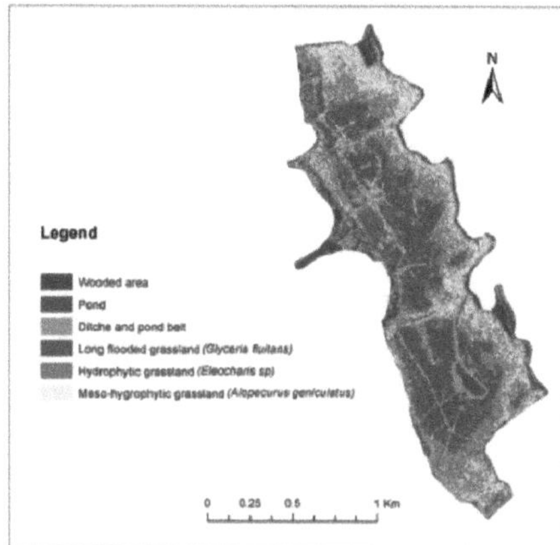

a)

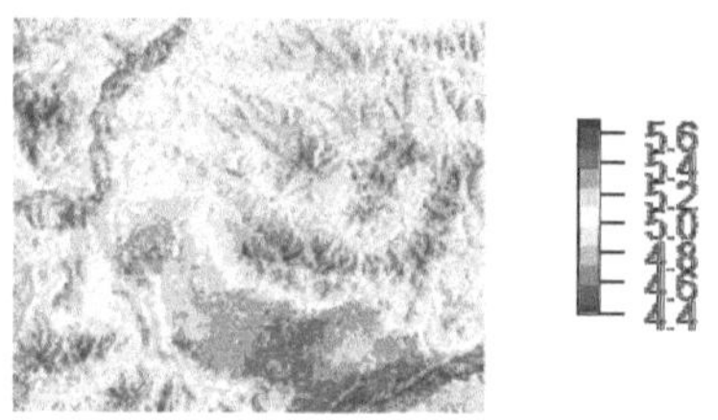

b)

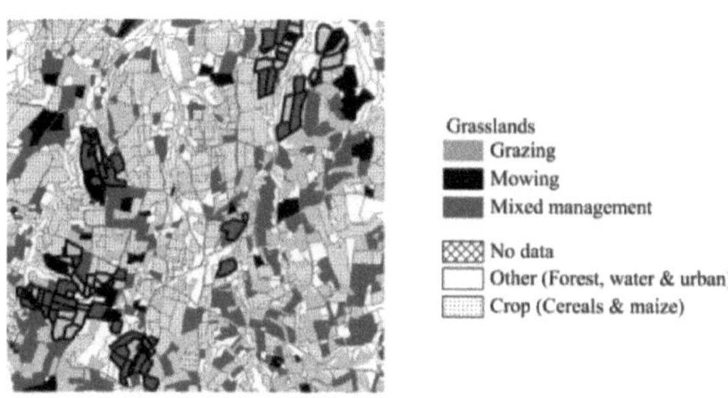

c)

Figure 2.33. *Examples of the results of the different techniques presented. a) Classification resulting from a temporal series of TerraSAR-X images using SVM and entropy on a territory in Brittany [BET 14]; b) estimation of a surface temperature map (in °C) in a wine-growing plot by SVR regression from a few measurement stations and satellite images [LER15]; c) identification of the different types of grasslands with SVM from time series of LAI (Leaf Area Index) estimated on images with very high spatial resolution on a watershed in Brittany, France [DUS 11]*

At present, expectations in the field of multispectral data analysis are diversifying and many subject areas are affected. So, fields as varied as urbanism, hydrology, agronomy, ecology and health are resorting to the analysis of satellite or airborne images and are requesting specific methods.

The statistical approaches constitute a possible response to these needs. In a unsupervised way, they allow the extraction of consistent sets from multi-dimensional data. In a supervised way, they define and extrapolate to the whole image links (sometimes complex) between the radiometry and the variables to be estimated (membership value or quantitative value of a parameter). This chapter develops, in particular, the principle of SVM and SVR techniques, which are widely used and offer interesting opportunities.

With the exception of a few geometric and radiometric preprocessing procedures, it is up to the readers to choose the methods which seem the best adapted with respect to their needs and capabilities (computing resources, execution time and expected). Indeed, the use of multispectral satellite images is always the result of a reflection conducted upstream. For example, a classification implies among other things building in advance a land cover nomenclature, to define a scale and a minimum unit of mapping. Before embarking on sometimes lengthy and complicated processing, and before acquiring images which can sometimes be expensive, it is necessary to identify the issues, the objectives and the expected results in order to correctly specify the input data, and the suitable tools and methods.

Multispectral image analysis methods have developed a lot in these last 20 years due to the increase in the available calculation means and significant progress in the areas of data analysis (signal and image processing, applied mathematics and statistics in particular). So the current problems are no longer limited to the recognition of “low level” entities in the image (road, urban area and vegetation) but the availability of more precise data now allows access to more “high level” information (distinction of vegetation type, habitats and identification of patterns of agricultural management). Some new data are now accessible (drones data, optical data, in particular those free of charge from the SENTINEL-2 sensor which has a very good spatial resolution – approximately 10 m – and temporal resolution – 5 days) and this requires a review of the theoretical foundations of the data analysis tools. The issues concerning dimension reduction and interpretation support tools are, among others, key points. It is a question of reducing the size of the data but, unlike a PCA, we are looking for “interpretable” bases related to a particular semantic (for example, a base can consist of the element "building" in an "agricultural plot"). The issues of clustering and classification are also crucial and the definition of fast algorithms to be run is an important research axis. Finally, we must state that the domain adaptation problems (how to adapt the training and learning phases when the acquisition conditions differ between the data sets and the test sets), active training (allowing the user to confirm in real time the classification results) or time series processing (how to properly address the fact that two time series can be shifted in time while representing the evolution of the same phenomenon) are points which are currently being studied in order to manage the significant and new flow of Earth observation data.

2.7. Key points

The satellite image is a complex product that requires a lot of preprocessing prior to the interpretation phase and information extraction. The automatic processing of images is more and more essential because of the increasing number of images. The easy access to very high spatial resolution optical data makes it necessary to use processing tools adapted to the presence of a strong spectral heterogeneity within the same object. Image processing software that are open-source/free of charge or even commercial ones are offering users more powerful and user-friendly algorithms. Today, there are many positive factors contributing to the democratization of the use of remote sensing data.

2.8. Bibliography

[AGU 13] AGUILAR M.A., SALDAÑA M., DEL M. *et al.*, "Assessing geometric accuracy of the orthorectification process from GeoEye-1 and WorldView-2 panchromatic images", *International Journal of Applied Earth Observation and Geoinformation*, vol. 21, pp. 427–435, 2013.

[BAR 10] BARALDI A., GIRONDA M., SIMONETTI D., "Operational two-stage stratified topographic correction of spaceborne multispectral imagery employing an automatic spectral-rule-based decision-tree preliminary classifier", *IEEE Transactions on Geoscience and Remote Sensing*, vol. 48, no. 1, pp. 112–146, 2010.

[BET 14] BETBEDER J., RAPINEL S., CORPETTI T. *et al.*, "Multitemporal classification of TerraSAR-X data for wetland vegetation mapping", in BERNARDO J.M., BAYARRI M.J., BERGER J.O. *et al.*, (eds), *Journal of Applied Remote Sensing,* vol. 8, no. 1, pp. 083648–083648, 2014.

[BIS 07] BISHOP C.M., LASSERRE J., "Generative or discriminative? Getting the best of both worlds", in BERNARDO J.M., BAYARRI M.J., BERGER J.O. *et al.*, (eds), *Bayesian Statistics 8,* Oxford University Press, pp. 3–23, 2007.

[BRE 84] BREIMAN L., FRIEDMAN J.H., OLSHEN R.A. *et al.*, *Classification and Regression Trees*, Wadsworth, Belmont, United States, 1984.

[BRI 16] BRIOTTET X., "Radiometry in the optical domain", in BAGHDADI N., ZRIBI M., *Microwave Remote Sensing of Land Surfaces*, ISTE Press Ltd, London and Elsevier Ltd, Oxford, 2016.

[CAL 01] CALOZ R., COLLET C., *Précis de Télédétection, Traitements numériques d'images de Télédétection*, vol. 3, Presses de l'université du Québec, Québec, 2001.

[CHA 88] CHAVEZ P.S., "An improved dark-object substraction technique for atmospheric scattering correction of multispectral data", *Remote Sensing of Environment*, vol. 24, pp. 459–479, 1988.

[CHA 96] CHAVEZ P.S., "Image-based atmospheric corrections – revisited and improved", *Photogrammetric Engineering & Remote Sensing*, vol. 62, no. 9, pp. 1025–1036, 1996.

[DUS 11] DUSSEUX P., HUBERT-MOY L., LECERF R. *et al.*, "Identification of grazed and mown grasslands using a time series of high-spatial-resolution remote sensing images", *6th International Workshop on the Analysis of Multi-Temporal Remote Sensing Images (Multi-Temp)*, pp. 145–148, 2011.

[FRE 99] FREUND Y., SCHAPIRE R.E., "Large margin classification using the perceptron algorithm", *Machine Learning*, vol. 37, no. 3, pp. 277–296, 1999.

[FRE 09] FREEDMAN D.A., *Statistical Models: Theory and Practice*, Cambridge University Press, 2009.

[GAO 96] GAO B.C., "NDWI – a normalized difference water index for remote sensing of vegetation liquid water from space", *Remote Sensing of Environment*, vol. 58, no. 3, pp. 257–266, 1996.

[HAJ 08] HAJJ M.E., BÉGUÉ A., LAFRANCE B. *et al.*, "Relative radiometric normalization and atmospheric correction of a SPOT 5 time series", *Sensors*, vol. 8, no. 4, pp. 2774–2791, 2008.

[JAW 13] JAWAK S., LUIS A., "A comprehensive evaluation of PAN-sharpening algorithms coupled with resampling methods for image synthesis of very high resolution remotely sensed satellite data", *Advances in Remote Sensing*, vol. 2, no. 4, pp. 332–344, 2013.

[KAU 92] KAUFMAN Y.J., TANRE D., "Atmospherically resistant vegetation index (ARVI) for EOS-MODIS", *IEEE Transactions on Geoscience and Remote Sensing*, vol. 30, no. 2, pp. 261–270, 1992.

[KER 00] KERGOMARD C., "Pratique des corrections atmosphériques en télédétection : utilisation du logiciel 5S-PC", *Cybergeo : European Journal of Geography*, document 181, available at http://cybergeo. revues.org/1679, 2000.

[KOH 95] KOHAVI R., "A study of cross-validation and bootstrap for accuracy estimation and model selection", *International joint Conference on Artificial Intelligence*, vol. 14, no. 2, pp. 1137–1145, 1995.

[LEF 08] LEFEBVRE A., CORPETTI T., HUBERT-MOY L., "Object-oriented approach and texture analysis for change detection in very high resolution images", *IEEE International Geoscience and Remote Sensing Symposium,* vol. 4, pp. IV–663, 2008.

[LEF 11] LEFEBVRE A., Contribution de la texture pour l'analyse d'images à très haute résolution spatiale : application à la détection de changement en milieu périurbain, PhD thesis, University of Rennes, 2011.

[LER 15] LE ROUX R., NEETHLING E., VAN LEEUWEN C. *et al.*, "Multi-scalar modelling of climate applied to European vineyard sites in the climate change context", *Procedia Environmental Sciences,* vol. 29, pp. 62–63, 2015.

[MAT 98] MATERKA A., STRZELECKI M., Texture analysis methods – a review, Technical University of Lodz, COST B11 report, Brussels, 1998.

[MER 03] MERCIER G., LENNON M., "Support vector machines for hyperspectral image classification with spectral-based kernels", *IEEE International Geoscience and Remote Sensing Symposium*, vol. 1, pp. 288–290, 2003.

[MIN 41] MINNAERT M., "The reciprocity principle in lunar photometry", *Astrophysical Journal,* vol. 93, pp. 403–410, 1941.

[MOR 92] MORAN M.S., JACKSON R.D., SLATER P.N. *et al.*, "Evaluation of simplified procedures for retrieval of land surface reflectance factors from satellite sensor output", *Remote Sensing of Environment*, vol. 41, pp. 169–184, 1992.

[OMR 07] OMRAN M.G., ENGELBRECHT A.P., SALMAN A., "An overview of clustering methods", *Intelligent Data Analysis*, vol. 11, no. 6, pp. 583–605, 2007.

[PAL 05] PAL M., "Random forest classifier for remote sensing classification", *International Journal of Remote Sensing*, vol. 26, no. 1, pp. 217–222, 2005.

[QUÉ 14] QUÉNOL H., "Méthodes d'analyse et de modélisation agroclimatique du changement climatique à l'échelle des terroirs viticoles", *Changement climatique et terroirs viticoles*, Lavoisier Tec&Doc, 2014.

[RIA 03] RIAÑO D., CHUVIECO E., SALAS J. *et al.*, "Assessment of different topographic corrections in Landsat-TM Data for mapping vegetation types", *IEEE Transactions on Geoscience and Remote Sensing*, vol. 41, no. 5, pp. 1056–1061, 2003.

[SAL 04] SALOMONSON V.V., APPEL I., "Estimating fractional snow cover from MODIS using the normalized difference snow index", *Remote Sensing of Environment*, vol. 89, no. 3, pp. 351–360, 2004.

[SCH 02] SCHÖLKOPF B., SMOLA A.J., *Learning with Kernels: Support Vector Machines, Regularization, Optimization, and Beyond*, MIT Press, p. 648, 2002.

[TEI 82] TEILLET P.M., GUINDON B., GOODENOUGH D.G., "One the slope-aspect correction of multispectral scanner data", *Canadian Journal of Remote Sensing*, vol. 8, no. 2, pp. 84–106, 1982.

[TOU 04] TOUTIN T., "Geometric processing of remote sensing images: models, algorithms and methods", *International Journal of Remote Sensing*, vol. 25, no. 10, pp. 1893–1924, 2004.

[VER 97] VERMOTE E.F., TANRÉ D., DEUZÉ J.L. *et al.*, "Second simulation of the satellite signal in the solar spectrum, 6s: an overview", *IEEE Transactions on Geoscience and Remote Sensing*, vol. 35, no. 3, pp. 675–686, 1997.

[WHI 04] WHITESIDE A., Some image geometry models, *Open Geospatial Consortium,* document 04-071, available at: http://www.opengeospatial.org/standards/dp, 2004.

[ZHA 02] ZHANG Y., "Problems in the fusion of commercial high-resolution satellites images as well as LANDSAT 7 images and initial solutions", *Proceedings of the ISPRS, CIG, and SDH Joint International Symposium on Geospatial Theory, Processing and Applications*, pp. 9–12, 2002.

[ZHA 08] ZHANG Y., "Pan-sharpening for improved information extraction", in LI Z., JUN C., BALTSAVIAS E. (eds), *Advances in Photogrammetry, Remote Sensing and Spatial Information Sciences*, CRC Press, pp. 185–204, 2008.

[ZHA 13] ZHANG N., HONG Y., QIN Q. *et al.*, "VSDI: a visible and shortwave infrared drought index for monitoring soil and vegetation moisture based on optical remote sensing", *International Journal of Remote Sensing*, vol. 34, no. 13, pp. 4585–4609, 2013.

3

Digital Terrain Models from Optical Images

3.1. Introduction: the main principles of spatial stereoscopy

Images, which are two-dimensional (2D), allow us to visualize a scene and locate each point of longitude and latitude within this scene. This 2D information is needed for many applications including mapping. Stereoscopy provides an estimate for the third dimension in this scene: altitude. To do this, in spatial stereoscopy, one can put two pairs of "eyes" on board a satellite, like on SPOT 5 (high-resolution spatial sensor (HRS)): one of the devices looks ahead while the other looks behind. The Pleiades satellite only uses one device but has the agility to achieve stereoscopy: it takes two images of the same area in a very short space of time and from two different angles. The ability to pinpoint the altitude of each point within a scene allows users to access highly accurate three-dimensional information at local to global scales. This additional dimension first and foremost improves image interpretation. Developments can thus be easily quantified using elevation change detection techniques once multi-temporal stereoscopic information is obtained. It is a fundamental tool for better understanding and decision support for problems such as defense, public safety, urban planning, environment, glaciology, seismology, etc.

How can we obtain this altitude information from a stereoscopic pair? By being aware that the relief distorts obliquely viewed images. Indeed,

---

Chapter written by Jean-Marc DELVIT and Julien MICHEL.

let us consider, for example, a tower on the ground: it finds itself "lying" on the image, with its summit shifted to the right of the base of the tower, as shown in Figure 3.1.

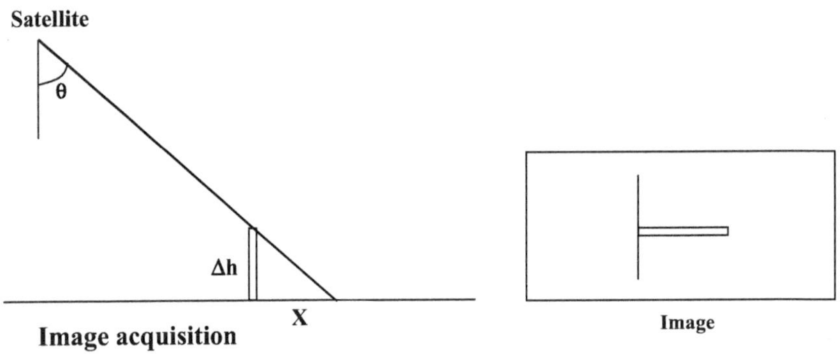

Figure 3.1. *Displacement of objects due to the relief*

Displacement ΔX of objects, due to the relief, depends on their elevation Δh and the viewing angle θ, according to the relationship $\Delta X = \Delta h.\tan(\theta)$ (see Figure 3.1).

Note that, from examining the above relationship, the greater the viewing angle, the greater the displacement will be. It is therefore possible to restore the altimetry of a scene based on these deformations. To do this, at least two images of the same area must be acquired under different angles of incidence. These images form one or more stereoscopic pairs. The measurement accuracy of the terrain, called vertical accuracy, is a function of the factor B/H, called the stereoscopic angle. B represents the distance between the two viewing centers (S_1 and S_2 in Figure 3.2) and is called the baseline. This can be two different carriers (two satellites), or could also be two detectors in the same focal plane, two cameras on the same satellite or two images acquired successively by an agile satellite, as is the case for Pleiades satellites. Height H represents the average altitude of the satellites. Let us consider a ground point of altitude Δh. We obtain the relationship:

$$\Delta X = \frac{B}{H}\Delta h \qquad [3.1]$$

By measuring ΔX and by perfectly mastering geometric modeling (characterized by the intersection of the target (S_1P_1) and (S_2P_2), Figure 3.2) which can be approximated by the ratio B/H, it is possible to deduce Δh. In this chapter, we will discuss geometric modeling (how to best intersect the targeted directions) and we will show how to measure disparity ΔX, which are both key performance elements [VAL 08].

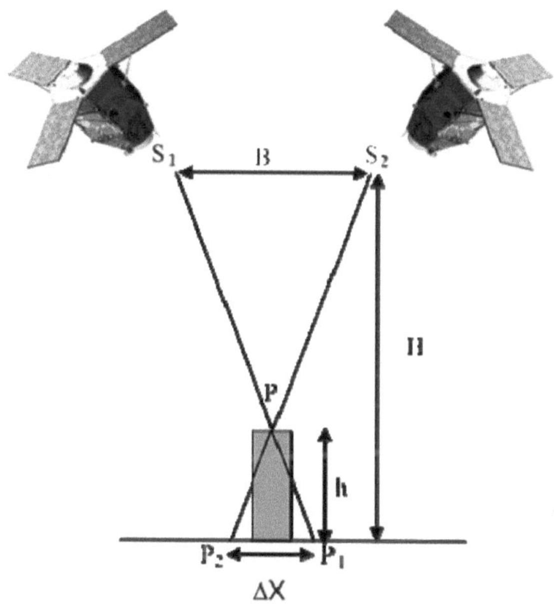

Figure 3.2. *Example of lateral stereoscopy*

3.2. Geometric modeling of the scene

3.2.1. *General principle*

Modeling the scene involves locating a pixel of the raw image on the ground, and conversely calculating which pixel of the raw image contains the detail of the ground. We therefore define the direct model and the inverse model as follows:

– the direct model, also called localization function, is the mathematical relationship that allows us to locate a pixel (l,c), where l is the index of the row and c is the index of the column from the ground image to altitude h [RAD 91]:

- $\left(x,y,h\right)_{\text{TerrestrialReference}} = F(l,c,h)$

with (x,y,h) map coordinates of the point on the ground;

- $\left(\lambda,\phi,h\right)_{\text{TerrestrialReference}} = F(l,c,h)$

with (λ,φ,h) geographical coordinates of point on the ground;

- $\left(X,Y,Z\right)_{\text{TerrestrialReference}} = F(l,c,h)$

with (X,Y,Z) Cartesian coordinates of the point on the ground.

– The inverse model is the mathematical relationship that allows us to know which pixel (l,c) contains detail on the ground of coordinates (x,y,h) or (λ,φ,h) or (X, Y, Z):

- $(l,c) = G\left(x,y,h\right)_{\text{TerrestrialReferenceFrame}}$;

- $(l,c) = G\left(\lambda,\phi,h\right)_{\text{TerrestrialReferenceFrame}}$;

- $(l,c) = G\left(X,Y,Z\right)_{\text{TerrestrialReferenceFrame}}$;

There are three major types of model:

– the physical model;

– the analytical model;

– the sampled model.

The physical model, also called the geometric viewing model, is calculated from the scene parameters provided by the system. These scene parameters are only known with a certain precision, which is insufficient for many applications. It is possible to refine the physical model, thus improving the accuracy of the scene parameters through external data.

The analytical model is used when the scene parameters are not known or to simplify exchanges between users. In this case, the direct and inverse models are modeled by polynomials or rational functions.

The sampled model is a model in the form of a localization grid: for each point on the grid, (l,c) is known in the raw image and for the ground coordinates of the ground point corresponding to (x,y,h), (λ,φ,h) or (X,Y,Z). This model is derived from either the physical model or the analytical model. It is used in certain downstream applications, such as geometric corrections.

3.2.2. *Physical model of the scene*

3.2.2.1. *Calculating the direct model*

The physical viewing model, also called the geometric model, requires knowledge of the scene parameters, namely:

– the time of acquisition of the pixel;

– the position of the satellite at that time;

– the orientation of the satellite at that time;

– the viewing direction of the detector relative to the satellite;

– the altitude of the target point.

These scene parameters are measured either on board or on the ground and are then provided in the auxiliary data of the image. The data for calculating the position, the satellite's orientation and the viewing directions are sampled. For example, satellite position is given at different times t in a ephemerides table. The first step in developing the physical model of image acquisition is therefore to model the various phenomena involved in a defined manner: satellite position can be expressed in the form of an arc of local orbit (during the acquisition) and the satellite's orientation can be expressed in the form of polynomials in terms of roll, pitch and yaw. The advantage of this approach is to achieve a comprehensive and consistent preprocessing of each scene segment (thus not reduced to a single scene), which then allows rapid calculation of the localization function and the possibility of global refinement of the model.

After this initialization step, the second step is the direct implementation of the physical viewing model, which involves calculating the intersection of the viewing direction of the sensor at the time of pixel acquisition, with the Earth, which is modeled as an ellipsoid or a digital terrain model (DTM). The viewing direction must be expressed in the terrestrial reference frame.

However, it is supplied as a reference linked to either the sensor, the instrument or the satellite. The scene parameters, such as the orientation of the sensor relative to the instrument, the orientation of the instrument relative to the satellite, the satellite attitude and the satellite position, will allow us, per set of rotations, to calculate the viewing direction in a terrestrial reference frame. Figure 3.3 illustrates the general principle of this calculation. It may be slightly different depending on the data and references used.

Once the viewing direction is calculated in the terrestrial reference frame, it just needs to be intersected with the Earth:

– if Earth is modeled by an ellipsoid and we know height h relative to the ellipsoid of the target point, this calculation is reduced to solving a simple quadratic equation;

– if Earth is modeled by a DTM, the calculation is a bit more complex and consists of searching through terrain elevation point iterations until there is convergence.

After this second step, we obtain the direct geometric model of image acquisition.

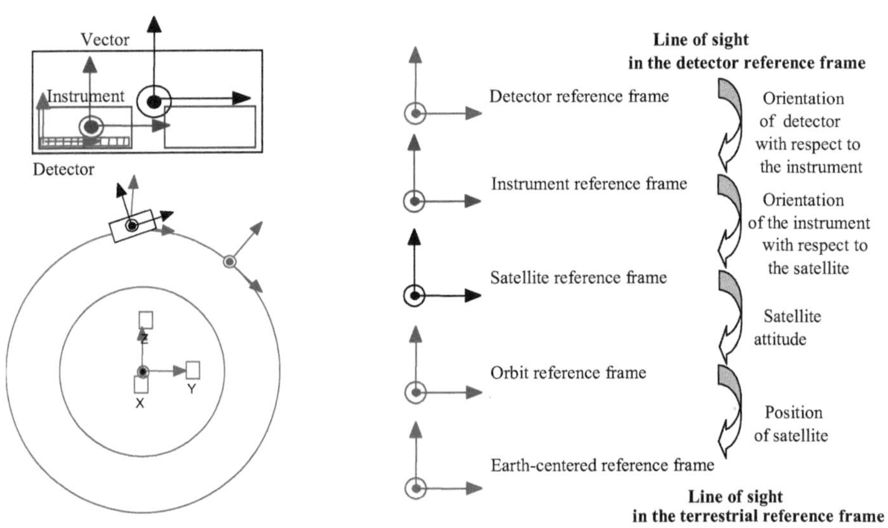

Figure 3.3. *General principle for calculating the viewing direction in the terrestrial reference frame*

3.2.2.2. *Calculating the inverse model*

In practical applications, we often use the inverse model rather than the direct model.

3.2.2.2.1. On a known elevation plane

This involves calculating (l, c) from ground coordinates (λ, φ, h).

This calculation is performed through an iterative process successively involving an inverse localization predictor and direct localization onto a plane.

The inverse localization predictor is a function that allows us to approximately connect the ground coordinates (λ, φ, h) to image coordinates (l, c).

This predictor takes the form of a pair of polynomials.

These polynomials are calculated by least squares from a direct localization grid of size (5 × 5) projected at elevation $h_{\min}$:

$$\begin{aligned} \tilde{l}_{\min} &= b_0 + b_1\lambda + b_2\varphi + b_3\lambda^2 + b_4\varphi^2 + b_5\lambda\varphi + b_6\lambda^3 + b_7\varphi^3 + b_8\lambda^2\varphi + b_9\lambda\varphi^2 \\ \tilde{c}_{\min} &= a_0 + a_1\lambda + a_2\varphi + a_3\lambda^2 + a_4\varphi^2 + a_5\lambda\varphi + a_6\lambda^3 + a_7\varphi^3 + a_8\lambda^2\varphi + a_9\lambda\varphi^2 \end{aligned} \quad [3.2]$$

They are also calculated by least squares from a direct localization grid of size (5 × 5) projected at elevation $h_{\max}$:

$$\begin{aligned} \tilde{l}_{\max} &= e_0 + e_1\lambda + e_2\varphi + e_3\lambda^2 + e_4\varphi^2 + e_5\lambda\varphi + e_6\lambda^3 + e_7\varphi^3 + e_8\lambda^2\varphi + e_9\lambda\varphi^2 \\ \tilde{c}_{\max} &= d_0 + d_1\lambda + d_2\varphi + d_3\lambda^2 + d_4\varphi^2 + d_5\lambda\varphi + d_6\lambda^3 + d_7\varphi^3 + d_8\lambda^2\varphi + d_9\lambda\varphi^2 \end{aligned} \quad [3.3]$$

For a given pair (λ, φ), these predictors give the approximate image localizations $(\tilde{l}_{\min}, \tilde{c}_{\min})$ and $(\tilde{l}_{\max}, \tilde{c}_{\max})$ of ground points $(\lambda, \varphi, h_{\min})$ and $(\lambda, \varphi, h_{\max})$. These values are then used to calculate the approximate localization of point (λ, φ, h) by inverse linear interpolation according to altitude.

This inverse localization predictor only allows us to initialize the iterative process.

The iterative process is performed until the projection onto a plane at altitude h of the image point calculated in this way is in the close vicinity (set value generally a few mm) of the starting point (λ,φ,h).

This method is quite effective: determining the polynomial coefficients in a (5×5) grid is immediate and the iterative process converges quickly after a few iterations.

3.2.2.2.2. On DTM

Consideration of the DTM is not a problem here, because from point (λ,ϕ) and the DTM, we can deduce the value of altitude h and return to the previous case.

However, two DTM operating methods exist:

– the conventional method: the altitude at a point (λ,ϕ) is calculated through a bilinear interpolation of the DTM at this point (the DTM and the coordinates of point (λ,ϕ) must be in the same representation system);

– the faceted method: the altitude at a point (λ,ϕ) is determined by intersecting the local standard at a point with the corresponding facet.

The above two methods can be used punctually or on grids.

3.2.3. *Analytical modeling of the image geometry*

In the case of analytical models, direct and inverse models are described by polynomials or rational functions. This type of model can be used for simplicity: it is generic in its form and very fast-calculating, especially for inverse localization. It can be supplied in addition to the physical model (as is the case for the Pleiades[1] or Worldview[2] satellites).

1 http://pleiades.cnes.fr and http://www.geo-airbusds.com/fr/237-pleiades-imagerie-satellite-tres-haute-resolution.

2 http://www.satimagingcorp.com/satellite-sensors/worldview-3/.

Rational functions are analytic functions for connecting an image pixel (defined by its row and column coordinates) with the value of its ground localization (defined by geographical coordinates). These functions depend on altitude.

The so-called "inverse" rational function is one that calculates the position of a pixel in the image based on its position on the ground (by analogy with the "inverse localization"). It is defined as follows:

$$\begin{array}{lccc} & \Re^3 & \to & \Re \\ F_C: & (\lambda,\varphi,h) & \to & \tilde{c} \\ F_L: & (\lambda,\varphi,h) & \to & \tilde{l} \end{array} \quad [3.4]$$

F_C and F_L are rational functions for which the most general form is written:

$$\begin{cases} F_C: & \Re^3 \to \Re \\ & (\lambda,\varphi,h) \to \tilde{c} = F_C(\lambda,\varphi,h) = \dfrac{\sum_{i=0}^{d}\sum_{j=0}^{d-i}\sum_{k=0}^{d-j-1} p_{ijk}\lambda^i\varphi^j h^k}{\sum_{i=0}^{d}\sum_{j=0}^{d-i}\sum_{k=0}^{d-j-1} q_{ijk}\lambda^i\varphi^j h^k} = \dfrac{P(\lambda,\varphi,h)}{Q(\lambda,\varphi,h)} \end{cases} \quad [3.5]$$

Function F_L is analogous.

Usually, we use functions of degree 3 $(d=3)$, that is to say that the polynomials $P(\lambda,\varphi,h)$ and $Q(\lambda,\varphi,h)$ are written:

$$\begin{aligned} P(\lambda,\varphi,h) = {} & p_0 + p_1\lambda + p_2\varphi + p_3 h + p_4\lambda^2 + p_5\lambda\varphi + p_6\lambda h + p_7\varphi^2 + \\ & p_8\varphi h + p_9 h^2 + p_{10}\lambda^3 + p_{11}\lambda^2\varphi + p_{12}\lambda^2 h + p_{13}\lambda\varphi^2 + \\ & p_{14}\lambda\varphi h + p_{15}\lambda h^2 + p_{16}\varphi^3 + p_{17}\varphi^2 h + p_{18}\varphi h^2 + p_{19}h^3 \end{aligned} \quad [3.6]$$

$$\begin{aligned} Q(\lambda,\varphi,h) = {} & 1+q_1\lambda+q_2\varphi+q_3h+q_4\lambda^2+q_5\lambda\varphi+q_6\lambda h+q_7\varphi^2+q_8\varphi h+ \\ & q_9h^2+q_{10}\lambda^3+q_{11}\lambda^2\varphi+q_{12}\lambda^2h+q_{13}\lambda\varphi^2+q_{14}\lambda\varphi h+ \\ & q_{15}\lambda h^2+q_{16}\varphi^3+q_{17}\varphi^2h+q_{18}\varphi h^2+q_{19}h^3 \end{aligned} \quad [3.7]$$

Every rational function thus comprises 39 coefficients.

One can also define the "direct" rational function for calculating ground coordinates of the pixel from its image coordinates (by analogy with the "direct localization"). In this case, the functions are defined as:

$$\begin{array}{lccc} & \Re^3 & \rightarrow & \Re \\ F_\lambda : & (l,c,h) & \rightarrow & \tilde{\lambda} \\ F_\varphi : & (l,c,h) & \rightarrow & \tilde{\varphi} \end{array} \quad [3.8]$$

Analogously to inverse rational functions, F_λ and F_φ are usually degree 3.

The main advantage of rational functions is to provide the user with an easy-to-use model that does not require downstream knowledge and use of geometric viewing parameters, and does not require complex software modeling. They also allow us to take altitude into account.

3.2.4. *Refining the viewing geometric model*

Refinement of one or more geometric models involves improving the performance of geometric or imaging systems. This means improving the performance of the absolute localization and internal consistency of images, using external data (ground control points (GCPs), maps, etc.) and/or characteristic points seen on each image.

In this section, we denote the direct geometric model as:

$$\begin{cases} x = F_x\left(l,c,h,P_1^0,\ldots,P_k^0,\ldots,P_n^0\right) \\ y = F_y\left(l,c,h,P_1^0,\ldots,P_k^0,\ldots,P_n^0\right) \end{cases} \quad [3.9]$$

where $\left(P_1^0,\ldots,P_k^0,\ldots,P_n^0\right)n$ scene parameters are provided by the system (such as orbit, attitude, dating, viewing directions, or coefficients of an analytic function, etc.).

3.2.4.1. *General principle*

The physical viewing model is imperfect, because of the inaccuracy of the viewing parameters P_k^0 provided by the system. We notice, for example, errors in orbit restitution, ranging from a few meters to a hundred meters and altitude restitution errors of up to a few milliradians on some satellites (e.g. 1 microradian equals one pixel for the Pleiades satellite), which are due to a misunderstanding of the sensor/satellite alignments.

On SPOT 4, for example, absolute localization performance (provided by the system, which therefore depends only on the accuracy of the sceneparameters) was about 350 m root mean square error. Indeed, the ground sensors on board SPOT4 were not able to obtain good attitude restitution accuracy, resulting in poor performance of absolute localization, unlike Pleiades, which has a very good geometrical performance.

These characteristics are insufficient for applications requiring high precision such as mapping or 3D restitution. To improve the model, we then use information that is external to the acquisition system, such as:

– GCPs, or landmarks, which are known ground and image coordinate points that ensure absolute recalibration of the image relative to the ground;

– homologous points, or tie points, which are unknown plane coordinate points but with known image coordinates of at least two images (even terrain detail seen on several images); these points are used to calculate the relative recalibration of multiple images.

Improving the geometric model through these ground control and homologous points is done by estimating the real value P_k of the image parameters, or at least some of them. This is called refining or adjusting the model. The principle of this calculation is as follows:

– introducing unknowns dP_k to the value of certain scene parameters P_k^0 provided by the system; these unknowns are called model adjustment parameters and it is they that we try to estimate;

– the GCPs and tie points used to establish the following equations:

- for a GCPi of image coordinates (l_i, c_i) and ground coordinates (x_i, y_i, h_i):

$$\begin{cases} x_i = F_x(l_i, c_i, h_i, P_1, \ldots, P_k, \ldots, P_n) \\ y_i = F_y(l_i, c_i, h_i, P_1, \ldots, P_k, \ldots, P_n) \end{cases} \quad [3.10]$$

where $P_k = P_k^0 + dP_k$;

- for a homologous point j of image coordinates (l_j, c_j) on image 1, of image coordinates $(l_j^{'}, p_j^{'})$ on image 2 and ground altitude h_j:

$$\begin{cases} F_x(l_j, c_j, h_j, P_1, \ldots, P_k, \ldots, P_n) = F_x^{'}(l_j^{'}, c_j^{'}, h_j, P_1^{'}, \ldots, P_k^{'}, \ldots, P_n^{'}) \\ F_y(l_j, c_j, h_j, P_1, \ldots, P_k, \ldots, P_n) = F_y^{'}(l_j^{'}, c_j^{'}, h_j, P_1^{'}, \ldots, P_k^{'}, \ldots, P_n^{'}) \end{cases} \quad [3.11]$$

where F is the direct model of image 1 and F'is the direct model of image 2.

– we obtain a system of equations that must be linearized and then solved using the least squares method; solving this system provides us with the values of unknowns dP_k and therefore the refined geometric model:

$$\begin{cases} x = F_x(l, c, h, P_1, \ldots, P_k, \ldots, P_n) \\ y = F_y(l, c, h, P_1, \ldots, P_k, \ldots, P_n) \end{cases} \quad [3.12]$$

3.2.4.2. *Choice of unknowns*

The choice of unknowns can be tricky. In fact, errors of various scene parameters can have similar impacts on the ground. For example, an error in pitch, in time-tagor on the longitudinal position of the satellite induces a shift in the ground point along the track. These parameters are then said to be correlated. In this case, how can we estimate the real value of these parameters?

First, we only introduce unknowns to the most inaccurate imaging parameters, those that cause the greatest error on ground positioning if ignored. By solving the system, we do not necessarily get the real value of the scene parameters but rather we obtain values that allow us to correct image defects.

Moreover, in estimation problems, additional "constraint equations" are generally also introduced to constrain the value of the unknowns to within a realistic range, especially when several images are simultaneously realigned. This serves to strengthen the models.

Detailed knowledge of defects in the scene parameters is important: it provides a high level of accuracy for a limited number of measurements (GCPs or homologues) and allows access to a realistic estimate of the real value these parameters.

3.2.4.3. *Choice of ground control points*

GCPs are known ground coordinate points that can be located precisely on the image. They are used to absolutely recalibrate the geometric imaging model (with respect to the ground).

The modeling accuracy and therefore quality of geometric corrections that ensue depend on:

– the number of GCPs;

– their distribution in the image;

– their accuracy.

Each GCP bears two equations (in two directions, one in x and one in y). We therefore need a minimum of nbinc/2 GCPs (with nbinc number of unknowns) to recalibrate the model. But redundancy is necessary in order to identify and eliminate false points during the calculation of unknowns. It is indeed possible to misrecognize an image detail or for the detail to change over time from the design of the map to the date of acquisition of the image: different road elements, agricultural limits, etc. The number of GCPs also depends on the characteristics of the image: its resolution and its readability (optical or radar image, etc.).

The distribution of points in the image is also important. It is best to choose the most widely distributed points throughout the image. The accuracy of GCPs is a determinant of modeling accuracy. The measurement accuracy of the point in the image can vary, depending on the nature of the point (manual or automatic correlation) and the characteristics of the image (resolution, readability, etc.). In addition, the accuracy of field coordinates of the GCP may also vary according to the nature of their measurement (manual measurement on a map, GPS measurement, etc.) and the accuracy of the reference used. Thus, if maps are used to measure the ground coordinates of GCPs, it is essential to know the accuracy of the maps used, which can vary significantly (from several meters to several tens of meters on older maps). Finally, the quality of the GCPs is also important, especially in a spatiotriangulation process, where it may be necessary to identify the same GCP on different images, sometimes several years apart. The GCPs must of course be clearly identifiable in both the image and the reference frame. The sharpness of detail depends on the characteristics of the image, of the method of measurement, etc.

The accuracy of physical modeling will depend on all these factors. With very good GCPs, modeling accuracy on SPOT5 images can be achieved to a few meters with a resolution of 2.5 m or within 1 m on Pleiades.

3.2.4.4. *Simultaneous refinement of several models: spatio-triangulation*

Spatiotriangulation is a method for simultaneously recalibrating, absolutely and/or relatively, a set of scenes grouped into a single block within an area. This method is based on a physical description of the acquisitions that is as close as possible to acquisition conditions and therefore limits the number of unknowns. This is an essential step for many applications including 3D restitution.

The scenes, acquired under the same conditions as the acquisition, are grouped into segments; it is the segment that is the modeling unit, and not the scene, meaning that the unknowns are linked to the segment. Indeed, it is logical that consecutive scenes, acquired under the same conditions (same instrument, same sensor, etc.), share the same unknowns. In addition, this method allows multiple segments to share certain unknowns (for example, segments that were simultaneously acquired by two different instruments of the same platform can share unknowns on orbit determination and attitude of the satellite). With this method, the geometry of the acquisition is respected. The number of unknowns is considerably reduced and the system to be solved is therefore less complex.

This method uses two types of points:

– GCPs, known planimetric and altimetric terrain coordinates, and known image coordinates in at least one image: these points ensure absolute calibration of the segments;

– tie points, still known as homologous points, with unknown planimetric coordinates but for which altimetric coordinates are generally known, and known image coordinates in at least two images: these points ensure relative matching between adjacent segments.

The choice of GCPs and connection points is done in the same way as the recalibration of a single scene. Figure 3.4 illustrates the recalibration of segments by spatiotriangulation.

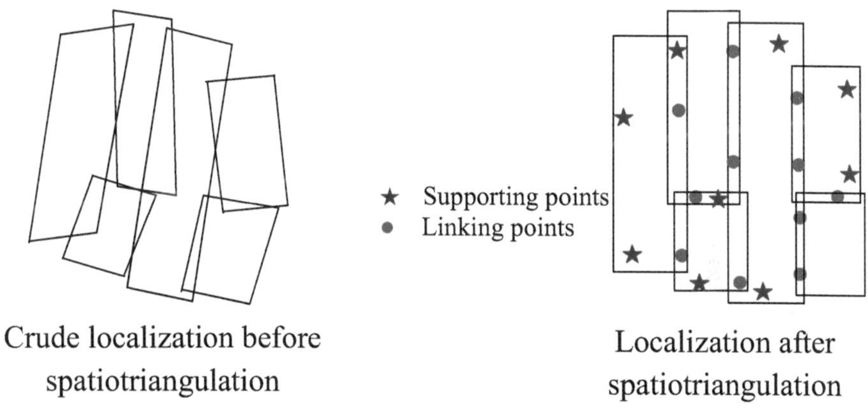

Figure 3.4. *Adjusting segments by spatiotriangulation*

There are two types of unknowns: unknowns in geometric models and unknown points. Spatiotriangulation therefore makes it possible to simultaneously estimate:

– unknowns in geometric models of segments;

– real ground coordinates of points;

– real image coordinates of points.

To calculate these unknowns, we must establish and solve a system of equations. This system consists of three types of equations:

1) image equations, which are observation equations of each measurement (l_{mes},c_{mes}) of the point M on the image I, weighted by the uncertainty of the image measurement σ_I :

– For a GCP, we obtain:

$$\begin{cases} \delta l = l_{mes} - l_{est} = l_{mes} - G_{I_l}(X_{est}, Y_{est}, Z_{est}, P_1^0, ..., P_n^0, dP_1, ..., dP_n) \\ \delta p = c_{mes} - c_{est} = c_{mes} - G_{Ip}(X_{est}, Y_{est}, Z_{est}, P_1^0, ..., P_n^0, dP_1, ..., dP_n) \end{cases} \quad [3.13]$$

– For a homologous point, we obtain:

$$\begin{cases} \delta l = l_{estI} - l_{estI'} = G_{I_l}(X_{est}, Y_{est}, Z_{est}, P_1^0, ..., P_n^0, dP_1, ..., dP_n) - \\ \qquad G_{I'_l}(X_{est}, Y_{est}, Z_{est}, P_1^0, ..., P_n^0, dP_1, ..., dP_n) \\ \delta c = c_{estI} - c_{estI'} = G_{I_p}(X_{est}, Y_{est}, Z_{est}, P_1^0, ..., P_n^0, dP_1, ..., dP_n) - \\ \qquad G_{I'_p}(X_{est}, Y_{est}, Z_{est}, P_1^0, ..., P_n^0, dP_1, ..., dP_n) \end{cases} \quad [3.14]$$

where G_I is the inverse geometric model of image I and $G_{I'}$ is the inverse geometric model of image I '.

2) the field equations, relationships between the measured point and the estimated point:

– $dX = X_{mes} - X_{est}$ and $dY = Y_{mes} - Y_{est}$ weighted by the uncertainty of the planimetric measurement σ_{XY};

– $dZ = Z_{mes} - Z_{est}$ weighted by the uncertainty of the altimetric measurement σ_Z.

3) the rigidity of equations, also known as constraint equations, on the unknowns dP_k, weighted by the accuracy of scene parameters σ_{Pk}.

The system thus formed is not linear. It must therefore be linearized in order to solve it by iterations, for example by using the least squares method. Figure 3.5 illustrates the general principle of spatiotriangulation.

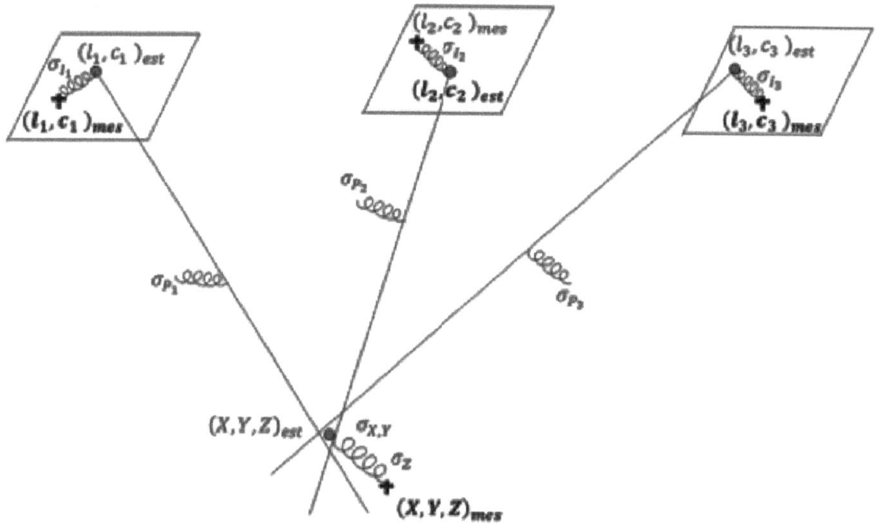

Figure 3.5. *General principle of spatiotriangulation*

3.2.5. *Concept of epipolar geometry*

Let us consider a regular surface S close to the actual land surface. Depending on the extent of the area to investigate and the main form of S, we show that S can be approximated locally by an ellipsoid. An epipolar curve E will be defined as an S-curve for which the tangent at any point M_0 on the curve is in the plane defined by the two homologous radii passing through M_0.

Let M be a point on the terrain represented by S.

Let m_1 and m_2 be image points representing M.

Let E_1 (and respectively, E_2) be the epipolar curve passing through m_1 (and respectively, m_2).

In aerial photography, perspective is conical and the acquisition is instantaneous. Epipolar curves are straight lines (D_1 and D_2 in Figure 3.6(a)).

For push-broom satellite images (or push-broom sensor line acquisition – principle of the photocopier), the perspective is subcylindrical-conical and image acquisition is not instantaneous so real epipolar curves no longer exist. Ignoring (m_2, E_2) of small distance, we consider that E_1 and E_2 are homologous curves (see Figure 3.6(b)). In this case, we measure a point m'_2 that defines an epipolar curve E'_2 at a slight distance from E_2.

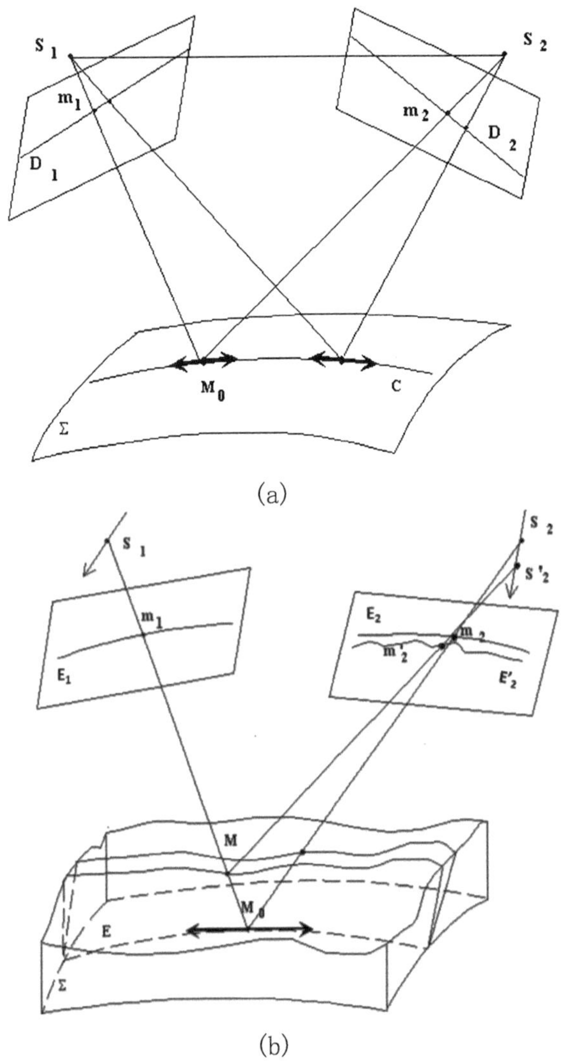

Figure 3.6. *a) Case for aerial photographs, b) case for satellite images*

This distance depends, among other reasons, on the height of M above S. By choosing a constant altitude for S, we can show that this distance is small, even at a high altitude. This allows us to assume that m_1 and m_2 are on the same epipolar curve. We can also show that the epipolar lines are almost straight by calculating their radius of curvature and maximum deflection. This observation is the basis of the one-dimensional correlation process used in the automatic restitution of the relief.

We can then either use the epipolar lines as guides for matching or resampling images along epipolar curves: the images are then said to be in epipolar geometry. In this case, the lines of the transformed images are parallel to the epipolar curves (similar to straight lines). Thus, the same epipolar curve is shown in each image by two lines of the same index.

This resampling, said to be "according to epipolar lines", is therefore such that the line coordinates of two homologous points are identical in each image. The homologue of line 1 of image 1 is the line 1 of image 2. We then simply need to measure their separation along the line to derive a measurement of the displacement of the point, and to determine its height. The measurement of this displacement, also known as disparity, is discussed in the next section.

3.3. Matching images

As we have seen in previous sections, if we know the positions of a feature in a scene for each element of a sequence of images (minimum of two images), of which the viewing angle varies, we can then restitute its position in space by simple spatiotriangulation. If one does not know these positions beforehand (which is usually the case), we can then estimate them: this is known as image matching. Moreover, to achieve the restitution of a complete digital surface model[3] (DSM), we seek to achieve dense matching (as opposed to matches for refining the geometric model as presented in section 3.2.4) along a regular grid, either in the image domain or directly in the geographical domain.

3 The DSM corresponds to the surface actually observed by the sensor, as opposed to the DTM in which the overground parts are not restituted.

3.3.1. *Some key principles*

If we restrict the general problem of matching to a pair of resampled images in epipolar geometry, we introduce the following formulation: if U_1 and U_2 are both resampled images in epipolar geometry, the one-dimensional field vector D, called *disparity*, is estimated such that the following condition is satisfied:

$$U_1(x,y) = U_2(x + D(x,y), y) \quad [3.15]$$

This formulation is easily generalized:

– in any geometry, making the field vector D two-dimensional;

– to a series of images, in which case the constraint [3.11] exists for each pair of images that can be formed from the sequence.

We note that disparity samples D(x, y) at the positions of pixels in image U_1, and that one will obtain a different disparity sample by inversing the images U_1 and U_2.

If we consider two images in epipolar geometry, provided that image U_1 is a Nadir view, field D is almost geometrically homothetic to a height map (in the same geometry) through a simple multiplication by baseline ratio $\frac{B}{H}$.

All the techniques described in this section estimate the disparity field D, as shown in Figure 3.7.

3.3.2. *Taxonomy of stereo-restitution algorithms*

This section builds on the excellent article by Schartein *et al.* [SCH 02], author of the renowned Middleburry[4] benchmark, which allows us to compare the algorithms for estimating disparity. Readers interested in an in-depth study of the concepts presented hereby may refer to this work. Finally, we note that the free library OpenCV[5] and the online journal Ipol[6] offer many implementations of disparity estimation algorithms.

4 http://vision.middlebury.edu/stereo/.

5 http://opencv.org/.

6 http://www.ipol.im/.

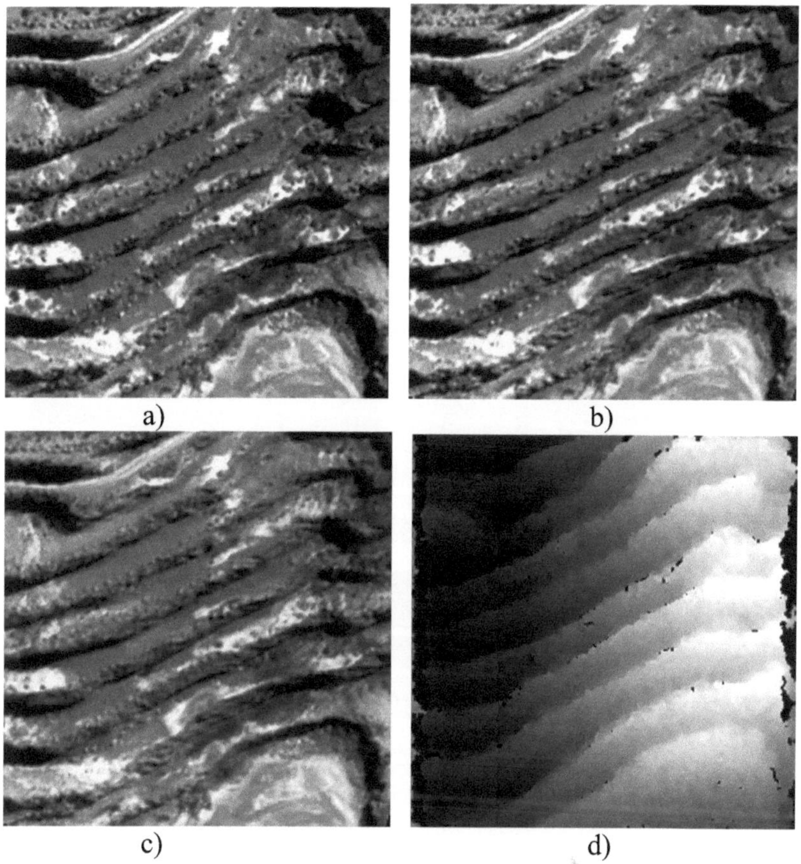

Figure 3.7. *Illustration of disparity estimation: (a and b) excerpts from a stereoscopic Pleiades pair resampled in epipolar geometry, c) anaglyph superposition of two images of the pair with complementary colors, allowing 3D visualization with "anaglyph" glasses (red on the right, blue on the left), d) disparity map estimated by the semi-global block matching (SGBM) algorithm [HIR 08]. The color codes the value of the measured disparity, from -13 pixels (black) to 10 pixels (white)*

3.3.2.1. *Hypotheses*

There are two strong hypotheses that are implicitly or explicitly incorporated into almost all stereo-restitution algorithms, and for which the validity deserves to be further elaborated. First, there is the hypothesis of a scene made of Lambertian surfaces, for which the response does not change with the viewing angle. This hypothesis ensures that the radiometry of images, and therefore the appearance of objects, is maintained from one viewpoint to another. These two viewpoints can be compared for identification purposes without introducing any knowledge on the nature of the observed material and an associated bidirectional reflectance distribution function (BRDF) model. Nevertheless, there are many situations where this hypothesis is faulted, such as in the effects of specular reflections on metal or glass surfaces, as shown in Figure 3.8, or as for water or waterlogged surfaces.

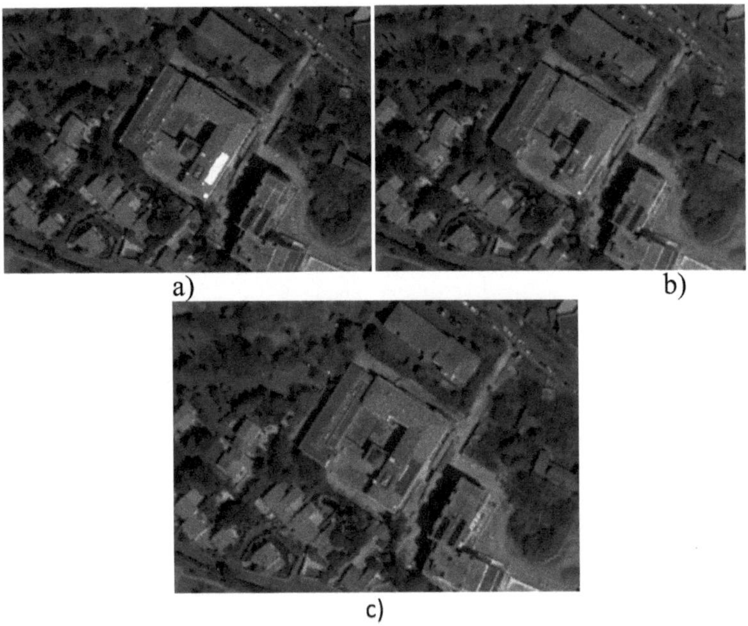

Figure 3.8. *Example of default layout of the Lambertian surfaces hypothesis. The roof successively shows: a) strong reflection (known as specular reflection); b) normal appearance and c) low reflection. A panchromatic extract of a stereoscopic Pleiades triplet on the city of Marseille © CNES (2013), distribution by Airbus DS/Spot Image*

The second hypothesis refers to the model of the observed scene: in order to constrain the problem of estimation of D, a certain regularity of disparity is assumed. Whether this assumption is implicit (for example, assuming constant disparity in the measurement window) or explicit (by integrating it into a cost function to be minimized), it simplifies the observed scene. The adherence effect presented in section 3.3.2 and even the erosion of isolated heavy disparities are typical consequences of the limited validity of this hypothesis.

3.3.2.2. *Cost functions*

The cost function, which is the basis of any stereoscopic restitution algorithm, measures the similarity between a pixel (x, y) of image U_1 and a pixel $(x + D(x, y), y)$ of image U_2. Given a set $A = \{(x_i, y_i, d_i), i \in [1, n]\}$ of candidate disparities, we can define the following traditional cost functions:

$$L_p(U_1, U_2, A) = \sum_{(x,y,d)\in A} |U_1(x, y) - U_2(x + d, y)|^p \qquad [3.16]$$

For this cost function L_p, two classical values of p are in the order of 2, called sum of squared difference (SSD) and in the order of 1, called sum of absolute difference (SAD).

Another commonly used cost function is the normalized cross-correlation:

$$NCC(U_1, U_2, A) = \sum_{(x,y,d)\in A} \frac{(U_1(x,y) - \overline{U_1^A})(U_2(x+d,y) - \overline{U_2^A})}{\sigma_{U_1^A} \times \sigma_{U_2^A}} \qquad [3.17]$$

where $\overline{U_1^A}$ and $\sigma_{U_1^A}$ are the mean and standard deviation U_1 on (x, y) such that $(x, y, d) \in A$ and $\overline{U_2^A}$ and $\sigma_{U_2^A}$ are the mean and standard deviation of U_2 on $(x + d, y)$ such that $(x, y, d) \in A$. Compensation of the mean of images and normalization by standard deviation makes this measurement robust toward linear changes of contrast between the two images.

Among the interesting cost functions, we also notice the presence of mutual information, which is expressed as follows:

$$\mathrm{IM}(\mathrm{U_1}, \mathrm{U_2}, \mathrm{A}) = \sum_{\mathrm{u_1,u_2}} \mathrm{P_{U_1,U_2,A}}(\mathrm{u_1}, \mathrm{u_2}) \times \log \frac{\mathrm{P_{U_1,U_2,A}(u_1,u_2)}}{\mathrm{P_{U_1,A}(u_1) \times P_{U_2}(u_2)}} \qquad [3.18]$$

where $P_{U_1,U_2,A}$ represents the joint probability density of images U_1 and U_2 on $\{((x, y), (x + d, y)), (x, y, d) \in A\}$ and P_{U_1}(respectively, P_{U_2}) is the probability density of U_1 on $(x, y), (x, y, d) \in A$ (respectively, U_2 on $(x + d, y), (x, y, d) \in A$). This cost function is attached to the local similarity of statistical distributions, and therefore provides a certain robustness to the default setting of the Lambertian surfaces hypothesis, and even to the operation of a pair of images from different modalities. For example, it is used in the semi-global block matching algorithm [HIR 08].

The definition of set A, for which cost functions are evaluated, depends on the type of approach.

3.3.2.3. *Local methods*

With local methods, set A is restricted to the local neighborhood of the pixel (x_0, y_0) of the image U_1 for which it seeks to restore disparity. The classic algorithm, called block matching, uses a spatial square neighborhood and constant disparity:

$$A_l(x_0, y_0, d) = \{(x, y, d), (x, y) \in [x_0 - l, x_0 + l] \times [y_0 - l, y_0 + l]\} \quad [3.19]$$

The estimated disparity at point (x_0, y_0) is that which minimizes the cost of A:

$$D^*(x_0, y_0) = argmin_{d \in [D_{min}, D_{max}]} \left(C\left(U_1, U_2, A_l(x_0, y_0, d)\right) \right) \quad [3.20]$$

Optimization is trivial, as it is exhaustive. To restitute all values of $D(x, y)$, the search is repeated for each coordinate (x, y). We note that the choice of constant disparity on A introduces an implicit constraint on disparity regularity. A more visual explanation of the algorithm is given in Figure 3.9.

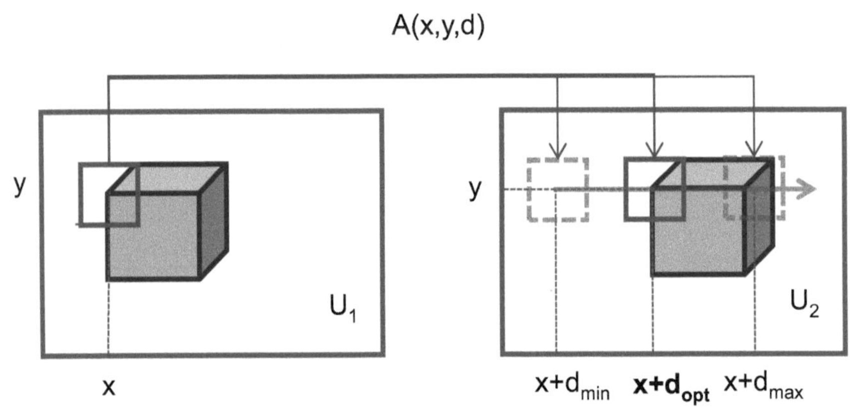

Figure 3.9. *The block-matching principle: the reference window centered at (x,y) in U_1 is compared successively to windows centered in(x+d_{min},y) to(x+d_{max},y) through the function cost. The retained disparity (d_{opt}) is that which minimizes the value of this cost function*

There are many variations of local algorithms:

– it is possible to vary d within set A in order to overcome the frontal-parallel hypothesis (the observed surface is parallel to the focal planes of cameras), and thus better manage the oblique surfaces;

– it is possible to change the shape of the window, to use multiple windows at the same time or even dynamically set the window size based on the context of the relevant pixel or a first disparity estimation [COR 02, BLA 11].

3.3.2.4. *Global methods*

Where local methods limit the extent of the cost function to the spatial neighborhood of the pixel, global methods instead measure this cost function over the entire image: $A_g = \{(x, y, D(x, y)\}$. This involves determining function $D(x, y)$ to minimize global energy in the classic form:

$$E(D) = C(A_g, U_1, U_2) + \lambda \cdot S(D) \quad [3.21]$$

The cost function plays the role of attachment term to the data, while the function $S(D)$, called the regularization term (one can also substitute it with $\sum C(A_l, U_1, U_2)$), forces D to be solved with some regularity among eligible solutions. S is therefore an explicit formulation of the regularity hypothesis assumed by the algorithm. A classically used constraint is:

$$S(D) = \sum_{x,y} \rho(D(x, y) - D(x + 1, y)) + \rho(D(x, y) - D(x, y + 1)) \quad [3.22]$$

where ρ is a monotonically increasing function. Some more complex regularization constraints seek to preserve the discontinuities of D while promoting its regularity.

According to the functional sets, the optimization problem is often NP-hard, but a wide range of tailored optimization methods exist in the literature, including graph-cuts or max-flow type methods, which have been acclaimed in recent years.

This global method is used in MicMac[7] software [PIE 06], which combines it with a multi-resolution approach.

---

7 http://logiciels.ign.fr/?-Micmac,3-.

Let us also note the methods based on dynamic programming, which maintain the global formulation while restricting A to a whole line of the image: $A_d(y_0) = \{(x, y_0, D(x, y_0)\}$.

Optimization of this problem, which is reduced to a single dimension, can indeed be carried out in polynomial time by searching for an optimal path in the matrix of costs between the two considered lines. Complete restitution of $D(x,y)$ is obtained by applying the optimization to each row independently, which decreases consistency between lines. However, one of the most powerful algorithms in current literature [HIR 08] is based on dynamic programming. To work around the problem of consistency between lines, this dynamic programming is performed several times in several directions in space, and the most frequently affected disparities are then retained.

3.3.2.5. *Refinement*

Once the disparity is estimated, there are several types of commonly used improvements to correct defects from different methods.

First, most methods produce sampled disparities in whole pixels, and thus lead to quantification of the DSM in the order of B/H. This quantification can hinder certain applications. To correct this, there are two solutions:

– oversampling the images prior to calculation of the disparity, which quantifies the disparity with a step less than the pixel and thus reduces quantification of the elevation. However, the complexity of the restitution step, which is already very significant, increases quadratically with respect to the reduction of the quantification step. This also allows a better sampling of the cost function [DEL 07b];

– using a second algorithm that will estimate the real value of disparity. Several techniques are available for this: iterative gradient descent by interpolating images, local regression of a curve on the sampled cost function in full disparity and search for the real extrema;

– once the coarse displacement is estimated, optical flow type techniques can be used. These techniques are quick and efficient under the hypothesis of low offsets [FLE 06].

Let us consider an image I(x,y,t) at time t+Δt (for example, a stereoscopic pair), in which image I is displaced:

$$I(x,y,t) = I(x+\Delta x, y+\Delta y, t+\Delta t) \quad [3.23]$$

If the displacements are small, a Taylor expansion gives:

$$I(x+\Delta x, y+\Delta y, t+\Delta t) = I(x,y,t) + \frac{\partial I}{\partial x}\Delta x + \frac{\partial I}{\partial y}\Delta y + \frac{\partial I}{\partial t}\Delta t + \varepsilon \quad [3.24]$$

We deduce:

$$\frac{\partial I}{\partial x}\Delta x + \frac{\partial I}{\partial y}\Delta y + \frac{\partial I}{\partial t}\Delta t = 0 \quad [3.25]$$

or even:

$$\frac{\partial I \Delta x}{\partial x \Delta t} + \frac{\partial I \Delta y}{\partial y \Delta t} + \frac{\partial I \Delta t}{\partial t \Delta t} = 0 \quad [3.26]$$

Thus

$$\frac{\Delta x}{\Delta t} = V_x, \frac{\Delta y}{\Delta t} = V_y, \frac{\partial I}{\partial x} = I_x, \frac{\partial I}{\partial y} = I_y, \frac{\partial I}{\partial t} = I_t \quad [3.27]$$

is the same as writing:

$$I_x V_x + I_y V_y + I_t = 0 \quad [3.28]$$

I_x is the gradient in x, I_y is the gradient in y and I_t is the difference between the two images.

The solution to this proposed problem, in particular for Lucas and Kanade [LUC 81], is simple. For all pixels $q = (x, y)$, we can write the following relationships:

$$\begin{cases} I_x(q_1)V_x + I_y(q_1)V_y = -I_t(q_1) \\ I_x(q_2)V_x + I_y(q_2)V_y = -I_t(q_2) \\ \therefore \\ I_x(q_n)V_x + I_y(q_n)V_y = -I_t(q_n) \end{cases} \quad [3.29]$$

The solution to this sytem is:

$$\begin{bmatrix} V_x \\ V_y \end{bmatrix} = \begin{bmatrix} \sum_i I_x(q_i)^2 & \sum_i I_x(q_i)I_y(q_i) \\ \sum_i I_x(q_i)I_y(q_i) & \sum_i I_y(q_i)^2 \end{bmatrix} \begin{bmatrix} -\sum_i I_x(q_i)I_t(q_i) \\ -\sum_i I_y(q_i)I_t(q_i) \end{bmatrix} \qquad [3.30]$$

This technique is simple in terms of computation time and allows independent work to be done on each sufficiently small image subblock such that the disparity amounts to pure translation.

Apart from calculation of subpixel disparity, many processes can be implemented to correct various defects:

– detection of potentially erroneous values: coherence test, weak signal areas, areas likely to be tainted by adherence, outliers, etc.;

– fixed non-restituted areas (occlusions or areas identified by the above methods as false), regularization, etc.

3.3.3. *Typical defects in disparity estimation*

The aim of this section is to give the readers a better understanding of the defects that are commonly encountered in DSM, which are automatically restituted from optical satellite imagery.

3.3.3.1. *Occlusions*

Because of the difference in viewpoints, sometimes points in an image have no homologues in any of the remaining images: these are occlusions. If a point is only visible in a single image, it is not possible to estimate its disparity, let alone its position in space. In the best case, the occlusion and disocclusion phenomena (point visible in one image only) are properly detected and we can indicate that for this point on the disparity map or DSM, no information could be restituted. But, it is also common that these points lead to mismatches and create visible defects in the DSM. Figure 3.10 shows this occlusion phenomenon.

3.3.3.2. *Adherence*

As we saw in the previous section, local methods of "block matching" are based on a finite neighborhood of the pixel used to define a similarity measure, which is then used to search for the homologous pixel in the second image. This consideration of the spatial context, which is essential for proper identification of homologous areas, introduces the phenomenon of adherence. This phenomenon results from a combination of two distinct effects:

– the measured disparity mainly comes from high gradients present in the window, as they bring a greater contribution to the measurement of similarity;

– the hypothesis of constant disparity in the measurement window is often invalid.

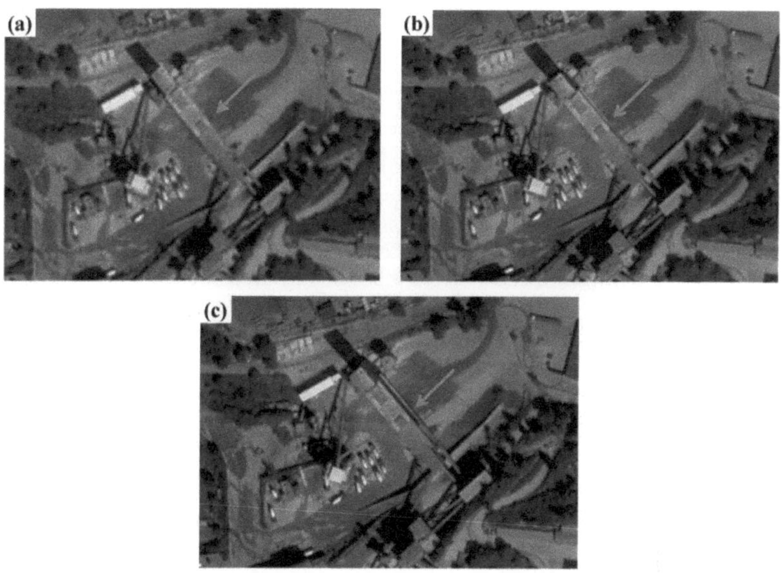

Figure 3.10. *Example of occlusion: the southwest side of the building in the center of the image, indicated by the red arrow, is never seen and hides the ground in images b) and c), while the northeast face, indicated by the blue arrow, is only visible in the last two pictures © CNES (2013), distribution by Airbus DS/Spot Image*

Consequently, if high gradients are present in the measurement window for pixels of which the disparity differs from that of the central pixel, the measured disparity will be biased toward the one carried by the pixels with high gradients. This is particularly noticeable for buildings, for which the influence is overstated due to the adherence induced by a combination of the breaking disparity and the significant gradient along the building edge. This effect is also known as fattening. However, adherence occurs to a lesser extent wherever disparity is not constant and the window is not uniform, and adds noise to the measurement of disparity. Figure 3.11 illustrates this adherence phenomenon.

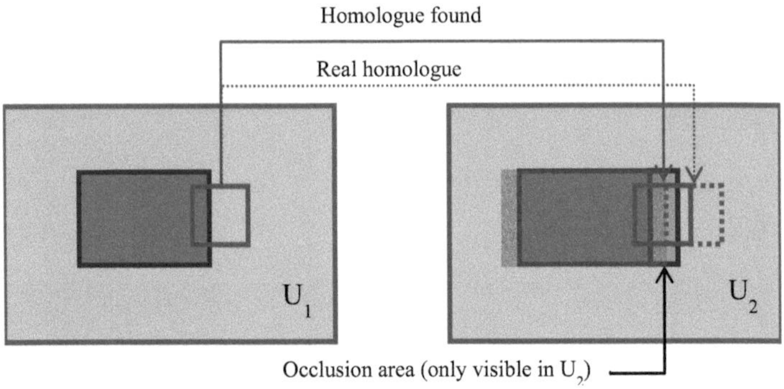

Figure 3.11. *Illustration of the adherence phenomenon. The green area shows high disparity, while the blue area corresponds to a small disparity. The gray area is occluded in U_1 but visible in U_2. The central pixel of the measurement window in red is given the disparity in the green area, although the center of the window is in the blue zone. Therefore, in the restituted disparity map, the green area is expanded. For a color version of this figure, see www.iste.co.uk/baghdadi/1.zip*

3.3.3.3. *Non-discriminative areas*

When using local methods, it is possible that set A is considered insufficiently informative (if it only contains homogeneous radiometric values) relative to the image content: for example, constant areas of the image will be sources of ambiguity in the search for extrema of a cost function. In the case of a search for an epipolar line, which is only horizontal, the neighborhood can also be non-discriminatory if it is constant in the horizontal direction.

3.3.3.4. *Pseudo-periodic structures and auto-similarity*

Another frequent case of failure involves periodic structures in images. These elements, which are not uncommon in urban areas, for example, cause ambiguities in the search for local extrema, in particular for local approaches. An auto-similarity test can be set up to limit these effects.

3.3.3.5. *Asynchrony*

Finally, a specificity of spatial stereoscopy is that it is never synchronous, meaning that acquisitions are spaced from a few seconds apart (when acquisitions are carried out consecutively from a single orbit) to several days apart (when acquisitions are done from different orbits). Consequently, some objects in the scene (for example, vehicles) move between shots, and so disrupt the estimation of disparity.

3.4. From disparity to the digital surface model

As we saw in the previous section, disparity is the correspondence between images, sampled regularly in the image space (i.e. native or possibly epipolar geometry). Each disparity or matching can restitute a point in space.

Let us consider a line Δ that passes through a summit S with direction vector $\vec{V}$, a point M and the distance δ of M to the line Δ. For a set of n targeted lines Δ_i with $(S_i, \vec{V}_i)$, we seek the point M that minimizes the sum of squared distances:

$$\sum_i \delta_i^2 = \sum_i \left\| \overrightarrow{S_i M} \wedge \vec{V}_i \right\|^2 = \sum_i \left[\overrightarrow{S_i M}^2 - \left(\overrightarrow{S_i M} . \vec{V}_i \right)^2 \right] \qquad [3.31]$$

By differentiating, the solution obtained is:

$$M = \left[\sum_i \left(Id - V_i V_i^t \right) \right]^{-1} \sum_i \left[\left(Id - V_i V_i^t \right) S_i \right] \qquad [3.32]$$

There are several advantages to this method: for each new line, we constitute each $Id - V_i V_i^t$ term then $\left(Id - V_i V_i^t \right).S_i$, which are combined on the fly; a single inversion of a 3 × 3 matrix; to eliminate a target that is too far away, we simply subtract its contribution in $\sum_i \left(Id - V_i V_i^t \right)$ and $\sum_i \left[\left(Id - V_i V_i^t \right) S_i \right]$ [DEL 07a].

The disparity map is thus translated into a pattern of points with *irregular* planimetric sampling, as shown in Figure 3.12. This is especially true as the geometry of the reference image (which determines sampling disparity) is far from a Nadir (vertical) view.

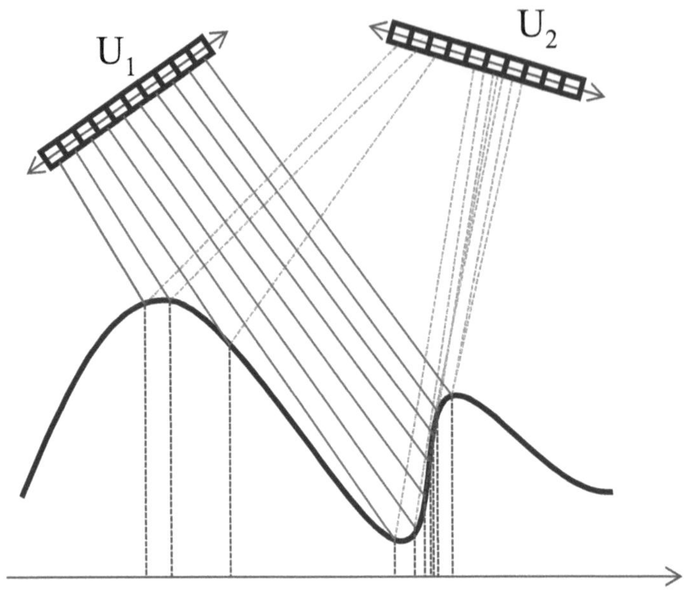

Figure 3.12. *Illustration of the sampling problem: the disparity between U_1 and U_2 is sampled regularly in the geometry of image U_1, but the points that are restituted in space are sampled planimetrically in an irregular manner*

The irregular pattern of points that correspond to the targeted intersections accurately reflects the localization information in three dimensions extracted from the image. However, many applications assume a regularly planimetrically sampled DSM. To overcome this problem, we can define a regular grid of planimetric cells and aggregate the elevation measurements (mean, median, maximum, etc.) in each cell. However, the derived DSM is a non-reversible transformation of the 3D information actually extracted from the images. In particular, empty cells (with no measurement) and cells aggregating a large number of points (for example, a building façade) may be found.

Regarding the assessment of restitution accuracy for the entire chain, the readers may refer to the works presented in [DUR 13] and [BER 14]. Figure 3.13 shows an example of restitution for an area of 2500 × 1385 m located near Marseille in France. It shows the Nadir image of the triplet used (the image with the most vertical angle of incidence), the estimated DSM in the geometry of the image and this same DSM projected onto the ground on a regular grid with a mesh of 1 m (31 N UTM projection). In addition to good restitution of the observed relief, one may note the presence of holes in the DSM image geometry, which correspond to areas that have not been restituted by the method due to the defects mentioned above, as well as additional holes in the DSM projected onto the ground. These result from the presence of occlusion.

Figure 3.14 shows a synthetic three-dimensional view of the same scene, represented as a cloud of unstructured points. Once the DSM is estimated, one can simulate a virtual view of the observed scene, colorized using radiometry from the image or with a color gradient that corresponds to the change in height.

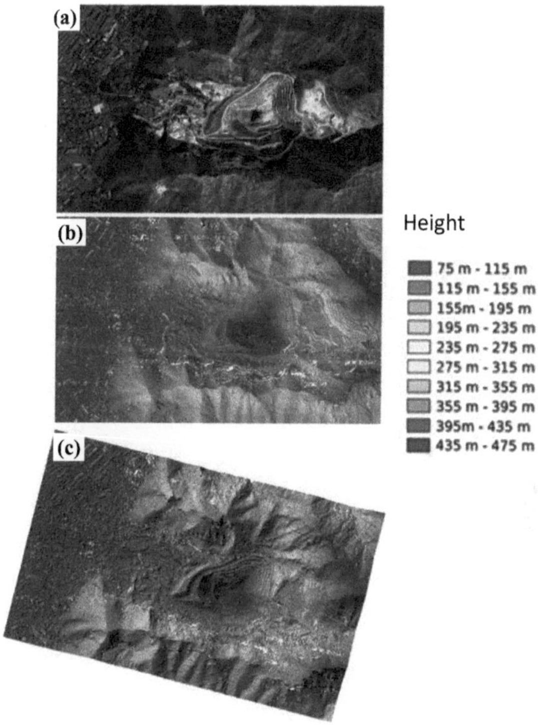

Figure 3.13. *a) Extract of an image acquired by the Pleiades satellite (mine in the hills of Marseille, 43°15′50.0"N , 5°26′08.5"E, 2500 m × 1385 m), © CNES (2013), distribution by Airbus DS/Spot Image. b) Triangulated disparity map calculated using the SGBM algorithm [HIR 08] and the Satellite Stereo Pipeline restitution chain (S2P) [DEF 14], sampled in the image geometry. c) Digital Surface Model generated at a resolution of 1 m. Rendered by: open-source Qgis software. For a color version of this figure, see www.iste.co.uk/baghdadi/1.zip*

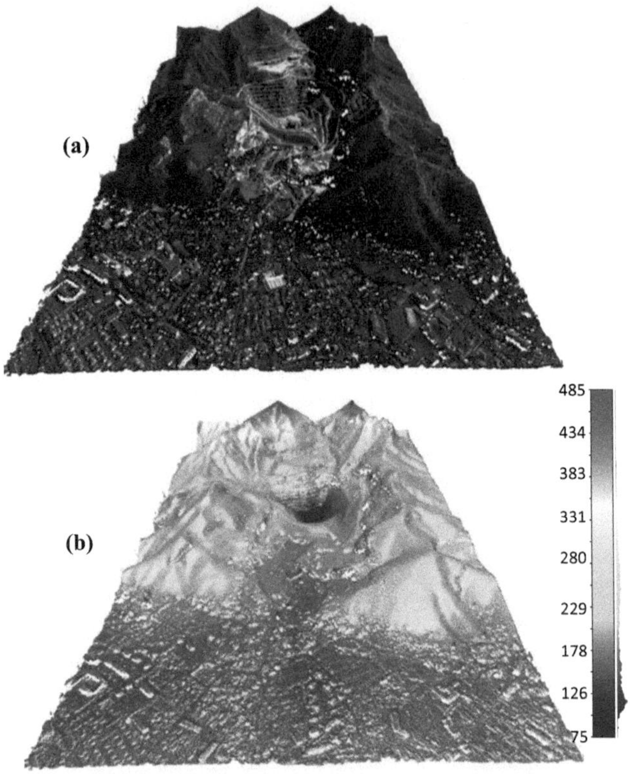

Figure 3.14. *a) Point cloud colorized with radiometries of the image. The numerous missing areas (white) correspond to occlusions of the scene (the target satellite being opposed to the view simulated here). b) Three-dimensional representation of the DSM (each point has a color corresponding to its height). Rendered by: open-source CloudCompare software. For a color version of this figure, see www.iste.co.uk/baghdadi/1.zip*

3.5. Conclusions and perspectives

Since the commoditization of multi-stereoscopic acquisitions, users have access to the algorithms and tools allowing for the restitution of the relief. This can be achieved through different imaging systems (land, aerial or spatial) like SPOT, Pleiades, GeoEye, Worldview or even Skybox for spatial.

Spatial resolution of the system allows access to different types of 3D restitution that are summarized in Table 3.1. The contribution of a spatial resolution is critical for certain applications.

System solution	*Natural terrain*	*Low-density urban (hangars, isolated houses, boats and airplanes)*	*Dense urban (dense downtown, slum and light vehicles)*	*Fine elements (antennas, air conditioning and pedestrians)*
Spatial to 70 cm	+++	++	+ (a few very isolated 3D items can be restituted, cars on bare land, etc.)	--- (elements are too fine for spatial resolution)
Spatial or aerial to 30 cm	++ (Some overground elements are difficult to filter)	+++	++	+ (depends on the size and neighborhood)
Aerial to 10 cm	+ (too many overground elements are difficult to filter)	+++	+++	++

Table 3.1. *Capacity of a system to restitute 3D elements*

3D restitution requires a large number of operations and task parallelism is desirable to ensure that the user obtains results in an acceptable computation time. Improving the accuracy of altimetric restitution within a reasonable computation time remains a challenge, as it is always better to control the geometric phenomena and matching algorithms described in this chapter. Interested readers may explore the concepts further by reading [HAR 03].

3.6. Key points

When we image a scene from different angles, a shift in the scene components is observed. This displacement is increasingly significant the closer the object is to the axis formed by the optical centers of cameras (or the orbit taken by the satellite). Restitution of relief by stereoscopy involves measuring this displacement for each pixel of the image through image matching algorithms. We then just need to calculate the intersection of the viewing directions generated by the pixels thus matched in order to restitute in space the position of the considered point. However, this step requires mastering the acquisition geometry perfectly, which can be improved by re-estimating the viewing parameters, which is also called refinement of sensor models.

Around these two central pillars, many elements are needed to achieve a complete chain of DSM restitution: consideration of a low resolution DTM (tens of meters) to reduce the search interval, postprocessing treatment for filtering outliers, projection onto a regular grid, etc. Finally, it is important to pinpoint the main sources of error: occlusions that can create areas that are not indicated in the DSM, low-contrast areas that are not restituted or even the adhesion or diachronism problem. In particular, we note that water surfaces are often not restituted because they are of low contrast, not Lambertian and often in movement.

3.7. Bibliography

[BER 14] BERTHIER E., VINCENT C., MAGNÚSSON E. *et al.*, "Glacier topography and elevation changes derived from Pleiades sub-meter stereo images", *The Cryosphere*, vol. 8, no. 6, pp. 2275–2291, 2014.

[BLA 11] BLANCHET G., BUADES A., COLL B. *et al.,* "Fattening free block matching", *Journal of Mathematical Imaging and Vision*, vol. 41, no. 1–2, pp. 109–121, 2011.

[COR 02] CORD M., JORDAN M., BELLI T. *et al.*, "Analyse d'images aériennes haute résolution pour la reconstruction de scènes urbaines", *Bulletin de la Société Française de Photogrammétrie et de Télédétection*, vol. 166, pp. 34–43, 2002.

[DEF 14] DEFRANCHIS C., MEINHARDT-LLOPIS E., MICHEL J. *et al.*, "An automatic and modular stereo pipeline for pushbroom images", *ISPRS Annals of Photogrammetry, Remote Sensing and Spatial Information Sciences*, vol. 1, pp. 49–56, 2014.

[DEL 07a] DELVIT J.M., FAVÉ P., GACHET R., "The geometric supersite of Salon de Provence", *Symposium of ISPRS Commission I*, vol. 181, pp. 19–24, 2007.

[DEL 07b] DELON J., ROUGÉ B., "Small baseline stereovision", *Journal of Mathematical Imaging and Vision*, vol. 28, no. 3, pp. 209–223, 2007.

[DUR 13] DURAND A., MICHEL J., DEFRANCHIS C. *et al.*, "Qualititive assesment of three DSM generation approaches using Pleiades-HR data", *Earsel Symposium*, Materra, Italie, 2013.

[FLE 06] FLEET D.J., WEISS Y., "Optical flow estimation", in *PARAGIOS N. et al.*, *Handbook of Mathematical Models in Computer Vision*, Springer, 2006.

[HAR 03] HARTLEY R., ZISSERMAN A., *Multiple View Geometry in Computer Vision*, Cambridge University Press, 2003.

[HIR 08] HIRSCHMULLER H., "Stereo processing by semiglobal matching and mutual information", *IEEE Transactions on Pattern Analysis and Machine Intelligence,* vol. 30, no. 2, pp. 328–341, 2008.

[LUC 81] LUCAS B.D., KANADE T., "An iterative image registration technique with an application to stereo vision", *Proceedings of Imaging Understanding Workshop*, pp. 121–130, 1981.

[PIE 06] PIERROT-DESEILLIGNY M., PAPARODITIS N., "A multiresolution and optimization-based image matching approach: an application to surface reconstruction from SPOT5-HRS stereo imagery", *International Archives of Photogrammetry, Remote Sensing and Spatial Information Sciences*, vol. 36, no. 1, p. w41, 2006.

[RAD 91] RADIX J.C., *Répertoire géodésique en vue de la navigation*, Cépaduès, 1991.

[SCH 02] SCHARSTEIN D., SZELISKI R., "A taxonomy and evaluation of dense two-frame stereo correspondence algorithms", *International Journal of Computer Vision*, vol. 47, no. 1–3, pp. 7–42, 2002.

[VAL 08] VALORGE C., LIER P. *et al.*, *Imagerie Spatiale Des principes d'acquisition au traitement des images optiques pour l'observation de la Terre*, Cépaduès, Collection CNES, 2008.

4

Processing Hyperspectral Images

4.1. Introduction

Hyperspectral imaging is a technology that has been used in remote sensing to study planetary surfaces for over 20 years. Hyperspectral cameras (or imaging spectrometers) have a high spatial resolution (~1–30 m) coupled with regular sampling (every ~4–15 nm) of a broad spectral range, which can cover wavelengths ranging from ultraviolet (~0.35 μm) to thermal infrared (~12 μm). With this dense spectral sampling, imaging spectrometers provide images with an extra dimension to conventional imaging: the spectral dimension. Indeed, each point of a hyperspectral image is associated with a vector called a "spectrum", which is formed by a few hundred spectrally adjacent values with a physical meaning (for example, radiance measured by the sensor or surface reflectance after an atmospheric correction step). In this chapter, we are primarily concerned with the processing of remote-sensing hyperspectral images to study surface materials, which is one of the most common applications. In this context, analysis of land scenes imaged by their associated spectra (not by their pixels) becomes interesting because radiance or reflectance spectra contain a wealth of information on the chemical and physical state of surface materials.

The main difficulty in hyperspectral imaging is the large amount of data to be processed. Indeed, the number of spectral bands often makes it difficult

Chapter written by Xavier CEAMANOS and Silvia VALERO.

to extract the relevant information for a given application. Also, hyperspectral images are relatively sparse (empty) due to a combination of relatively invariant spectral responses of most materials and dense wavelength sampling of imaging spectrometers. Therefore, there is a need for efficient (accurate and fast) tools:

– for visualization and representation, to understand and explore hyperspectral data;

– for analysis, to carefully study the images in order to identify and extract the relevant information.

This chapter aims to provide a succinct but comprehensive overview on processing of remote-sensing hyperspectral images. First, the characteristics of hyperspectral images, their visualization and the most commonly applied preprocessing techniques are presented in section 4.2. Compression of this type of large image is also discussed. Subsequently, the main tools for extracting relevant information from the images are presented. Approaches are organized by type of processing:

– *spectral matching* techniques in section 4.3, which compare observed spectra to a database of reference spectral measurements, which may contain hundreds of materials;

– *spectral segmentation* in section 4.4, which seeks to group similar and spatially adjacent spectra;

– *spectral classification* in section 4.5, which aims to assign a label to each spectrum of the image from a list of classes that may or may not be predefined;

– *spectral unmixing* in section 4.6, which goes down to the subpixel scale to decompose each spectrum into the components that form it;

– detection of anomalies, targets or objects in section 4.7.

The approaches that are most commonly used in the literature are presented for all previous types of processing. In this chapter, remote-sensing hyperspectral images that are freely available on the Internet are processed to illustrate the usefulness of such techniques. Technical tools that have been used are ENVI[1], MATLAB[2], Orfeo ToolBox[3] software and C++ executables.

This chapter concludes with a summary of the latest trends in hyperspectral image processing and with a summary of the key points.

4.2. Hyperspectral images, representation and preprocessing

4.2.1. *What is a hyperspectral image?*

A hyperspectral image is a compilation of several hundred images of the same scene, each of which contains information for a given wavelength. In remote sensing, and after a calibration step, this information is the radiant flux that reaches the sensor (or radiance). Stacking these images (called spectral bands) produces a hyperspectral image, which is formed by two spatial dimensions and one spectral dimension. This third dimension finely and uniformly samples the spectral range covered by the imaging spectrometer. In this configuration, each spatial unit of the image is characterized by a radiance spectrum, which is a vector containing the evolution of radiant flux measured by the camera according to wavelength.

The utility of hyperspectral imaging is based on the fact that every material absorbs solar irradiance at specific wavelengths, which are determined by the physicochemical properties of the material. The low values of a reflectance spectrum occur at wavelengths where the observed material strongly absorbs incident irradiance, thus creating absorption bands. The convex hull of the spectrum is called the continuum. These spectral attributes uniquely characterize each material, shaping its characteristic reflectance spectrum (or its spectral signature). Figure 4.1 shows the main elements of a remote-sensing hyperspectral image. The reflectance spectrum of two different materials (crops and bare ground) is also displayed. The dotted line of each spectrum represents the continuum, while

1 http://www.harrisgeospatial.com/IntelliEarthSolutions/GeospatialProducts/ENVI.aspx.

2 http://www.mathworks.com/products/matlab/index.html.

3 https://www.orfeo-toolbox.org/.

the arrows point to the absorption bands. The spectral differences between the two materials are notable; for example, the characteristic increase in reflectance of vegetation at 0.65 μm, which is only observed in the crops spectrum.

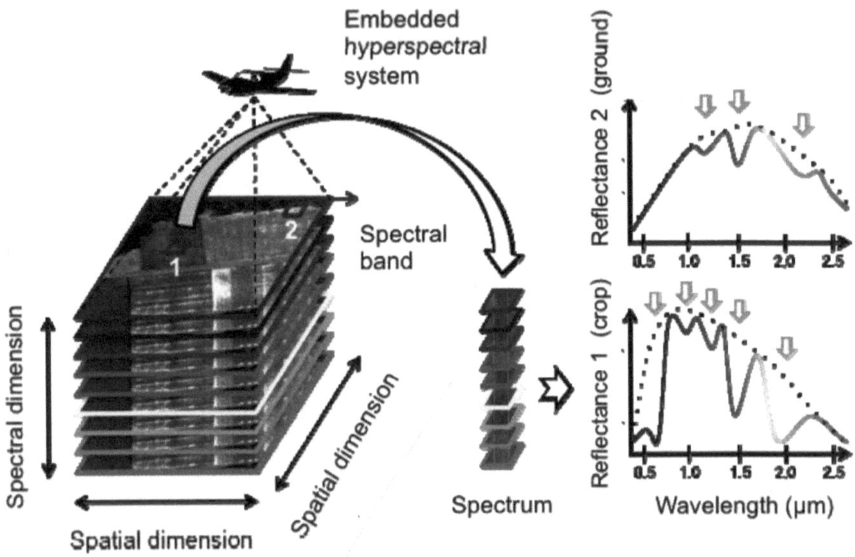

Figure 4.1. *Diagram of a remote sensing hyperspectral image. The reflectance spectrum of the two materials (crops and bare ground) is shown with its continuum (dotted line) and its absorption bands (arrows). For a color version of this figure, see www.iste.co.uk.baghdadi/1.zip*

4.2.2. *Noise reduction and retrieval of surface properties*

Any remote-sensing image must be preprocessed before accessing information related to the surface. This preliminary step is intended to eliminate signals other than those of the surface. This includes noise from electronic, optical and thermal artifacts. In addition, images in radiance units must be systematically corrected for effects of the atmosphere in order to retrieve the surface reflectance. The most commonly applied techniques in these two fields are:

– *noise reduction*: different types of noise from different origins affect hyperspectral sensors. For example, the intrinsic temperature of camera components gives rise to random (and often not Gaussian) noise, which particularly affects the spectral bands associated with larger wavelengths. Hyperspectral images are also affected by instrumental artifacts of electronic or other origin, known as "spikes"; pixels with outliers; or "stripes", a parasite bias for which the magnitude varies depending on the image column. We also include the "smile" effect, which makes the spectral characteristics of the camera (for example, spectral resolution) dependent on the column number. This artifact of optical origin leads to strong column-dependent bias in some bands of hyperspectral images. Methods have been proposed in the literature to correct these and other types of noise and artifacts [GOM 08, LET 08, CEA 10];

– *atmospheric correction*: in remote sensing, the surface signal is mixed with the atmospheric contribution. Retrieval of surface reflectance from radiance images is done by atmospheric correction methods. These approaches are often based on a physical model of the radiative transfer that takes place in the surface/atmosphere coupled system. More information on these approaches and the physics that intervene is available in Chapter 1 of this book. Correction methods for hyperspectral data must work at a high-spectral resolution (equivalent to that of the hyperspectral sensor at least). A detailed review of the most used correction methods for hyperspectral images is available in [GAO 09].

4.2.3. *Dimensionality reduction*

Hyperspectral data processing is affected by the "curse of dimensionality" [BEL 61], which describes the complexity of extracting useful information from datasets of a larger size. Furthermore, hyperspectral images are "sparse" datasets because of the high correlation among consecutive spectral bands. This spectral redundancy is due to the dense spectral sampling of imaging spectrometers combined with the low variations in the spectral reflectance of most materials. These characteristics of hyperspectral images have led to preprocessing techniques that aim to reduce their size by decreasing the number of spectral bands while maintaining the significant information. These approaches are also useful for reducing the computation time associated with hyperspectral processing. Three groups of methods are available:

– *feature selection*: techniques that select only the spectral bands containing information that is discriminative. This selection may be supervised if the spectral signature of the object to be detected is already known (only keeping the bands associated with characteristic spectral attributes, for example). If this information is not available, there are techniques that allow automatic selection of bands by maximizing the separability of the different classes of the image [BAC 05]. To illustrate feature selection in hyperspectral image processing, we use the "Beltsville[4]" image, which corresponds to an agricultural scene with several crop parcels surrounded by trees (Figure 4.2). This image was acquired using a SpecTIR hyperspectral sensor at 2 m spatial resolution and wavelength coverage from 390 to 2450 nm. Figure 4.2(a) shows the three selected spectral bands containing different information: the band at 548 nm related to the intensity of the color green, the band of near infrared at 807 nm that highlights vegetation and the band at 2091 nm that highlights the presence of bare ground. The information provided by these three spectral bands alone may be sufficient for some applications;

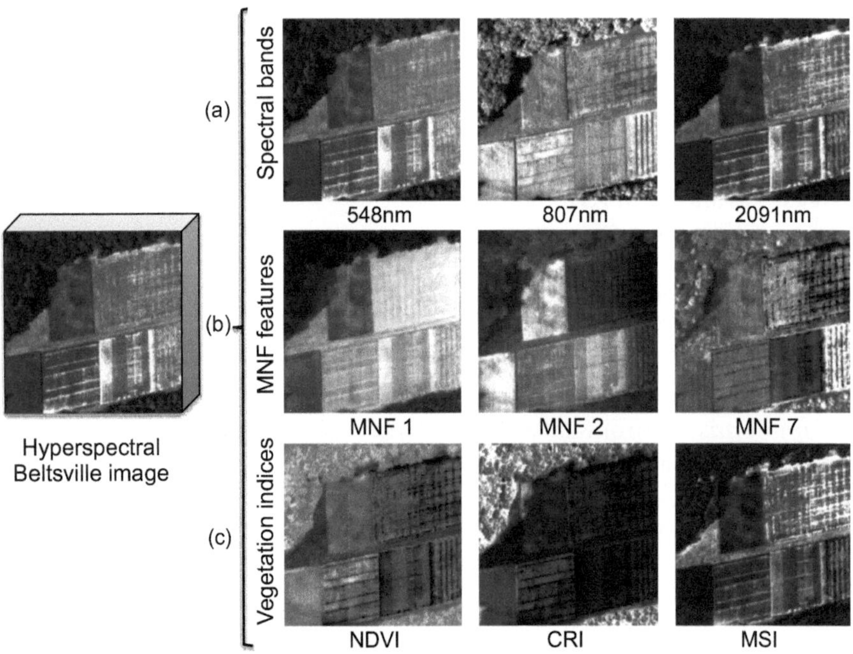

Figure 4.2. *Dimensionality reduction of the Beltsville hyperspectral image by a) feature selection, b) extraction of MNF features and c) calculation of physical vegetation indices*

4 http://www.spectir.com/free-data-samples/

– *feature extraction*: techniques that transform hyperspectral images in a new space where we can define a reduced dataset containing only the essential information. Among the best-known techniques are principal components analysis (PCA), which transforms spectral bands – correlated between themselves – into new uncorrelated features (or components) ordered in terms of variance percentage [ROD 02] or Maximum Noise Fraction (MNF), which sorts the features by their signal-to-noise ratio [GRE 88]. A simple way to reduce the size of a hyperspectral image is to work with the first components of respective PCA or MNF transformations only. For the PCA, this reduction amounts to withholding a certain percentage of the accumulated data variance. The major drawback of these techniques is the difficulty in automatically determining the number of significant components that should be kept. Figure 4.2(b) shows three features of the MNF transformation applied to the Beltsville image: MNF-1 is related to the brightness of the image, MNF-2 highlights some crops and MNF-7 emphasizes the presence of bare ground in the right part of the image with low values (dark pixels);

– *calculation of physical indices* that generate features with physical meaning from the combination of spectral bands. Physical indices can provide information on vegetation, such as the normalized difference vegetation index (NDVI), which determines the presence and health status of plants [ROU 73]; soils, like the $dD_{1300,1970}$[5] that quantifies soil moisture [WAN 11]; and oceans, like the Floating Algae Index (FAI)[6], which measures the concentration of suspended algae [HU 09]. The NDVI is calculated as follows:

$$NDVI = (r_{NIR} - r_{RED}) / (r_{NIR} - r_{RED}) \quad [4.1]$$

where r_{NIR} is the reflectance value of near infrared (NIR) between 800 and 850 nm and r_{RED} is the reflectance corresponding to the color red (between

5 $dD_{1300,1970} = d(r_{1300})/d\lambda - d(r_{1970}) - d\lambda$ where $d(r_{1300})/d\lambda$ and $d(r_{1970})/d\lambda$ are the first derivatives of reflectance at 1,300 and 1,970 nm, respectively. The $dD_{1300,1970}$ term is proportional to soil moisture, reaching near-zero values for dry soils.

6 $FAI = r_{NIR} - [r_{RED} + (r_{SWIR} - r_{RED}).(\lambda_{NIR} - \lambda_{RED}) / (\lambda_{SWIR} - \lambda_{RED})]$ where r_{SWIR} is the reflectance in the shortwave range (Shortwave InfraRed or SWIR) at 1,240 nm. The λ_i terms correspond to the wavelengths used, in nanometers. The FAI parameter is proportional to the presence of algae (greater than about 0.10 for the detection of algae under conditions of good atmospheric visibility).

645 and 670 nm). The NDVI is correlated with plant health status and is normally higher than 0.3 for green vegetation. Figure 4.2(c) shows three indices calculated on the Beltsville image containing information about vegetation: the NDVI, the Carotenoid Reflectance Index (CRI)[7] [GIT 02] that quantifies the presence of carotenoids linked to chlorophyll, which protects plants against the undesirable effects of excess light and the Moisture Stress Index (MSI)[8] [CEC 01], which detects areas under high water stress.

4.2.4. *Hyperspectral image compression*

The large size of hyperspectral images, typically in the order of megabytes or gigabytes, can cause significant problems for data storage and processing. In this case, data compression may be a solution. The main difficulty in achieving this goal is the impossibility of adapting traditional 3D compression methods because radiance and reflectance values vary differently according to two spatial dimensions than they do it according to the spectral dimension. In this context, specific approaches have been proposed in the literature:

– *prediction methods*, which result from the hyperspectral adaptation of this set of approaches that are lossless and are typically used for 2D images. The data are uncorrelated by a predictor so an entropic encoder then encodes the associated error [AIA 09];

– *vector quantification approaches* that break down a hyperspectral image into a set of vectors [QIA 04]. A training set of vectors is selected to generate a codebook using an iterative clustering algorithm. Finally, each of the image vectors is quantized to the closest representative vector in the codebook based on the minimization of an appropriate distortion measure. This set of approaches introduces distortion into the reconstructed images but not in the classification results;

– *transformation methods*, which work in two stages. First, the data are transformed in a space where they are represented more compactly [PEN

7 CRI = $1/r_{510} - 1/r_{550}$ where r^{510} and r^{550} are the reflectances at 510 and 550 nm, respectively. CRI can range from 0, when the abundance of carotenoids is zero, to over 15, when their concentration is significant.

8 MSI = r_{1599} / r_{819} MSI gives an index of the abundance of water in leaves from the absorption of water at 1,599 nm. It can have values ranging from 0, for a very high water content, to more than 3 for severe water stress.

07]. The Karhunen–Loeve transform (KLT) or the wavelet decomposition is used at this stage due to their compression efficiency. However, KLT causes high computation times. Second, the compacted information is coded, with or without loss of information, using image-coding methods such as Embedded Zerotrees of Wavelet (EZW) transforms [CHR 08].

We recommend the book by [CHR 11] to gain further insights on the subject of compression of hyperspectral images.

4.2.5. *Other representations of hyperspectral data*

Typically, hyperspectral images are represented in a classical 2D image space with an additional third dimension, which may correspond to the wavelength or the space defined by the PCA transform, for example. Other alternatives exist such as representations based on regions, which allow a high level of abstraction in comparison to the classical array representation defined by the pixel or spectrum. This type of representation provides semantic information on the content of the hyperspectral image. Representations by regions establish connections between the spectra in order to distinguish shapes and objects with richer spatial information than individual pixels.

An example of a hierarchical representation of hyperspectral images by region is the binary partition tree (BPT) [VAL 13]. The BPT turns a hyperspecral image into a set of hierarchical regions (a region being a group of adjacent spectra) stored in a tree structure (see Figure 4.3). This representation succeeds in:

– decomposing the image in terms of coherent regions;

– exploring the different relationships of inclusion of regions in the scene.

The BPT is obtained from an initial partition of the image, which is often the one in which each spectrum of the image corresponds to a region. If the initial partition implies n hyperspectral regions, the BPT will correspond to a structure containing $2n - 1$ nodes. The BPT must be created so that the regions in the image with semantic content become nodes of the tree. One way to do this is to create the tree through a hyperspectral region-merging algorithm (more details in section 4.4). Figure 4.4 gives an example of construction for an image with four initial regions that leads to a BPT with seven nodes.

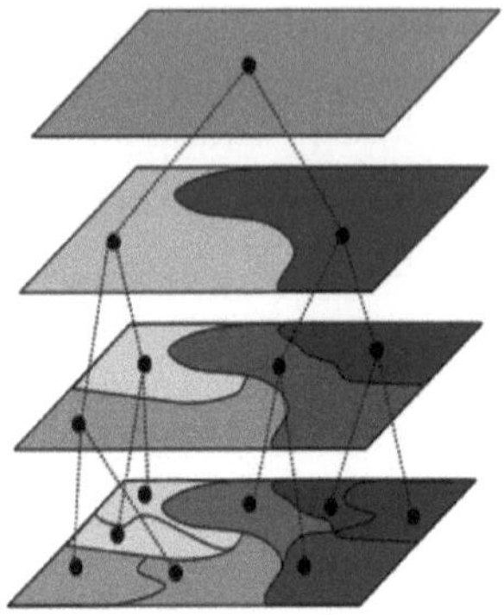

Figure 4.3. *Diagram of the binary partition tree (BPT)*

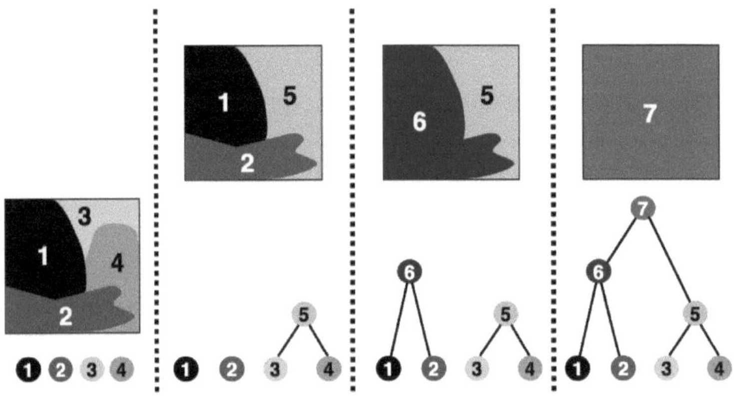

Figure 4.4. *Example of construction of a BPT from an image with four initial regions*

The representation of a hyperspectral image as a tree allows a large number of processing techniques to be defined [ALO 13]. These approaches are normally based on the tree pruning, which extracts the most interesting hyperspectral regions for a given application. Pruning can be seen as the extraction of "sub-trees" that consist of homogeneous nodes under a certain criterion (for example, spectral homogeneity, intensity, shape, texture, etc.). While construction of the tree is a generic operation, the pruning criterion is driven by the application. Two applications of the BPT are discussed in this chapter: image segmentation in section 4.4 and object detection in section 4.7.3.

4.3. Spectral matching

The spectral characterization of a site (detection, discrimination and/or identification of surface elements) by hyperspectral remote sensing is possible due to the analysis of the spectral signature of observed materials. Spectral matching techniques exploit this discriminant information by comparing the observed spectra with a set of reference spectra for known materials: spectral libraries.

4.3.1. *Spectral libraries*

Spectral databases arise from the need for reference measurements to perform spectroscopic imaging studies of the Earth and other planets. Spectral libraries may contain hundreds of reflectance spectra covering the spectral range from ultraviolet to far infrared. Measurements of emissivity or transmission are also possible. Spectra obtained from spectroscopic measurements of various materials such as minerals, vegetation types, synthetic materials and others are acquired in different ways: in the laboratory (when mastered samples are available), in the field (useful during airborne campaigns) or from an airborne sensor (when it is difficult to make measurements any other way). A well-known example is NASA's ASTER Spectral Library[9], which includes measurements of over 2000 materials found on the Earth. There is also the MEMOIRES database[10], which contains a large number of measurements on reflectance, emissivity and other units acquired in the lab or field.

4.3.2 *Spectral matching methods*

Spectral libraries are used to identify materials on the surface or for mapping the presence of a material of interest. For this, spectral matching techniques are used to calculate the degree of similarity between an unknown spectrum (from the hyperspectral image) and a known spectrum (from the spectral library). We distinguish between two sets of spectral matching techniques: deterministic methods such as the Euclidean distance, the Mahalanobis distance or the spectral angle mapper (SAM) and stochastic methods such as the spectral information divergence (SID). All these

9 http://speclib.jpl.nasa.gov/.

10 https://memweb.onecert.fr/sitools/.

methods are designed to detect the presence of materials at the pixel scale. For example, the spectral angle between two spectra $\mathbf{s}_i$ and $\mathbf{s}_j$, with norms$||\mathbf{s}_i||$ and $||\mathbf{s}_j||$, is calculated as follows:

$$SAM(\mathbf{s}_i,\mathbf{s}_j) = \cos^{-1}[<\mathbf{s}_i,\mathbf{s}_j> / (||\mathbf{s}_i|| \, ||\mathbf{s}_j||)] \qquad [4.2]$$

where < > is the scalar product. A lower SAM value indicates a small spectral angle and therefore strong similarity between the shapes of the two spectra. More information on spectral similarity measurements is available in [CHA 03].

Spectral characterization of a site by applying spectral matching techniques to a hyperspectral image may be done using all the available spectral bands or only the most discriminant ones (see section 4.2.3 for feature selection). Also, for materials with strong absorption bands, it is interesting to carry out spectral matching after eliminating the continuum of the spectra to be compared. This is done by dividing each spectrum by its convex hull. Spectral matching in the continuum-removed spectra space gives the full weight of the comparison to the absorption bands. In this way, the importance of the continuum can be minimized, as it is often linked to factors that are independent of the surface material (for example, solar irradiance, sensor viewing conditions, etc.).

An example of spectral matching is shown in Figure 4.5. For this, we use the Reno[11] hyperspectral image, which corresponds to an urban scene of the city of Reno in the United States. This image was acquired with a SpecTIR hyperspectral sensor with 2 m spatial resolution and a spectral range between 390 and 2450 nm. The Reno image is analyzed by the SAM method and two spectra from the ASTER database corresponding to the materials: "white coating" and "asphalt" (see Figure 4.5(a)). Spectral matching is done in the space with continuum and without continuum (CR for continuum removed). For the white coating, the use of SAM in the classical spectral space with continuum highlights a few buildings due to their association with small angles (cyan and light green circles in Figure 4.5(c)). However, if the spectra of the image and those from the ASTER database are transformed in the CR space (see Figure 4.5(b) for ASTER spectra), the detection of buildings is refined. For example, the building that was previously circled in cyan is no longer detected as white coating because it is now associated with large

11 http://www.spectir.com/free-data-samples/

SAM values (Figure 4.5(d)). The reason for this change is that the roof of this building is not coated in white but in a material with a similar spectral continuum. This improved discrimination in the CR space is possible through the strong absorption bands of the white coating spectrum. However, for asphalt, the SAM only works properly in space spectra with continuum, where roads and parking areas get the lowest spectral angles (Figure 4.5(e)). In the CR space, the SAM is not a good discriminant for asphalt because of the absence of absorption bands in the spectral signature (Figure 4.5(f)). In general, the CR space is only discriminant for characterizing materials with strong absorption bands.

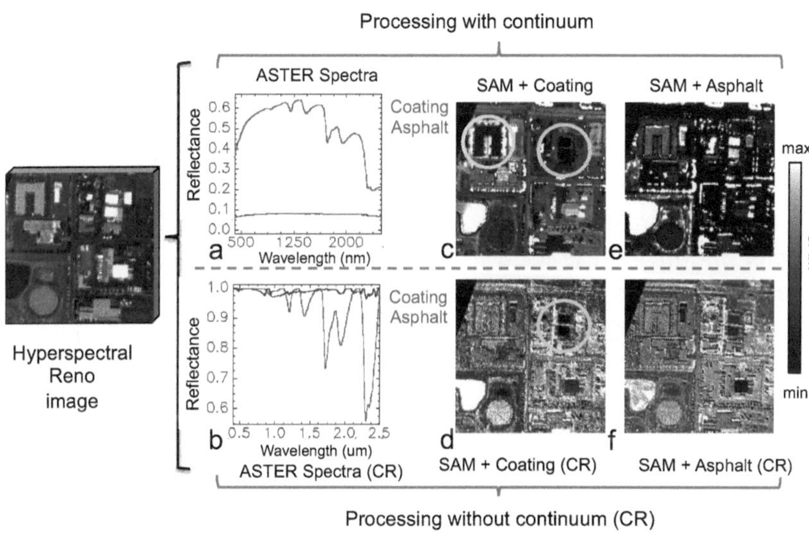

Figure 4.5. *Spectral characterization of the Reno image with the SAM method and ASTER classes "asphalt" and "white coating". Spectral matching between the Reno image and the two ASTER materials is first done in the conventional spectral space (a, c, and e), while a comparison in the continuum removed space (CR) is also carried out (b, d, and f). For this second comparison, spectra of the Reno image also lack their continuum. The SAM values indicate the presence of spectra near ASTER classes by low values (dark colors). For a color version of this figure, see www.iste.co.uk/baghdadi/1.zip*

4.4. Spectral segmentation

The segmentation of a hyperspectral image aims to partition it into spatially connected non-overlapping regions satisfying certain homogeneity criteria (spectral, shape, textural, etc.). The extraction of regions representing real objects of the image is possible due to the homogeneity of the spectra forming the same object. However, the definition of a homogeneity criterion within a group of spectra is not straightforward because of high spectral variability within the region and strong correlation between the spectral bands of the image. In addition, the large volume of data to be processed often requires a long computation time.

In the literature, studies have been proposed for segmentation of hyperspectral images from conventional image segmentation techniques. This is the case for the watershed method [TAR 10]. To process large amounts of hyperspectral data, some studies propose reducing the size of the images (for example, to three channels, see section 4.2.3) as a preliminary step to the segmentation using conventional techniques.

An alternative approach is to expand the definition of conventional segmentation techniques to this new multi-dimensionality. To this end, various methods for merging iterative regions have been proposed in recent years for the segmentation of hyperspectral images [GUA 02, VAL 13]. For example, one method is used to build the BPT presented in section 4.2.5. This is based on a bottom-up approach: the algorithm starts with small elements of the image (spectra), which will be grouped into elements with a semantic direction (regions). Each merging iteration involves searching for and merging the two most similar regions among all pairs of regions. Iterative algorithms are based on the definition of a region model and a merging criterion for hyperspectral data. The former represents the set of spectra in the region, while the latter is a measurement of similarity between the two models of regions. A simple solution is to use the spectra average as a region model and a measurement of similarity between two spectra (see section 4.3.2) as a criterion for fusion between the average spectra of two neighboring regions. Other alternative models have been proposed for better modeling of hyperspectral regions [VAL 13].

The result of an iterative region-merging algorithm is a sequence of hierarchical partitions called multi-scale segmentation. In this case, different levels of segmentation with differently sized regions are created. Figure 4.6 shows the results of multi-scale segmentation by the region-merging algorithm that is used to build the BPT applied to an area of the "Pavia Center"[12] image. This urban image was acquired by the ROSIS

12 http://www.ehu.eus/ccwintco/index.php?title=Hyperspectral_Remote_Sensing_Scenes.

hyperspectral sensor, which covers the spectral range of the visible and near infrared with 102 spectral bands and at a resolution of 1.3 m. In this example, the multi-scale segmentation was performed using the 102 spectral bands of the image. Figure 4.6 shows three different levels of the multi-scale segmentation as three segmentation results containing a number of different regions, allowing the user to choose the desired spatial detail.

(a) (b) (c) (d)

Figure 4.6. *Example of multi-scale segmentation with a region-merging algorithm of a) the "Pavia Center" image. Several segmentation results with b) 39 regions, c) 24 regions and d) 13 regions are shown. For a color version of this figure, see www.iste.co.uk/baghdadi/1.zip*

In a typical multi-scale approach [GUA 02], a stopping criterion allows us to choose from one of the hierarchy levels of partitions as a result of the segmentation. However, the use of a region-merging algorithm for creating a BPT allows us to define a pruning method to extract regions at different levels of the multi-scale segmentation in order to form a better segmentation result.

4.5. Spectral classification

4.5.1. *Introduction*

Classification of hyperspectral images involves assigning each spectrum of the image to a class (or label) from a list of classes that are known or not. By definition, the classes are categories of interest that need to be identified in the images, for example, different types of crops, tree species, man-made materials, etc. The result of a classification process is a set of labels assigned to all pixels, which creates a classification map. The peculiarity of spectral classification compared to traditional image classification is the access provided to the spectral dimension for achieving a more accurate assignment of classes.

There are two distinguishable categories in hyperspectral image classification techniques: parametric methods and non-parametric methods [LAN 03]. A classifier is said to be parametric if it combines a probabilistic model to all spectra within a class. In this category, we find the method of maximum likelihood, which assumes a Gaussian mixture model for the data. However, faults in parametric methods are often highlighted when applied to hyperspectral data due to the curse of dimensionality (see section 4.2.3). However, non-parametric classifiers are not based on a probability distribution. This is the case of support vector machines (SVMs), neural networks or the k-means classifier. The remainder of this chapter focuses on non-parametric methods, which are the most used in the literature.

A second division exists in spectral classification methods. If the spectral behavior of the class is unknown, we speak of *non-supervised classification*. In this case, classes are either set automatically or according to a list of classes that is defined independently of the image. However, if information on the spectral signature of the classes is available, we speak of *supervised classification*. In this case, the reference spectra for each class is required, which may be provided by:

– delineation by an expert of the areas in the image corresponding to classes of interest;

– acquired measurements in the field or in the laboratory;

– independent thematic maps.

In some cases, the supervised classifier is built from some labeled spectra in the image (called training samples) to then predict the class of the remainder spectra of the image. In a classification context, the set of reference data is called the ground truth (GT).

In addition to the importance of GT for training in supervised methods, GT is also essential in validation classification results. Therefore, GT is often divided into two subsets: the first is used for training and the second is used for testing. This validation is performed by comparing the result obtained by the classifier with the GT testing samples. A confusion matrix, which is a table for assessing the classification of all spectra, is often used to express this comparison. The confusion matrix is a widely used classification tool because it facilitates the understanding of results and details the effects of incorrect predictions [ING 16].

Figure 4.7 shows an example of a confusion matrix, where *Nc* is the number of classes. The rows of the matrix correspond to the values predicted by the classifier, while the columns represent the GT testing samples. The values on the diagonal of the matrix are the number of correctly classified pixels for each class. The elements of the matrix outside of the diagonal show different confusions. For example, the value n_{ij} is the number of pixels corresponding to the class *j* that have been misclassified as class *i*.

Reference data

Classification results	**1**	**2**	**...**	**Nc**
1	n_{11}	n_{12}	...	n_{1Nc}
2	n_{21}	n_{22}	...	n_{2Nc}
...	...	...	...	n_{3Nc}
Nc	n_{Nc1}	n_{Nc2}	...	n_{NcNc}

Figure 4.7. *Example of a confusion matrix for classification with Nc classes*

The confusion matrix allows us to calculate several quality measurements for classification results:

– the *overall accuracy* (OA) between 0 (no spectrum correctly classified) and 1 (100% correctly classified spectra), which is the ratio between the spectra that have been satisfactorily classified and the total number *Np*:

$$OA = \frac{\sum_{i=1}^{N_c} n_{ii}}{NP = \sum_{i=1}^{N_c}\sum_{j=1}^{N_c} n_{ij}} \quad [4.3]$$

– the *kappa* coefficient (κ), also between 0 and 1, which is a measure of accuracy that reflects the positive results that occur due to chance:

$$k = \frac{OA - P_c}{1 - P_c} \qquad [4.4]$$

where OA is the overall accuracy and P_c is the accuracy owed to chance defined as:

$$P_c = \frac{1}{Np^2} \sum_{i=1}^{Nc} (n_{+i} n_{i+}) \qquad [4.5]$$

where n_{+i} is the total of the observations in column i and n_{i+} is the same value for row i.

Evaluation of the individual accuracy of each class may be done through the following steps:

– *accuracy* (p_i), which is the ratio between the number of spectra correctly assigned to class i and the total number of spectra assigned to class i:

$$p_i = \frac{n_{ii}}{\sum_{j=1}^{Nc} n_{ij}} \qquad [4.6]$$

– *recall* (r_i), which is the ratio between the number of spectra correctly assigned to class i and the total number of spectra belonging to class i:

$$r_i = \frac{n_{ii}}{\sum_{j=1}^{Nc} n_{ji}} \qquad [4.7]$$

4.5.2. *Non-supervised classification methods*

In the case of non-supervised classification, the image is partitioned into several classes without using any prior knowledge. Thus, it is not necessary for the spectra to be labeled already. The non-supervised classifiers analyze the data distribution within their area of representation in order to identify natural groupings that correspond to high-density areas. At the end of the classification process, classes extracted from the image have to be interpreted by the user.

There are two types of approach: classification by partitioning, called "clustering", and bottom-up or top-down hierarchical classification algorithm. In the field of hyperspectral processing, partitioning methods such as the k-means, expectation-maximization or ISODATA are most prevalent. The principle of these methods is to consider that the representation space is divisible into a finite number of disjointed groups. Each group is defined such that the spectra contained within each group are as similar as possible and the groups are the most dissimilar possible between them.

The k-means algorithm is one of the most used methods of hyperspectral processing in this category [RIC 99]. Its objective is to partition all spectra of the image into several groups (classes), where each group is defined by a centroid that minimizes the distance between itself and the spectra belonging to its group. An example describing the various stages of the k-means algorithm is shown in Figure 4.8, where the number of classes has been set to three. Initially, the centroids of the three classes (triangles) are chosen randomly from the image spectra (circles). The latter are associated with the closest centroid, which is then updated. These last two steps are repeated until stability is achieved.

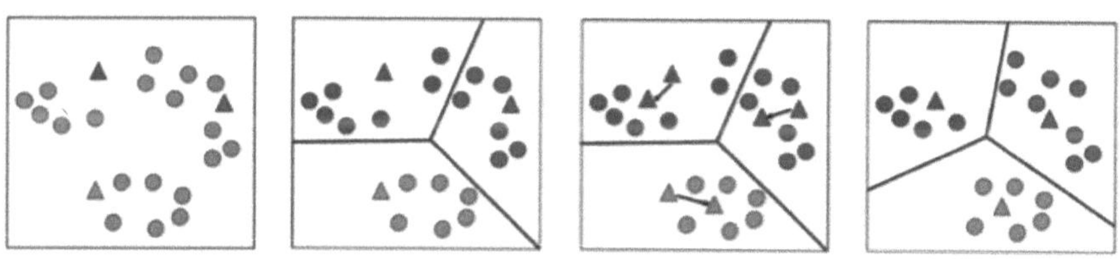

(a) Initialization (b) Allocation (c) Update (d) Stability

Figure 4.8. *Different stages of the k-means algorithm. For a color version of this figure, see www.iste.co.uk/baghdadi/1.zip*

The k-means algorithm has been widely used because of its speed and simplicity. However, it has significant limitations, such as needing to set the number of classes beforehand or the high dependency of results on the choice of initial centroids. To circumvent these limitations, alternatives are available such as the ISODATA method [BAL 65]. This classifier is used to automatically determine the number of classes due to thresholds that determine the minimum standard deviation within each cluster and the maximum distance between cluster centers.

However, non-supervised classification methods remain marginal in the field of hyperspectral data processing. Indeed, supervised methods have shown better performances, particularly since they allow for information that may be available to the user to be considered in the classification (for example, classes of interest, prior information, etc.).

4.5.3. *Supervised classification methods*

Supervised classification requires previously classified reference samples (the GT) in order to train the classifier and subsequently classify unknown data. In the field of hyperspectral image classification, supervised methods are divided according to their training system.

The simplest methods do not go through training but rather use notions of proximity between GT spectra and unknown spectra. These methods rely on the use of similarity measures like those set out in section 4.3.2. For example, the spectral angle (SAM) can be used to obtain a classification map by assigning each unknown spectrum to the GT class with the lowest SAM value. These spectral matching methods are useful when the GT is a spectral library that is independent of the image. In the case of classification by similarity measurements, one reference spectrum is sufficient to define each class.

A second set of methods is formed by those for which training is based on searches in a hypothesis space. In this category, there are methods such as the decision trees and neural networks [GOE 03]. However, the most widely used methods among these strategies for classification of hyperspectral images are those based on SVMs [VAP 98]. The strength of these methods lies in its low sensitivity to the Hughes phenomenon, which links processing a large number of spectral bands to significant decreases in classification performance for a fixed number of training samples [HUG 68]. The SVMs have a strong ability to produce good results with small training sets, as well as a discriminative power with similar classes [MEL 04].

The SVM classifier seeks to separate two classes by a hyperplane of which the minimum distance (called the margin) between the training samples of the two classes is the biggest it can be. The closest spectra, which are used for determining the hyperplane, are called support vectors. Figure 4.9(a) shows an example of binary classification for which the hyperplane appears in red. Sometimes, a separator hyperplane does not exist, or if it does exist then it is not the best solution. One solution is to relax the separation constraint and allow errors. In this case, we use soft margin SVM such as those illustrated in Figure 4.9(b). We note that although the SVMs were designed for binary classification, various methods exist to extend the SVM to multi-class classification of type "one versus all" or "one versus one" [MUL 01].

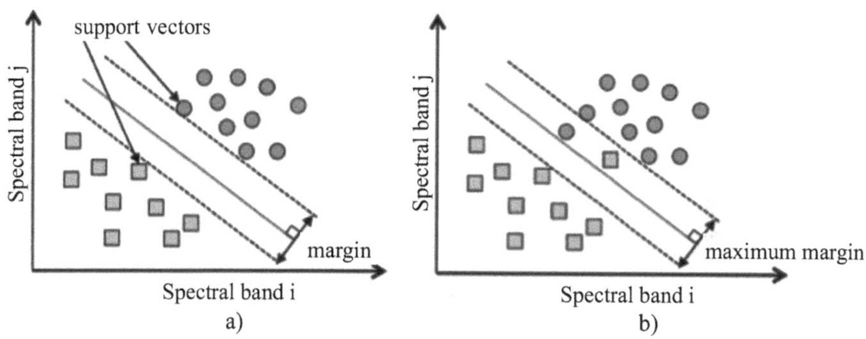

Figure 4.9. *a) Illustration of the concepts of margin and support vectors in an example linear discrimination that is separable in two dimensions; b) Example of soft margin SVM that allows errors. For a color version of this figure, see www.iste.co.uk/baghdadi/1.zip*

Kernel methods are used when the SVM decision function (the hyperplane) cannot be approximated by a line [CAM 05]. In this case, a function (the kernel) is used to transform data into a larger space (feature space) in which the data can be linearly separated. The kernel function ϕ, which can take many forms (for example, linear, polynomial, Gaussian, etc.), calculates a scalar product as a measure of similarity between two spectra to be processed. Figure 4.10 illustrates the use of a ϕ kernel to linearize the data space in order to define the SVM hyperplane. Then, the inverse transformation ϕ^{-1} is used to return to the original space.

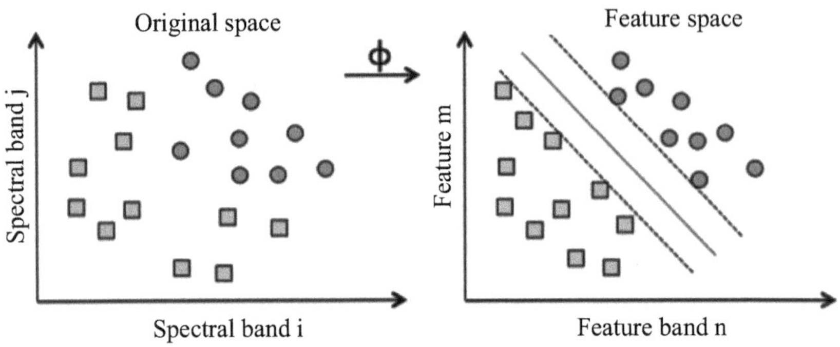

Figure 4.10. *Illustration of the use of a kernel ϕ in the SVM method. On the left is the original space of attributes and on the right is the projection through a kernel function ϕ. For a color version of this figure, see www.iste.co.uk/baghdadi/1.zip*

An example of supervised classification is shown in Figure 4.11, where the hyperspectral image "Indian Pines[13]" was classified using several methods. This image of a rural scene was acquired by the airborne AVIRIS sensor using 220 spectral bands covering the visible and near infrared. The "Indian Pines" image has been widely used in the literature to test hyperspectral classification methods. In addition to an overview of this dataset, Figure 4.11 shows the GT available for the scene. Each color represents one of the 16 classes defined by experts on the area. The bottom of Figure 4.11 shows the results of three different classification processes. First, the non-supervised k-means classifier was applied by determining the number of classes at 16. The resulting classification map is difficult to interpret because classes are not labeled (hence the different colors for GT classes). Subsequently, a supervised classification is performed by the SAM spectral angle, which compares each spectrum in the image with the average spectrum of each class defined by the GT. The class with the smallest SAM value is assigned to the given spectrum. The SAM classification map is more readable than the one obtained by the k-means method, although some inaccuracies can be observed. Finally, the classification map obtained using an SVM classifier with a Gaussian kernel is more consistent when compared to the GT. In this case, 25% of spectra of the image labeled by the GT were used to train the SVM.

13 http://www.ehu.eus/ccwintco/index.php?title=Hyperspectral_Remote_Sensing_Scenes.

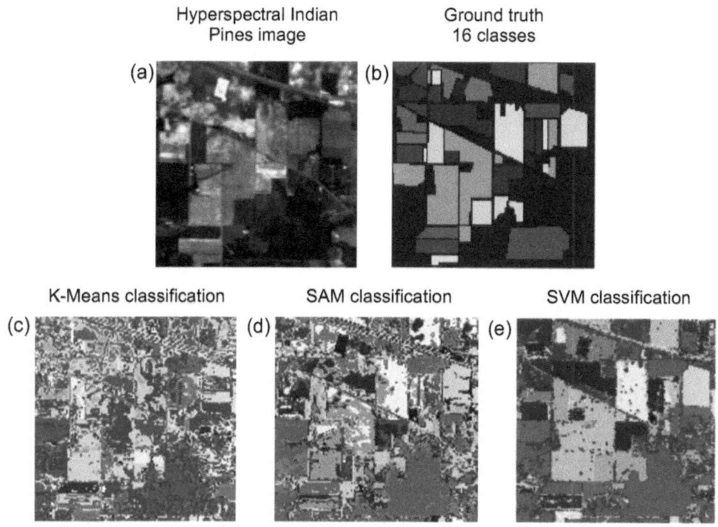

Figure 4.11. *a) "Indian Pines" hyperspectral image. b) Ground truth containing 16 classes. c) Classification map obtained with the non-supervised method of k-means. d) Classification map obtained with the SAM supervised method. e) Classification map obtained with a supervised SVM classifier with Gaussian kernel. For a color version of this figure, see www.iste.co.uk/baghdadi/1.zip*

4.5.4. *Spatial-spectral classification*

The supervised methods presented so far (and in particular, the kernel methods) obtain good classification performances by focusing only on the spectral dimension. However, not taking spatial information into account is a major limitation of these techniques. Consequently, these classification techniques usually produce results containing the so-called "salt and pepper" noise represented by misclassified individual pixels (Figure 4.11(e)).

To address this problem, many studies have proposed including spatial information in the classification process. To this end, each pixel is simultaneously defined in the spectral domain using its spectrum, and in space by applying a localized feature extraction approach in its spatial neighborhood. Three different approaches are identified for this purpose [PLA 09]:

– one solution is to include spatial information in the classifier decision rule. For example, the joint use of spatial and spectral information by defining a spatial-spectral kernel [FAU 08]. Another possibility lies in the adaptation of a Markov fields model by adding contextual information into the energy function [REL 04]. However, the Markov fields neighborhood system can cause the computation problem to become untreatable due to the large size of hyperspectral data [MOS 13];

– a second solution is to add a feature vector to each spatial spectrum of the image in addition to the spectral information. Then, feature vectors are used as input data for the classifier in addition to the regular spectral bands [BEN 05];

– a third solution proposes using spatial information for postprocessing of spectral classification results. This regularization step attempts to improve classification results by making the classification map elements more uniform. For example, Tarabalka *et al.* [TAR 09] use spectral segmentation results (see section 4.4) to apply majority votes on regions.

4.6. Spectral unmixing

4.6.1. *Types of subpixel mixing and the mixing model*

The concept of spectral unmixing goes beyond the previously detailed image analysis techniques by exploiting the spectral information at the subpixel scale. This family of techniques has its origin in blind source separation, which considers a signal to be a combination of different sources. By extension to hyperspectral imaging, each spectrum is broken down into a combination of spectral signatures that correspond to the surface materials in the associated pixel [KES 02]. In spectral unmixing, sources forming the image spectra are called endmembers, while the proportion of each of these endmembers in a spectrum is called the abundance.

The main reason behind mixed spectra in remote sensing is insufficient spatial resolution for separating adjacent materials at the surface. In Figure 4.12(a), this issue is illustrated through a mixture of three endmembers. This lack of spatial resolution is the origin of the so-called geographic mixtures, which make the observed spectrum equal to the weighted sum of the spectral signatures of each endmember in the pixel. In this case, the abundances correspond to the fraction of surface area occupied by each endmember (one third for each endmember in Figure 4.12(a)). This particular case is considered by the linear mixing model (LMM), which is used by most spectral unmixing methods because of its simplicity. Other mixture types also exist such as mixtures of different materials on the granular scale (Figure 4.12(b)) or laminated surfaces of which the upper layers are optically thin enough to allow photons to pass through (Figure 4.12(c)). Both these reasons are at the origin of the so-called intimate or non-linear mixtures, which should be processed by nonlinear unmixing techniques. Today, these approaches are far less developed than those based on the LMM [DOB 14].

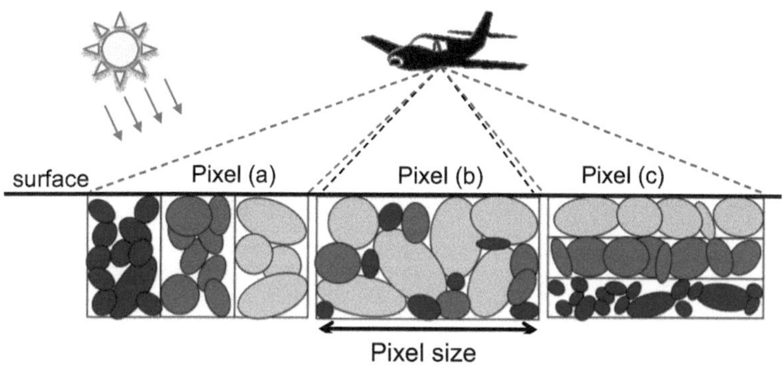

Figure 4.12. *Types of mixtures corresponding to different arrangements of three different materials (represented by grains of three different colors): a) geographical mixture, b) intimate granular mixture and c) intimate laminate mixture*

In the case of LMM, each $\mathbf{x}_i$ spectrum in a hyperspectral image $\mathbf{X}$ with P pixels ($\mathbf{X}=\{\mathbf{x}_1,\ldots,\mathbf{x}_P\}$) is modeled as a linear combination of N spectral signatures $\mathbf{M}=\{\mathbf{m}_1,\ldots,\mathbf{m}_N\}$ (one spectral signature for each endmember):

$$\mathbf{X} = \mathbf{M} \cdot \mathbf{S} + \mathbf{e} \qquad [4.8]$$

where $\mathbf{e}$ is the additive noise and $\mathbf{S}=\{\mathbf{s}_1,\ldots,\mathbf{s}_N\}$, where $\mathbf{s}_k=\{s_{k,1},\ldots,s_{k,P}\}$, gives the abundances of each endmember for each pixel.

Analysis of hyperspectral images through spectral unmixing can be supervised if information about the endmembers is available (for example, spectral signatures from a spectral library). Otherwise, we speak of non-supervised spectral unmixing in which there are three steps:

– determining N number of endmembers;

– extracting $\mathbf{M}$ spectral signatures of N endmembers;

– estimating $\mathbf{S}$ abundances associated with each endmember.

In this process, two constraints are often imposed: non-negativity of **S** values (abundances are always positive) and the sum of abundances $s_{k,I}$ for a given spectrum must be equal to the unit (the sum-to-one constraint).

4.6.2. *Estimation of the number of endmembers*

Spectral unmixing begins with an estimation of the number of endmembers N that exist in a hyperspectral image. This is often called the virtual dimensionality (VD) [CHA 04]. If *a priori* information on the scene is not available, we use:

– simple supervised techniques such as the one by [MOU 08], which sets the VD at 7 after finding that over 98% of the hyperspectral image variance in the study was concentrated in the first seven PCA components (see section 4.2.3);

– semi-automatic techniques such as in [HAR 93a], which set the VD equal to the number of components for which the difference between values of the covariance and correlation matrices of the hyperspectral image is not zero. The major drawback of this technique is the need to set a threshold parameter;

– completely unsupervised methods such as HySime [BIO 08], which, after estimating the correlation matrices of signal and noise, selects the values that best represent the signal space in a least squares sense.

4.6.3. *Extraction of endmembers and estimation of abundance maps*

Spectral unmixing continues by extracting a set of N spectral signatures, one for each endmember. Many algorithms are available to accomplish this under the assumption of a linear mixed model:

– *geometric methods* that exploit the parallelism between the LMM and the geometric distribution of data in a multidimensional space (the so-called simplex). This category is divided into two groups according to whether they adopt the pure pixel assumption or not. In fact, some methods seek the endmembers in the image and thus require the existence of their associated spectra. An example is the vertex component analysis (VCA) method [NAS 05]. Other methods aim to extract the spectra of endmembers without them existing in the image. Such is the case of the minimum volume simplex analysis (MVSA) approach [LI 08], which is useful for highly mixed data;

– *methods based on the concept of non-negative matrix factorization (NMF)*, which aim to factorize a matrix **X** into two matrices that only contain positive values. Approaches based on the NMF have been used for processing hyperspectral remote-sensing data [HUC 10];

– *statistical methods*, such as the PCA, the MNF (section 4.2.3) or the independent component analysis (ICA) that breaks the image down by assuming that the sources are independent and non-Gaussian [ERA 09]. A projection of hyperspectral data on two (or three) components obtained by these transformations can be used to identify endmembers and extract their characteristic spectra.

A complete review of endmember extraction methods by Bioucas-Dias *et al.* [BIO 12] is available, while a case study of a comparison of several methods is done by Ceamanos *et al.* [CEA 11].

After estimating the **M** spectral signatures of endmembers, each spectrum of the original image is reconstructed from the LMM linear model. The abundances are typically obtained by minimizing the reconstruction error by a least squares technique, generating an abundance map for each endmember.

4.6.4. *Example of spectral unmixing*

Figure 4.13 summarizes a simple spectral unmixing process applied to the Beltsville hyperspectral image. Here, we study the areas most affected by erosion among crops, i.e. those that contain bare soil rather than vegetation. To this end, we use the MNF transformation to generate a data cloud from the first and seventh components (see Figure 4.2). The choice of these two components produces a simplex with three extremes (Figure 4.13(a)), which correspond to three endmembers: trees, crops and bare soil (Figure 4.13(b)). We set N to three and consider the endmember spectra as the average spectrum of simplex extremes (Figure 4.13(c)). The three abundance maps are estimated from the LMM by minimizing the reconstruction error and imposing the sum-to-one constraint of the abundances for each spectrum, as well as their positivity (Figures 4.13(d–f)). From the abundance maps, we can identify the areas affected by erosion by, for example, looking at results associated with the bare ground endmember (Figure 4.13(d)). Indeed, an abundance value close to 1 indicates that the observed spectrum is close to the endmember of bare ground, thus where there is a significant lack of vegetation.

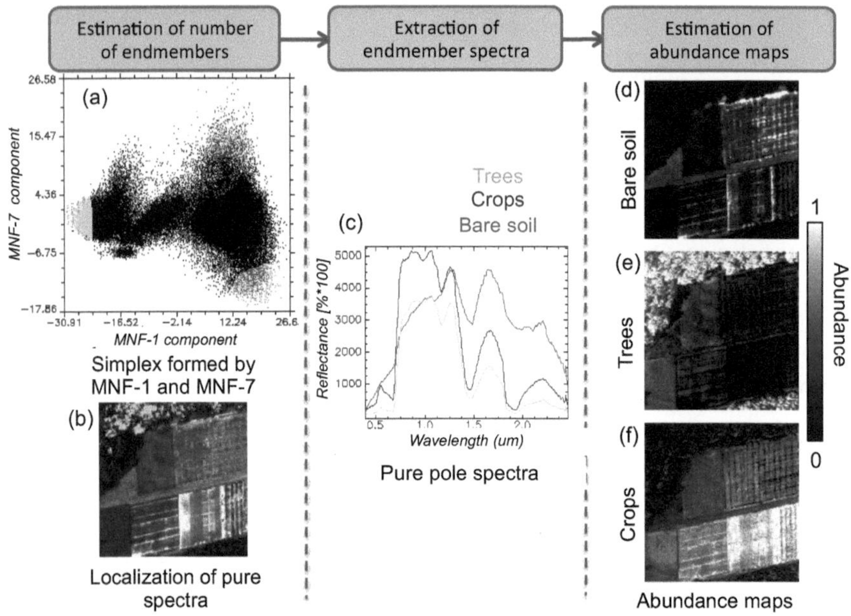

Figure 4.13. *Example of linear unmixing for the Beltsville image with three endmembers: trees, crops and bare ground. In this case, a combination of two MNF components is used to a) determine the number of endmembers and b) to spatially locate them. Subsequently, simplex extremes are used c) to define the spectra of endmembers prior to estimation of the associated abundance maps d–f). For a color version of this figure, see www.iste.co.uk/baghdadi/1.zip*

4.7. Detection of targets, anomalies and objects

4.7.1. *Target detection*

In some applications, the user may be interested in detecting a particular material in the hyperspectral image. Target detection techniques are available to meet this need where the spectrum of the desired target is known. In a detection problem, the image is divided into spectra corresponding to the *target* (if it exists) and the rest of the image, called the *image background.* Target detection techniques aim to extract the target from the image background.

Most detection methods are based on the principle of matched filtering, which aims to detect a known signal from an unknown signal by correlating the two of them. We recommend works by Chang [CHA 03] and Manolakis *et al.* [MAN 03] for a complete review of this category of methods. One of the most common techniques is the constrained energy minimization (CEM), which seeks a linear operator to maximize the correlation with the spectral signature of the target and to minimize it with the background spectral signature [HAR 93b]. For each unknown spectrum $\mathbf{r}$ of the image, the CEM method calculates a score from 0 (complete absence of the target spectrum) to 1 (detection of a spectrum equal to the target):

$$\delta_{CEM}(\mathbf{r}) = (\mathbf{d}^T\mathbf{R}^{-1}\mathbf{d})^{-1}(\mathbf{R}^{-1}\mathbf{d})^T\mathbf{r} \quad [4.9]$$

where $\mathbf{d}$ is the known spectral signature of the target and $\mathbf{R}$ is the correlation matrix of the image. If *a priori* information is available (for example, we seek man-made materials), preprocessing steps can be applied to improve the detection results (for example, masking vegetation).

Figure 4.14(a) illustrates the analysis using the CEM method on a section of the "Pavia Center" hyperspectral image (see section 4.4 for details on this image). In this case, we use the spectral signature of red tile from the ASTER database (see section 4.3.1) as a reference spectrum (d). The detection map obtained by the CEM assigns high values (orange and white pixels) to spectra that are likely to correspond to red tile or a spectrally similar material.

4.7.2. *Detection of anomalies*

If information on targets of interest is not available, we use anomaly detection techniques. An anomaly corresponds to an unknown but interesting target that is typically small in size and characterized by a spectral signature that is different from the background of the image.

A well-known anomaly detection method is the Reed–Xiaoli detector (RXD), which uses the observed spectrum as the reference spectral signature

[REE 90]. For each spectrum **r** in an image with covariance matrix **K** and mean μ, the RXD calculates the score by:

$$\delta_{RXD}(\mathbf{r}) = (\mathbf{r}-\mu)\mathbf{K}^{-1}(\mathbf{r}-\mu) \qquad [4.10]$$

The RXD method produces a score map in which the spectra that are most different from the background of the image are assigned large values, while small values are given to spectra that are similar to the background. A variant of this method is the local RXD, which applies the original technique several times on the image using a sliding window in order to overcome the potential background heterogeneity. More information on non-supervised detection methods is available in [CHA 03].

Figure 4.14(b) shows the results of the "Pavia Center" image obtained with the RXD method. In this case, the cars on the left of the image (circled in red) are detected as anomalies because of their unique spectral signature.

4.7.3. *Detection of objects*

As part of many current applications, research into the hyperspectral field attempts to develop techniques that approach the capabilities of the human visual system. The aim is to construct methods that recognize objects in the same way that the human brain does. Most automatic methods need to know at least some prior information about the object to be detected because a single object may appear differently depending on observation conditions.

Object detection in the hyperspectral field has mainly been treated as a target detection problem (see section 4.7.1). This is explained by the fact that the objects to be detected often occupy a small number of pixels in the image. However, an approach on the pixel level is not adapted for processing new hyperspectral images with a greater spatial resolution (therefore making objects to be made of several pixels) or for detecting larger objects. The main problem of a pixel-wise detection strategy is that objects are not only characterized by their spectral signature, but also by spatial characteristics such as shape, area or orientation [AKA 08].

A possible solution for integrating spatial and spectral information in the detection of objects is to use the representation provided by the BPT (see section 4.2.5). The proposed strategy analyzes a set of descriptors that has been calculated for each node of the tree. Subsequently, studying these descriptors allows us to recognize the object sought as a node of the BPT [VAL 15]. This analysis opens the door to a large number of techniques that capitalize on the very high spatial resolution in remote-sensing images such as those acquired by plane or drone.

For example, Figure 4.14 shows the results of a BPT pruning after introducing a shape descriptor measuring the squareness of regions into each node, as well as a spectral descriptor with a rooftop spectral signature. Thus, the majority of rooftops in the image are extracted with little confusion with other structures in the image.

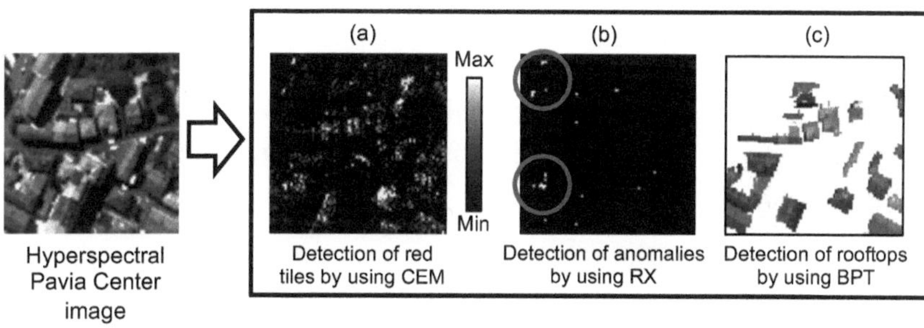

Figure 4.14. *Example of different detections on the "Pavia Center" image: a) Detection map obtained using the CEM method and a red tile spectrum as a reference; b) anomaly map obtained using the RXD method; c) extraction of rooftops from Binary Partition Tree constructed based upon the hyperspectral image. For a color version of this figure, see www.iste.co.uk/baghdadi/1.zip*

4.8. New directions in hyperspectral image processing

The wealth of information contained within hyperspectral data and the emergence of new hyperspectral sensors providing larger images require a continuous development of processing methods.

An example is the search for IT solutions that accelerate processing of large images. Indeed, the computation time and the manner of organizing data in the memory are major problems in the analysis of hyperspectral images. The implementation and parallelization on GPUs (graphics processors) of hyperspectral algorithms are a solution that is starting to be widely used. Methods have already been implemented on GPU systems with promising results in terms of reduction of the computation time [BER 13].

During recent years, there has also been a tendency to supplement the lack of reference data in supervised image processing through semi-supervised methods. These approaches are often designed to handle scenarios with hyperspectral data that are large in size and with limited prior information. In supervised classification, for example, semi-supervised training strategies use a large number of non-annotated data to address the lack of training data [BRU 06]. To do this, additional assumptions are initially made to classify non-annotated data (for example, the continuity hypothesis that states that if two points are close together in a high-density area, the same goes for their labels). Still concerning spectral classification, active learning is also an extension of semi-supervised methods that is increasingly used. Rather than exploiting non-annotated data, the idea is to allow the user to refine classification results [TUI 09]. For more information on advanced topics in spectral classification, we recommend the article by Camps-Valls *et al.* [CAM 14].

The high spatial resolution of future airborne hyperspectral images will allow a more detailed analysis of the elements that feature in a scene. Consequently, the incorporation of spatial information into processing will become more important. In addition to spatial-spectral classification methods (see section 4.5.4), studies have appeared in the field of spectral unmixing, for example using spectral and spatial information jointly [ECH 11]. Regarding spectral unmixing, the scientific community is making an effort to move toward nonlinear unmixing in order to bypass the linear mixing hypothesis, which is not fair in many situations [HEY 14].

Moreover, the temporal dimension in the hyperspectral field has not been considered much in the literature, as hyperspectral images acquired on the same site are rare. However, the appearance of remote sensing systems with low revisit time (for example, drones, satellite constellations, etc.) will open up this path in the coming years. Therefore, change detection techniques to process series of hyperspectral images are beginning to be available [LIU 15].

Lately, there is a tendency to use methods based on the transformation of hyperspectral images into the wavelet space. This mathematical tool has proven its effectiveness compared to other functions through its joint processing of spatial and spectral dimensions, among other benefits.

Wavelets are used in many other applications than just the compression of hyperspectral images (see section 4.2.4), such as image restoration [RAS 14] or classification [TAN 14].

Finally, hyperspectral imaging is increasingly combined with LiDAR (light detection and ranging) active imaging to characterize scenes with high relief like forest ecosystems or urban areas. Indeed, LiDAR images provide structural information on 3D objects on the surface [BAR 16], which are not accessible by hyperspectral imaging and which may enable, for example, better classification of different plant species in forests [TOR 14].

4.9. Key points

In the field of remote sensing, hyperspectral imaging provides a wealth of information on planetary surfaces through access to the spectral dimension. After presenting the preprocessing steps usually undergone by these data in order to retrieve surface reflectance, this chapter summarizes various existing processes for extracting the information of interest from hyperspectral images. This is the case for characterization, classification, unmixing, segmentation or detection methods to identify, classify, break down, extract or detect classes, materials or interesting objects on the surface. All these spectral processing methods provide improved results over traditional image processing due to the replacement of the pixel with the spectrum as the spatial unit. The major drawback of hyperspectral images is their large size, which can make it difficult to extract information. To face this, this chapter also details size reduction techniques, image compression and representations of alternative images for better use of hyperspectral data.

4.10. Bibliography

[AIA 09] AIAZZI B., BARONTI S., ALPARONE L., "Lossless compression of hyperspectral images using multiband lookup tables", *IEEE Geoscience Remote Sensing Letters*, vol. 16, no. 6, pp. 481–484, 2009.

[AKA 08] AKAY H., AKSOY S., "Automatic detection of geospatial objects using multiple hierarchical segmentations", *IEEE Transactions on Geoscience and Remote Sensing*, vol. 46, no. 7, pp. 2097–2111, 2008.

[ALO 13] ALONSO-GONZALEZ A., VALERO S., CHANUSSOT J. *et al.*, "Processing multidimensional SAR and hyperspectral images with binary partition tree", *Proceedings of the IEEE*, vol. 101, no. 3, pp. 723–747, 2013.

[BAC 05] BACKER S.D., KEMPENEERS P., DEBRUYN W. *et al.*, "A band selection technique for spectral classification", *IEEE Geoscience Remote Sensing Letters*, vol. 2, no. 3, pp. 319–323, July 2005.

[BAL 65] BALL G.H., HALL D.J., ISODAT A., A novel method of data analysis and pattern classification, Technical Report, Stanford Research Institute, Menlo Park, California, 1965.

[BAR 16] BARET F., "Estimation of biophysical variables from satellite observations", in BAGHDADI N., ZRIBI M. (eds), *Land Surface Remote Sensing in Urban and Coastal Areas*, ISTE Press Ltd, London and Elsevier Ltd, 2016.

[BEL 61] BELLMAN R., *Adaptive Control Processes: A Guided Tour*, Princeton University Press, 1961.

[BEN 05] BENEDIKTSSON J.A., PALMASON J.A., SVEINSSON J.R., "Classification of hyperspectral data from urban areas based on extended morphological profiles", *IEEE Transactions on Geoscience and Remote Sensing*, vol. 43, no. 3, pp. 480–491, March 2005.

[BER 13] BERNABE S., LOPEZ S., PLAZA A. *et al.*, "GPU implementation for an automatic target detection and classification algorithm for hyperspectral image analysis", *IEEE Geoscience and Remote Sensing Letters*, vol. 10, no. 2, pp. 221–225, 2013.

[BIO 08] BIOUCAS-DIAS J.M., NASCIMENTO J.M., "Hyperspectral subspace identification", *IEEE Transactions on Geoscience and Remote Sensing*, vol. 46, no. 8, pp. 2435–2445, August 2008.

[BIO 12] BIOUCAS-DIAS J.M., PLAZA A., DOBIGEON N. *et al.*, "Hyperspectral unmixing overview: geometrical, statistical, and sparse regression-based approaches", *IEEE Journal of Selected Topics in Applied Earth Observations and Remote Sensing*, vol. 5, no. 2, pp. 354–379, April 2012.

[BRU 06] BRUZZONE L., CHI M., MARCONCINI M., "A novel transductive SVM for the semisupervised classification of remote sensing images", *IEEE Transactions on Geoscience and Remote Sensing*, vol. 44, pp. 3363–3373, 2006.

[CAM 05] CAMPS-VALLS G., BRUZZONE L., "Kernel-based methods for hyperspectral image classification", *IEEE Transactions on Geoscience and Remote Sensing*, vol. 43, pp. 1351–1362, 2005.

[CAM 14] CAMPS-VALLS G., TUIA D., BRUZZONE L. *et al.*, "Advances in Hyperspectral image classification: earth monitoring with statistical learning methods", *IEEE Signal Processing Magazine*, vol. 31, no. 1, pp. 45–54, 2014.

[CEA 10] CEAMANOS X., DOUTÉ S., "Spectral smile correction of CRISM/MRO hyperspectral images", *IEEE Transactions on Geoscience and Remote Sensing*, vol. 48, no. 11, pp. 3951–3959, 2010.

[CEA 11] CEAMANOS X., DOUTÉ S., LUO B. *et al.*, "Intercomparison and validation of techniques for spectral unmixing of hyperspectral images: a planetary case study", *IEEE Transactions on Geoscience and Remote Sensing*, vol. 49, no. 11, pp. 4341–4358, November 2011.

[CEC 01] CECCATO P., FLASSE S., TARANTOLA S. *et al.*, "Detecting vegetation leaf water content using reflectance in the optical domain", *Remote Sensing of Environment*, vol. 77, pp. 22–33, 2001.

[CHA 03] CHANG C.I., *Hyperspectral Imaging: Techniques for Spectral Detection and Classification*, Kluwer Academic/Plenum Publishers, NY, 2003.

[CHA 04] CHANG C.I., DU Q., "Estimation of number of spectrally distinct signal sources in hyperspectral imagery", *IEEE Transactions on Geoscience and Remote Sensing*, vol. 42, no. 3, pp. 608–619, March 2004.

[CHR 08] CHRISTOPHE E., MAILHES C., DUHAMEL P., "Hyperspectral image compression: adapting SPIHT and EZW to anisotropic 3D wavelet coding", *IEEE Transactions on Image Processing*, vol. 17, no. 12, pp. 2334–2346, 2008.

[CHR 11] CHRISTOPHE E., "Hyperspectral data compression tradeoff" in PRASAD S. *et al.* (eds), *Optical Remote Sensing*, Springer-Verlag, Berlin Heidelberg, pp. 9–29, 2011.

[DOB 14] DOBIGEON N., TOURNERET J.Y., RICHARD C. *et al.*, "Nonlinear unmixing of hyperspectral images: models and algorithms", *IEEE Signal Processing Magazine*, vol. 31, no. 1, pp. 82–94, January 2014.

[ECH 11] ECHES O., DOBIGEON N., TOURNERET J.Y., "Enhancing hyperspectral image unmixing with spatial correlations", *IEEE Transactions on Geoscience and Remote Sensing*, vol. 49, no. 11, pp. 4239–4247, 2011.

[ERA 09] ERARD S., DROSSART P., PICCIONI G., "Multivariate analysis of visible and infrared thermal imaging spectrometer (VIRTIS) venus express nightside and limb observations", *Journal of Geophysical Research (Planets)*, vol. 114, no. E00B27, 2009.

[FAU 08] FAUVEL M., BENEDIKTSSON J.A., CHANUSSOT J. *et al.*, "Spectral and spatial classification of hyperspectral data using SVMs and morphological profiles", *IEEE Transactions on Geoscience and Remote Sensing*, vol. 46 no. 11, pp. 3804–3814, November 2008.

[GAO 09] GAO B.C., MONTES M.J., DAVIS C.O. *et al.*, "Atmospheric correction algorithms for hyperspectral remote sensing data of land and ocean", *Remote Sensing of Environment*, vol. 113, no. 1, pp. S17–S24, 2009.

[GIT 02] GITELSON A., ZUR Y., CHIVKUNOVA O.B. *et al.*, "Assessing carotenoid content in plant leaves with reflectance spectroscopy", *Photochemistry and Photobiology*, vol. 75 no. 3, pp. 272–281, 2002.

[GOE 03] GOEL P.K, PRASHER S.O., PATEL R.M. *et al.*, "Classification of hyperspectral data by decision trees and artificial neural networks to identify weed stress and nitrogen status of corn", *Computers and Electronics in Agriculture*, vol. 39, no. 2, pp. 67–93, May 2003.

[GOM 08] GOMEZ-CHOVA L., ALONSO L., GUANTER L. *et al.*, "Correction of systematic spatial noise in push-broom hyperspectral sensors: application to CHRIS/PROBA images", *Applied Optics*, vol. 47, no. 28, pp. F46–F60, October 2008.

[GRE 88] GREEN A.A., BERMAN M., SWITZER P. *et al.*, "A transformation for ordering multispectral data in terms of image quality with implications for noise removal", *IEEE Transactions on Geoscience and Remote Sensing*, vol. 26 no. 1, pp. 65–74, 1988.

[GUA 02] GUALTIERI J.A., TILTON J.C., "Hierarchical segmentation of hyperspectral data", *AVIRIS Earth Science and Applications Workshop*, Pasadena, CA, March 5–8, 2002.

[HAR 93a] HARSANYI J., FARRAND W., CHANG C.I., "Determining the number and identity of spectral endmembers: an integrated approach using Neyman–Pearson eigenthresholding and iterative constrained RMS error minimization", *Proceedings of 9th Thematic Conference on Geologic Remote Sensing*, February 1993.

[HAR 93b] HARSANYI J.C., Detection and classification of subpixel spectral signatures in hyperspectral image sequences, PhD Dissertation, University of Maryland, 1993.

[HEY 14] HEYLEN R., PARENTE M., GADER P., "A review of nonlinear hyperspectral unmixing methods", *IEEE Journal of Selected Topics in Applied Earth Observations and Remote Sensing*, vol. 7, no. 6, pp. 1844–1868, 2014.

[HU 09] HU C., "A novel ocean color index to detect floating algae in the global oceans", *Remote Sensing of Environment*, vol. 113, pp. 2118–2129, 2009.

[HUC 10] HUCK A., GUILLAUME M., BLANC-TALON J., "Minimum dispersion constrained nonnegative matrix factorization to unmix hyperspectral data", *IEEE Transactions on Geoscience and Remote Sensing*, vol. 48, no. 6, pp. 2590–2602, June 2010.

[HUG 68] HUGHES G.F., "On the mean accuracy of statistical pattern recognizers", *IEEE Transactions on Information Theory*, vol. 14, no. 1, pp. 55–63, 1968.

[ING 16] INGLADA J., "Land-cover mapping from optical images", in BAGHDADI N., ZRIBI M. (eds), *Land Surface Remote Sensing in Agriculture and Forest*, ISTE Press Ltd, London and Elsevier Ltd, Oxford 2016.

[KES 02] KESHAVA N., MUSTARD J.F., "Spectral unmixing", *IEEE Signal Processing Magazine*, vol. 19, pp. 44–57, 2002.

[LAN 03] LANDGREBE D.A., *Signal Theory Methods in Multispectral Remote Sensing*, John Wiley & Sons, New Jersey, 2003.

[LET 08] LETEXIER D., BOURENNANE S., "Noise removal from hyperspectral images by multidimensional filtering", *IEEE Transactions on Geoscience and Remote Sensing*, vol. 46, no. 7, pp. 2061–2069, July 2008.

[LI 08] LI J., BIOUCAS-DIAS J.M., "Minimum volume simplex analysis: a fast algorithm to unmix hyperspectral data", *IEEE International Geoscience And Remote Sensing Symposium*, Boston, pp. 250–253, July 2008.

[LIU 15] LIU S., BRUZZONE L., BOVOLO F. *et al.*, "Hierarchical unsupervised change detection in multitemporal hyperspectral images", *IEEE Transactions on Geoscience and Remote Sensing,* vol. 53, no. 1, pp. 244–260, January 2015.

[MAN 03] MANOLAKIS D., MARDEN D., SHAW G.A., "Hyperspectral image processing for automatic target detection applications", *Lincoln Laboratory Journal*, vol. 14, no. 1, pp. 79–116, 2003.

[MEL 04] MELGANI F., BRUZZONE L., "Classification of hyperspectral remote sensing images with support vector machines", *IEEE Transactions on Geoscience and Remote Sensing*, vol. 42, no. 8, pp.1778–1790, 2004.

[MOS 13] MOSER G., SERPICO S.B., "Combining support vector machines and Markov random fields in an integrated framework for contextual image classification", *IEEE Transactions on Geoscience and Remote Sensing*, vol. 51, no. 5, pp.2734–2752, May 2013.

[MOU 08] MOUSSAOUI S., HAUKSDÓTTIR H., SCHMIDT F. *et al.*, "On the decomposition of Mars hyperspectral data by ICA and Bayesian positive source separation", *Neurocomputing*, vol. 71, no. 10–12, pp. 2194–2208, June 2008.

[MUL 01] MULLER K.R., MIKA S., RATSCH G. *et al.*, "An introduction to kernel-based learning algorithms", *IEEE Transaction on Neural Networks*, vol. 12, pp. 181–202, 2001.

[NAS 05] NASCIMENTO J.M.P., DIAS J.M.B., "Vertex component analysis: a fast algorithm to unmix hyperspectral data", *IEEE Transactions on Geoscience and Remote Sensing*, vol. 43, no. 4, pp. 898–910, April 2005.

[PEN 07] PENNA B., TILLO T., MAGLI E. *et al.*, "Transform coding techniques for lossy hyperspectral data compression", *IEEE Transactions on Geoscience and Remote Sensing*, vol. 45, no. 5, pp. 1408–1421, 2007.

[PLA 09] PLAZA A., BENEDIKTSSON J.A., BOARDMAN J.W. *et al.*, "Recent advances in techniques for hyperspectral image processing", *Remote Sensing of Environment*, vol. 113, pp. 110–122, 2009.

[QIA 04] QIAN S.E., "Hyperspectral data compression using a fast vector quantization algorithm", *IEEE Transactions on Geoscience and Remote Sensing*, vol. 42, no. 8, pp. 1791–1798, 2004.

[RAS 14] RASTI B., SVEINSSON J.R. ULFARSSON M.O., "Wavelet-based sparse reduced-rank regression for hyperspectral image restoration", *IEEE Transactions on Geoscience and Remote Sensing*, vol. 52, no. 10, pp. 6688–6698, 2014.

[REE 90] REED I., YU X., "Adaptive multiple-band CFAR detection of an optical pattern with unknown spectral distribution", *IEEE Transactions Acoustics Speech and Signal Processing*, vol. 38, pp. 1760–1770, 1990.

[REL 04] RELLIER G., DESCOMBES X., FALZON F. *et al.*, "Texture feature analysis using a gauss-Markov model in hyperspectral image classification", *IEEE Transactions on Geoscience and Remote Sensing*, vol. 42, no. 7, pp.1543–1551, July 2004.

[RIC 99] RICHARDS J.A., JIA X., *Remote Sensing Digital Image Analysis*, Springer-Verlag, Berlin, 1999.

[ROD 02] RODARMEL C., SHAN J., "Principal component analysis for hyperspectral image classification", *Surveying and Land Information Science*, vol. 62, no. 2, pp. 115–22, 2002.

[ROU 73] ROUSE W., HAAS R.H., SCHELL J.A. *et al.*, Monitoring the vernal advancement and retrogradation (green wave effect) of natural vegetation, Report, RSC 1978-1, Remote Sensing Center, Texas A&M University, 1973.

[TAN 14] TANG Y.Y., LU Y., YUAN H., "Hyperspectral image classification based on three-dimensional scattering wavelet transform", *IEEE Transactions on Geoscience and Remote Sensing*, vol. 53, no. 5, pp. 2467–2480, 2014.

[TAR 09] TARABALKA Y., BENEDIKTSSON J.A., CHANUSSOT J., "Spectral-spatial classification of hyperspectral imagery based on partitional clustering techniques", *IEEE Transactions on Geoscience and Remote Sensing*, vol. 47, no. 8, pp. 2973–2987, August 2009.

[TAR 10] TARABALKA Y., CHANUSSOT J., BENEDIKTSSON J.A., "Segmentation and classification of hyperspectral images using watershed transformation", *Pattern Recognition*, vol. 43, no. 7, pp. 2367–2379, 2010.

[TOR 14] TORABZADEH H., MORSDORF F., SCHAEPMAN M.E., "Fusion of imaging spectroscopy and airborne laser scanning data for characterization of forest ecosystems – A review", *ISPRS Journal of Photogrammetry and Remote Sensing*, vol. 97, pp. 25–35, November 2014.

[TUI 09] TUIA D., RATLE F., PACIFICI F. *et al.*, "Active learning methods for remote sensing image classification", *IEEE Transactions on Geoscience and Remote Sensing*, vol. 47, no. 7, pp. 2218–2232, July 2009.

[VAL 13] VALERO S., SALEMBIER P., CHANUSSOT J., "Hyperspectral image representation and processing with Binary Partition Trees", *IEEE Transactions on Image Processing*, vol. 22, no. 4, pp.1430–1443, 2013.

[VAL 15] VALERO S., SALEMBIER P., CHANUSSOT J., "Object recognition in hyperspectral images using binary partition tree representation", *Pattern Recognition Letters*, vol. 56, pp. 45–51, 2015.

[VAP 98] VAPNIK V., *Statistical Learning Theory*, Wiley, 1998.

[WAN 11] WANG Q., LI P., PU Z. *et al.*, "Calibration and validation of salt-resistant hyperspectral indices for estimating soil moisture in arid land", *Journal of Hydrology*, vol. 408, no. 3–4, pp. 276–285, 13 October 2011.

5

Principle and Physics of the LiDAR Measurement

5.1. Introduction

In this chapter, a brief overview of the LiDAR measurement used for Earth and space observation is first presented (section 5.2). This is subsequently followed by an overview of the LiDAR instrument, describing its various optical constituent elements (section 5.3). The LiDAR equation is then described with emphasis on its key parameters (section 5.4). Finally, the main causes of LiDAR error on surface observations are specified, with a particular emphasis on airborne or spaceborne measurements.

5.2. General aspects concerning the LiDAR instrument

5.2.1. *Definition and background*

LiDAR is an acronym for Light Detection And Ranging. This remote sensing technique may also be found under different designations, such as light radar, Laser Detection And Ranging (LaDAR) or laser scanner. Although mainly used for the purpose of laser ranging (distance measurements), LiDARs also facilitate other types of measurements: the evaluation of ozone [MEG 89] and volcanic aerosol [CHA 95] concentrations in the stratosphere

Chapter written by Patrick CHAZETTE, Julien TOTEMS, Laurent HESPEL and Jean-Stéphane BAILLY.

and the evaluation of benthic seabeds, for instance. Following the example of RaDAR, LiDAR denotes both an active optical remote sensing technique and a sensor. LiDAR is an active remote sensing technique since it uses an artificial light source. This artificial light comes from a laser emitting in the optical spectrum ranging from approximately 0.25 to 11 µm [WEI 05].

Originally designed for a ranging measurement (using known properties of light velocities in air or water), many LiDAR systems used for Earth surface remote sensing are referred to as (simple or multiple) "echo systems". In these systems, the echoes correspond to the instants when high-intensity levels of backscattered light are detected (see Figure 5.1). Other systems referred to as profilers or full waveform systems record the profile of the temporal power of a light pulse backscattered by the media encountered by the laser beam. The information required in the latter case is represented by the profile form based on the properties of the investigated media (see Figure 5.1).

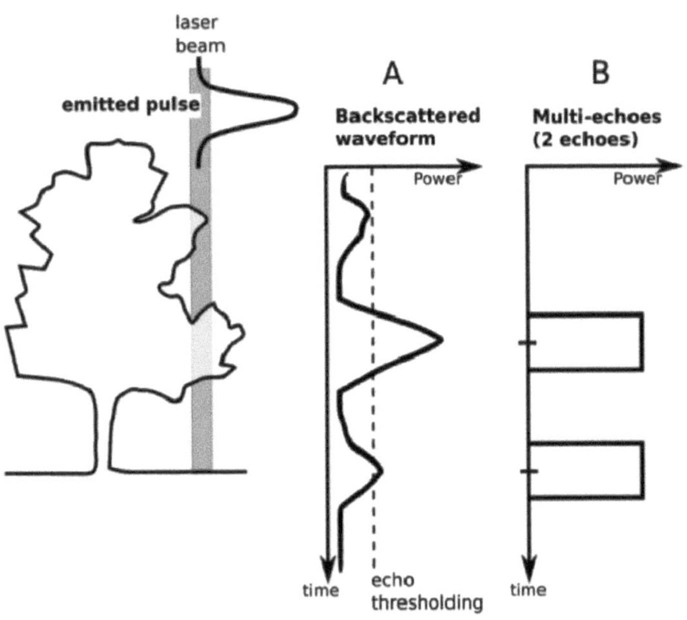

Figure 5.1. *Principle of profiler a) or echo b). LiDARs illustrated using vegetation [BAI 12]*

The laser represents one of the basic components of the LiDAR. Lasers originate from the stimulated transmission concept introduced by Albert Einstein in 1917, but the first laser system was designed only after advances in technology. It was (in particular, after the development of optical pumping by Albert Kastler in 1950). Two years after the "maser" which was based on the emission of radio waves, Charles Townes, Arthur Schawlow and Theodore Maiman invented the laser in 1960. The first LiDAR system was built by L.D. Smullin and G. Fiocco using the ruby laser invented by T. Maiman. This system made it possible to measure the distance between the Earth and the Moon [SMU 62]. Subsequently, the same type of LiDAR made possible the measurement of the distance between elements independently of the terrestrial geodetic network. This latter capacity was effective after the launch of the S-66-B satellite by NASA, which comprised 360 reflective prisms at the level of which laser light could be backscattered toward ground stations. One of these ground stations was the LiDAR station of the Haute-Provence Observatory built in 1962. It housed the first LiDARs using a switched ruby laser [CHA 08]. On the initiative of the Aeronomy Service of the CNRS, a geodetic junction between Europe, Africa and South America was created for the first time by means of laser ranging and space triangulation. This announced future uses of LiDARs in topographic applications. Nowadays, LiDAR systems are largely used outside the scientific sphere. Some applications are derived from sensor miniaturizing. In the industrial sphere, LiDAR systems will be incorporated into the next generation of vehicles in order to ensure autonomous vehicles and to increase road safety.

5.2.2. *Applications in geophysics*

LiDAR systems can be classified into four categories [FLA 05]:

– the first category refers to the most common LiDARs based on the elastic scattering principle, namely, without wavelength changes. These are used in order to carry out aerosol and cloud measurements in the atmosphere (Rayleigh or Mie scattering), but may also be used in ranging (surface backscattering). Such systems may potentially be polarized [CAI 99, CHA 12];

– the second category includes inelastic scattering LiDAR systems, which can be of the Raman [ROY 11], fluorescence [HOG 83] or Brillouin

[FRY 97] type and which can be used in order to determine the physico-chemical composition of the tested medium [MEG 85]. In all of these cases, the backscattering of light is affected by a wavelength offset with respect to the emitted light. This offset is conditioned by the molecules that form the targetted medium. For example, water molecules excited by light at 355 nm emit a Raman radiation at approximately 407 nm [CHA 14];

– the third category refers to differential absorption systems (DIAL), such as the LEANDRE-2 LiDAR system for instance [BRU 01]. In these DIAL systems, light is emitted at two very adjacent wavelengths. One of these wavelengths is strongly absorbed by the target component. Taking account of the adjacent wavelength, the ratio of backscattered energies between adjacent wavelengths is thus proportional to the molecular concentration of the target component;

– finally, the last category is constituted by heterodyne systems. They represent the majority of Doppler LiDARs used to measure wind speed, an essential component in operational meteorology. The principle of Doppler LiDARs consists of recording the wavelength offset of a backscattered laser pulse. This offset is induced by the speed of the particles encountered by the light beam during its propagation.

A large portion of LiDAR applications designed to observe terrestrial surfaces use elastic scattering and therefore fall under the first category of instruments. Precise sensor geolocation and attitude instruments are associated with the LiDAR measurement on terrestrial surface. The measuring platforms are therefore usually equipped with a differential GPS (DGPS) in order to accurately locate the instrument, and with an inertial unit whose role is to record the instrument attitude. The main applications and types of LiDAR instrumentation are indicated in Table 5.1. In this table, the measured variables are indicated, and the main references are provided for each application.

5.2.3. *Onboard LiDAR systems*

LiDAR systems can be carried onboard various platforms: satellites or shuttles (spaceborne LiDARs), aircraft, helicopters, ULM or drones. The first space LiDARs were designed to study the atmosphere or to explore the universe (altimeters). The first altimeter LiDAR used in space exploration was carried onboard Appolo 11 in 1969. It was followed by Clementine instruments in 1994 (topography of the Moon), NEAR Laser Ranging Investigation (NLRI, topography of the 433 Eros Asteroid) in 1996, Mars Orbiter Laser Altimeter (MOLA, topography of Mars) in 1996, Mercury Laser Altimeter (MLA, topography of Mercury) in 2004, Laser ALTimeter (LALT, topography of the Moon) as part of the KAGUYA mission in 2009, Lunar Orbiter Laser Altimeter (LOLA, topography of the Moon) in 2015 and recently BEpiColombo Laser Altimeter (BELA, topography of Mercury).

The first system deployed in orbit in order to complete a short Earth observation mission was the LITE (LiDAR In-Space Technology Experiment) instrument onboard the Discovery space shuttle in 1994 [BER 06]. This mission was followed by the Franco-American program Geoscience Laser Altimeter System (GLAS), [ZWA 02] using the Cloud-Aerosol LiDAR with Orthogonal Polarization (CALIOP), [WIN 02]) LiDAR. Prior to the CALIPSO mission, the ICESat satellite carrying the Geoscience Laser Altimeter System (GLAS), [ZWA 02], [ZWA 02]) instrument was launched in 2003. The main objective of the ICESat mission was the mapping of ice levels. The high vertical resolution (~15 cm) of the GLAS instrument also allowed studies on frost-free surfaces, especially on forest canopies.

LiDARs carried onboard space platforms collect samples along the ground track of the satellite. However, it may be noted that in the case of the LOLA instrument, a diffraction of the beam into five subbeams makes it possible to sample the surface not only along a transect. Airborne LiDAR systems are used in order to gather high resolution information at ground level. In addition, hydrographic LiDARs are all profilers (full waveform) and operate with green laser wavelengths, airborne LiDARs are often of the multi-echo type and operate in the near infrared. However, this configuration is developed toward full waveform systems. They are generally used within the context of scientific research as demonstration and/or validation tools for spatial observations. As an example, various iconic onboard LiDAR systems, applications and platforms are presented in Table 5.2.

LiDAR category	Type of data	Application	Measured variable(s)	Type of platform	Operational/ research	Historical reference(s)
Elastic scattering	Profiler	Bathymetry	Ocean floor topography, water depth,	Airborne	Operational	[GUE 78]
Polarized elastic scattering	Profiler	Bathymetry and oceanography	Water depth, phytoplankton concentration, water turbidity and optical properties of subsurface waters	Airborne	Research	[KRE 98]
Elastic scattering	Profiler	Mapping of ecological aquatic habitats	Local relief and turbidity	Airborne	Research	[BRO 06]
Elastic scattering	Multi-echo	Topography	3D topographic points and digital elevation models	Airborne/ Satellite/ Ground	Operational	[BAL 99]
Elastic scattering	Mono/ Multi-echo	Archaeology	3D topographic points and digital elevation models → 3D objects	Airborne/ Ground	Operational	[AHM 83]
Elastic scattering	Multi-echo	Fluvial, coastal geomorphology Glaciology	Eroded and deposited volumes	Ground/ Airborne/ Satellite	Research	[YOU 06]
Elastic scattering	Multi-echo/Profiler	3D urban mapping	3D topographic points → 3D objects	Ground/ Airborne	Operational/ Research	[YOU 03]
Elastic scattering	Multi-echo/Profiler	Forest management	Canopy height and basal surface	Airborne/ Satellite	Operational	[SOL 77]
Elastic scattering	Profiler	Biomass mapping	Canopy height and biomass	Airborne/ Satellite	Research	[NEL 88]
Elastic scattering	Profiler	Water altimetry (continental waters – oceans)	3D topographic points	Satellite	Research	[BRA 01]
Fluorescence	Profiler	Restoration of historic buildings	Type of mineral substrate and biological growth	Ground	Research	[LOG 03]
Fluorescence	Profiler	Detection of hydrocarbons at sea	Thickness of the hydrocarbon layer	Airborne	Operational	[HOG 83]
Fluorescence	Profiler	Detection of submarine strata	Thickness and composition of the stratum	Airborne	Research	[CHU 09]
Fluorescence	Profiler	Oceanography	Optical properties of subsurface waters	Airborne	Research	[CHU 98]
Raman	Profiler	Oceanography	Temperature of the subsurface water	Airborne	Research	[LEO 79]
Fluorescence	Profiler	Monitoring of vegetation	Chlorophyll content	Airborne/ Ground	Research	[SVA 95]
Brillouin	Profiler	Oceanography	Water temperature and speed of sound	Airborne	Research	[FRY 97]

Table 5.1. *LiDAR applications for Earth observations*

LiDAR system	Platform	Wavelength/laser power	Main objective	Secondary objectives	Scan or transect	Footprint (m) /distance at ground level between 2 successive measurements or point density per m^2	Reference(s)
GLAS (profiler)	IceSAT satellite	1,064 nm/ 75 mJ	Monitoring the altimetry of polar ice caps	Biomass mapping – water surface altimetry	Transect	70/170	[ZWA 02]
Hawkeye II (profiler)	Aircraft	532 nm/ 3 mJ	Coastal and riverine bathymetry	Mapping of ecological aquatic habitats	Scan	2.5/0.9	[AXE 10]
Research LiDAR ULICE	ULM/ aircraft	355 nm/ 7 mJ	Forest structure and LAI	Carbon balance Forest/atmosphere interaction	Transect	1–20/1–10	[SHA 14]
Yellowscan (multi-echo)	Drone	905 nm/ 375 nJ	Topography	Forest management	Scan	0.1/70	[ESP 14]
Riegl VZ-400 (multi-echo or profiler)	Ground	1,550 nm/ 3 mJ	3D objects and infrastructure	Forest topography	Scan	$0.0017/10^6$	[FOW 11]

Table 5.2. *Examples of LiDAR systems on satellite, airborne, ULM and ground platforms*

5.3. General aspects of LiDAR optical systems

In this section, the constitutive elements of the optical architecture of a LiDAR system dimensioned for forest canopy measurements will be presented.

We will hereafter consider a direct detection LiDAR, as opposed to LiDARs based on a heterodyne detection, also referred to as a "coherent detection". Only the amount of power backscattered by the targeted medium is recorded in time, in a resolute manner. The description spans the most compact airborne systems as well as existing and envisaged space systems.

5.3.1. *Architecture and functional chain*

The elements constituting the optoelectronic architecture of a LiDAR are represented in Figure 5.2, along with a diagram of the corresponding functional chain. A pulse is emitted by the laser source for each LiDAR measurement. Once the laser beam has been shaped (expanded and collimated), it passes through scanning optics and travels through space toward the targeted scattering medium. Following its interaction with the target, an echo of this pulse is backscattered toward the LiDAR. The optical flux is collected by an optical imaging system, often of large dimensions, such as a telescope. Stray flux is separated from the signal by means of spatial filtering in the image focal plane as well as spectral filtering. The former restricts the field of view to the space traveled by the laser pulse, whereas the latter is used to reduce the spectral bandwidth of the optical system around the emitted wavelength. Note that wavelength selection requires the beam to be collimated (parallel) upstream of the spectral filtering optical element. At the end of the optical chain, a photodetector converts the detected photon flow into an electrical signal, which is in turn digitized prior to its recording and processing.

The optical elements of this architecture are described in more detail in the remainder of this section.

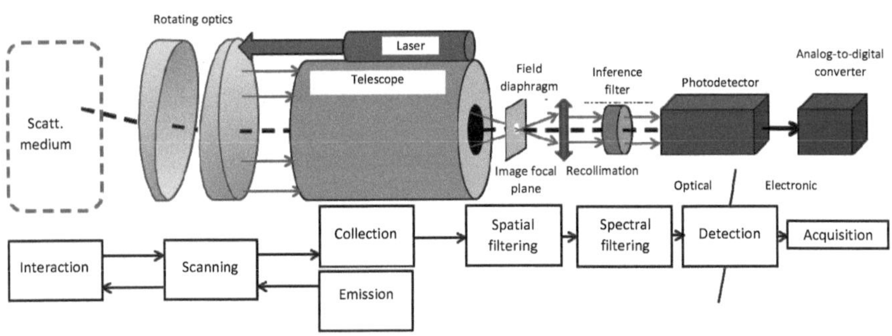

Figure 5.2. *Conventional architecture (top) and functional chain (bottom) of a direct detection LiDAR*

5.3.2. *Emission*

5.3.2.1. *Laser*

The transmission is derived from a pulsed laser source emitting in the near-infrared (IR, wavelength 780–3,000 nm), visible (390–780 nm) or near-ultraviolet domain (UV, 200–390 nm). A wide variety of laser sources are built around amplifying media which can be gas cells, crystals, diodes or fibers. In the case of space or airborne LiDARs for forestry applications, the selection of the emitted wavelength is conditioned by atmospheric transmission and vegetation reflectance. Other requirements in terms of compactness, stability and emitted energy must also be taken into consideration, especially in the case of pulse durations of the order of a few nanoseconds (required in the case of a vertical resolution lower than 1 m), and high repetition frequency. These various constraints restrict the selection to a range of basic sources summarized in Table 5.3. The table is ordered by the type of solid amplifying medium used.

Laser diodes, the most robust and compact laser sources, may be found in pulsed mode in the near IR medium. The most commonly used and favorable wavelength in terms of energy is 905 nm. These diodes have become a feature of laser range-finders and speed control LiDARs. However, their energy remains limited (a few microjoules), which restricts the range of the LiDAR to a few hundred meters.

Also in the near IR domain, fiber lasers at 1,064 nm (ytterbium) and 1,550 nm (erbium) are sufficiently compact and emit energy pulses of a few hundred microjoules.

Technology	Wavelength	Energy	Repetition frequency	Pulse duration
Pulsed laser diode	905 nm	< 10 µJ	several kHz	30–200 ns
Er fiber laser	1,550 nm	< 400 µJ	several 10 kHz	5–100 ns
Yb fiber laser	1,064 nm	< 200 µJ	several 10 kHz	5–100 ns
DPSS Nd:YAG	1,064,532, 355 nm	< 60, 30, 10 mJ	< 100 Hz	5–10 ns

Table 5.3. *Commonly used lasers and conventional characteristics categorized by increasing overall dimensions. Er: erbium, Yb: ytterbium, Nd:YAG: neodymium-doped yttrium aluminum garnet matrix, DPSS: diode-pumped solid-state lasers*

Finally, the Q-switched emission ("Q-Switch") and the quasi-continuous pumping (QCW) of Nd:YAG crystals (neodymium-doped yttrium aluminum garnet) make it possible to obtain energetic (several mJ) and short (nanosecond) pulses, best suited for the direct detection at long-range. The fundamental emission frequency corresponds to a 1,064 nm wavelength. It can be doubled or tripled in order to achieve wavelengths of 532 nm (visible) or 355 nm (UV), to the cost of a division of the transmitted energy. These systems are also more sizable and cannot dispense with active cooling. Diode-pumped solid-state lasers are more compact and robust but less powerful than flash-lamp-pumped lasers. Nevertheless, they provide pulse energies that are largely sufficient for the study of forest canopy.

The five wavelengths mentioned above have already been used by various LiDAR systems for surface applications. The spectral criteria to be taken into consideration when selecting one of these wavelengths are summarized in Figure 5.3:

– the energy link budget (power of the signal received for a given emitted power) is more favorable in the near IR than in the UV as a result of light scattering in the atmosphere (Rayleigh and Mie scattering) and the absorption capacity of vegetation. Note that the selection of a wavelength with a strong reflectance on the forest canopy can cause multiple scattering phenomena, which bias the observations of a full waveform LiDAR. A wavelength outside of the 600–1,300 nm range may be desirable in the latter case;

– the stray signal due to direct and scattered contributions of sunlight, referred to as "background noise", is not negligible except in the case of very narrow spectral filtering. Given the fact that the ambient heat radiation is low in the near IR, the 1,550 nm wavelength is the least affected by background noise;

– the limit laser radiance values permissible for the eye, which greatly vary as a function of wavelength, are set by norm EN 60825-1. In the case of LiDAR applications, the total exposure criterion applicable to systems without rapid scanning (10 s) has to be considered on top of the pulse exposure criterion. In both cases, these criteria are much more permissive between 310 and 400 nm and above 1,400 nm, with a ratio of up to 10^4 in the case of maximum permissible exposures (MPEs). These domains are referred to as "eye-safe" domains, even though a specific risk calculation has to be carried out for each case.

5.3.2.2. *Transmitted beam shaping*

At the laser output, the beam usually has to be slightly divergent and its diameter increased, in order to meet the criteria pertaining to eye safety. If the laser beam is not eye-safe at emission, the aim is to reduce the nominal ocular hazard distance (NOHD) below the flight altitude or values provided by aviation specifications (a few hundred meters). Beam divergence is also adjusted as a function of the measuring method selected: a small footprint (~cm) in the case of a discrete echo LiDAR or a scanning LiDAR, or a large footprint in the case of a full waveform LiDAR. An afocal optical system carries out the beam shaping. It is made up of two lenses or mirrors having merged focal points and a resulting magnification higher than 1, as shown in Figure 5.4.

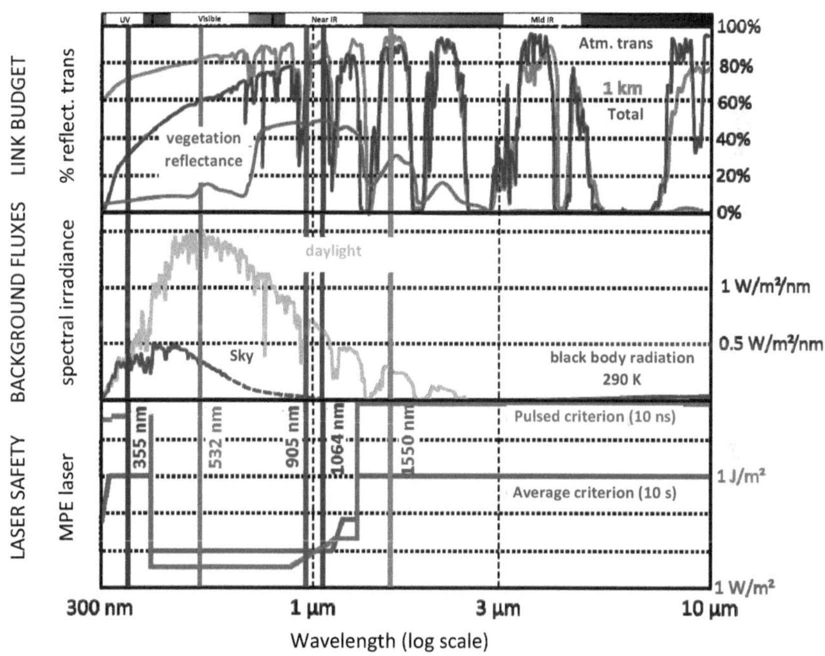

Figure 5.3. *Common LiDAR wavelengths and spectral parameters to be considered: vegetation transmission and reflectance, background spectral irradiance and maximum permissible exposure (MPE) to laser radiation in order to ensure the ocular safety of subjects on the ground (as per NF/EN 60825-1). The MPE is expressed in W/m² for continuous exposure and in J/m² for pulsed exposure. For a color version of this figure, see www.iste.co.uk/baghdadi/1.zip*

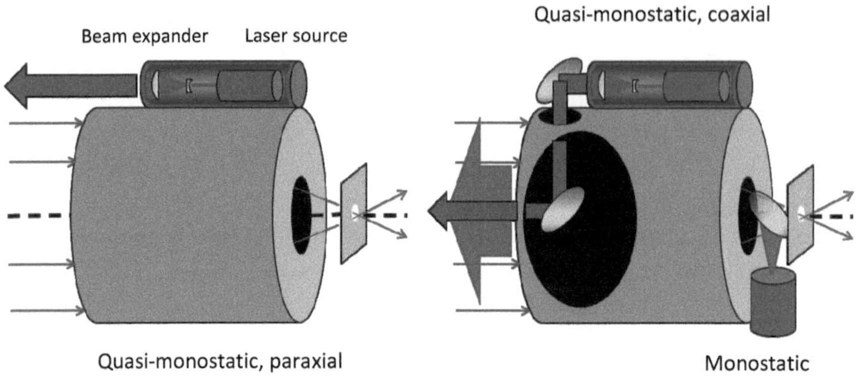

Figure 5.4. *Possible emission configurations*

A LiDAR configuration is determined by the layout of the emitting and receiving paths, as presented in Figure 5.4. If the two components of an optronic system have the same optical axis, said system is referred to as "monostatic", whereas if the axes are distinct and spatially separated, it is referred to as "bistatic". In our case, the proximity between the emitter and the receiver results in the LiDAR sensor being referred to as a "quasi-monostatic" system, which represents an intermediate case. Systems in surface applications can adopt the so-called paraxial configuration, shown on the right. Nevertheless, in order to allow the overlap of emission and reception over a greater distance, a coaxial or fully monostatic configuration can be preferred, shown on the left. The latter configuration also facilitates the passage of the emitting and receiving paths through the same scanning optics.

5.3.3. *Scanning*

Various mobile optical elements may be used in order to implement an airborne scanning LiDAR system capable of quickly sampling large surfaces. Four such solutions are illustrated in Figure 5.5, along with the associated scanning patterns. It is clear that the consecutive positions of laser footprints on these paths are conditioned by the pulse repetition frequency (see section 5.3.2.1). It can still be inferred from these patterns that a uniformly distributed sampling cannot be achieved using basic rotating optics. Another severe constraint associated with the implementation of a scanning system is that the latter has to accommodate both emitting and receiving paths, with a minimum of cross-talk. However, the fastest systems capable of providing uniform scanning have small pupils. This imposes reduced sensor dimensions and therefore restricts the levels of power received.

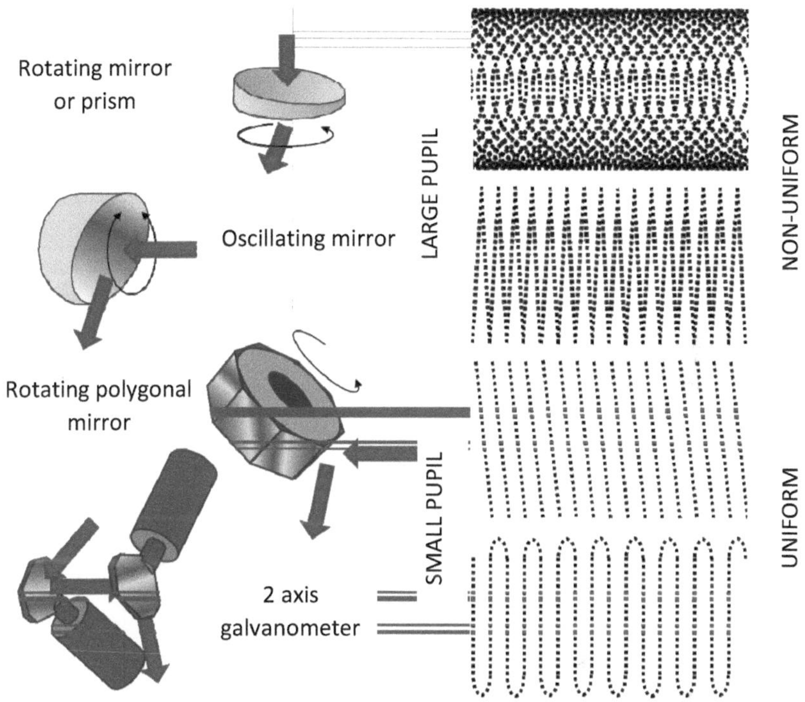

Figure 5.5. *Scanning devices and associated patterns from a moving carrier*

5.3.4. *Reception*

5.3.4.1. *Flux collection*

In order to collect the backscattered light flux, an optical system is used to create an image of the scene. Its field of view is restricted by a field diaphragm positioned in its image focal plane (see section 5.3.4.2).

Various receiving optical systems are used depending on application. It may be noted that the imaging resolution is generally low in the case of a flux-collecting device such as LiDAR. The ensuing constraints in terms of optical quality are loose except in the case of a spaceborne LiDAR system; in contrast, the compactness and stability requirement remains high in the case of an onboard instrument.

Figure 5.6 provides a few examples of optical collection systems:

– catoptric and catadioptric systems include mirrors and are more commonly referred to as reflective telescopes. These are large diameter systems that do not suffer from chromatic aberration, which reduces resolution when observations are conducted over a wide spectrum. Part of the pupil is obstructed by a secondary mirror except in the case of off-axis or fiber-coupled systems. The obstruction results in a loss of throughput but facilitates a coaxial emission (from behind the secondary mirror). In order to ensure good optical quality, the aperture (ratio between the diameter D and the focal length f of the telescope) and the field of view of the most basic off-the-shelf telescopes (Newton design) are low (D = f/10 and field-of-view around 1°). This implies a large focal length, not desirable in the case of a short range system, and significant overall dimensions. Nevertheless, the constraint imposed on image quality is largely loosened at the level of the LiDAR, especially if more developed systems are used, such as variations of the Cassegrain design. As a result, specially designed telescopes are preferred. A spaceborne LiDAR system generally uses a Cassegrain variation of up to 1,500 mm in diameter (allowing an effective collecting surface around 1 m^2) for a focal length of several meters;

– dioptric systems (refractive telescopes or refractors), solely made up of lenses, can rarely have the same diameter as reflective telescopes without additional costs or an inadmissible weight and overall dimensions. In the case of monochromatic LiDARs and reasonable diameter values, these systems can still provide improved compactness and throughput. Another major advantage is their high stability against vibrations, which facilitates their implementation on an airborne platform. In addition to a basic small-diameter system sufficient in the case of short range measurements, Figure 5.5 provides the example of two refractors dimensioned for full waveform LiDAR systems with a dual application to the study of vegetation and atmosphere. These LiDARs, designated as LAUVA and ULICE in Figure 5.6, are laboratory instruments. They are characterized by a resolution of the order of ~65 µrad and a field of view of 0.32° (respectively, 6 mrad) and 1.5° (respectively, 30 mrad).

Once the light flux has been collected, it can be injected into an optical fiber placed in the focal point of the reception system in order to dissociate the detection from the optical head of the LiDAR. This also makes it possible to limit telescope obstruction by placing the fiber at the focus of the primary mirror, or to combine the flow of several telescopes, a solution which results in reduced costs compared to a single larger telescope. An optical fiber is made up of a core and clad made of silica with slightly different refractive indices, which allows the assembly to guide light. Its diameter is limited to a few hundred microns in order to maintain flexibility. If used, it becomes the field diaphragm of the reception system, and reduces the field of view of the LiDAR to a great extent. The field of view can be expanded by using a bundle of fibers and a field lens, but this generates insertion losses which can amount to up to 60% of the incident light.

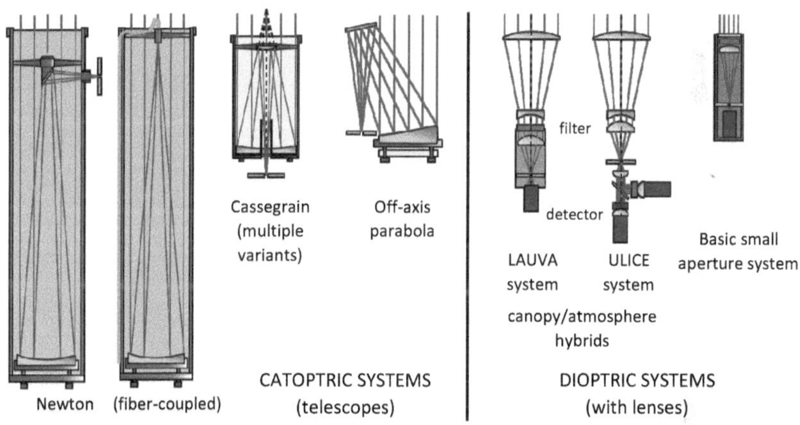

Figure 5.6. *Receiving optical systems (yellow: field diaphragm, green: spectral filter, blue: detector, see sections 5.3.4.2 and 5.3.4.3). For a color version of this figure, see www.iste.co.uk/baghdadi/1.zip*

In the case of a good quality optical system, well adjusted in terms of focal length, the only dimensioning parameters of the receiver are its field of view and its effective area A_{eff} which is connected to its actual area A and expressed as:

$$A_{eff} = \underbrace{\pi\left(\frac{D}{2}\right)^2 \times (1-O)}_{A} \times T_{rec}(\lambda) \quad [5.1]$$

where D is the pupil diameter of the optical system, O is the obstruction ratio and $T_{rec}(\lambda)$ is the transmittance of the receiver for laser wavelength λ. Table 5.4 provides orders of magnitude for the occultation and transmission of different types of receivers. It is more common for dioptric systems to have an effective collecting area in the vicinity of their actual area. Although in the case of short range measurements(<300 m), a small area optical system (A~10 cm^2, a few centimeters in diameter) with short focal length may suffice, large area systems (A~100 cm^2, of the order of 100 mm in diameter) with a long focal length are preferred at ranges over a few kilometers. In the case of a spaceborne LiDAR, a large telescope (A~1 m^2, diameter >200 mm) with a high optical quality is indispensable.

	Type of reception			
	Dioptric/Off-axis	Fiber catoptrics	Newton catoptrics	Cassegrain catoptrics
Obstruction ratio	**0 %**	**2 %**	**6 %**	**12 %**
	Dioptric		Catoptric	
Transmittance	**85–90 %**	**75 %(wide spectral band systems)** **95 % (narrow spectral band systems)**		

Table 5.4. *Conventional obstruction and transmittance values for different types of receivers*

5.3.4.2. *Filtering*

Due to the high luminance of daytime environment from near-ultraviolet to near-infrared, it is necessary to filter the incoming light flux both in a spatial and a spectral manner. This process is illustrated in Figure 5.7. In the image focal plane, a field diaphragm restricts the field of view of the system precisely to the space traveled by the laser beam. The optimum field diaphragm is rectangular in the case of a paraxial configuration, as represented in Figure 5.6.

A consequence of this limitation of the field is the existence of a minimum measurement distance, knowing that:

– in the case of non-monostatic configurations, the beam can only enter the field in a progressive manner;

– since the light rays backscattered in the near-field are focused behind the image focal plane, a large proportion is occulted by the diaphragm.

In the pupil plane (for example, on a recollimated beam after the first focus), a filter with a great spectral finesse and centered at the emitted wavelength is placed, in order to significantly reduce the bandwidth of the LiDAR system. The spectral width of the emission being less than a nanometer except in the case of laser diodes, an interference filter characterized by a width ranging from 0.5 to 10 nm can be used. Such an optical component is built using thin film stacks of the Bragg and Fabry–Pérot type in order to quench a broad spectrum and let through a narrow line.

Such a filter is designed for a perfectly collimated beam, at normal incidence to the filter. If the angle of incidence θ is different from 0, the central wavelength of filter $\lambda_{c,0}$ becomes $\lambda_c(\theta_i) = \lambda_{c,0} \cos \theta_i$. For the received flux at the laser wavelength, the effect is that of a limited acceptance angle for the filter, with $\Delta\theta_i = 2\ (\Delta\lambda/\lambda_c)^{1/2}$ in width, if $\Delta\lambda$ represents the spectral width of the filter. Therefore, for a filter of 1 nm in width, this acceptance only represents +/–1.5° in the near IR (1,064 nm), or 0.3 nm in the UV (355 nm). Furthermore, this also sets a minimum measurement distance for LiDARs with a large depth of field, given the fact that said acceptance restricts the field of view in the same manner as the field diaphragm.

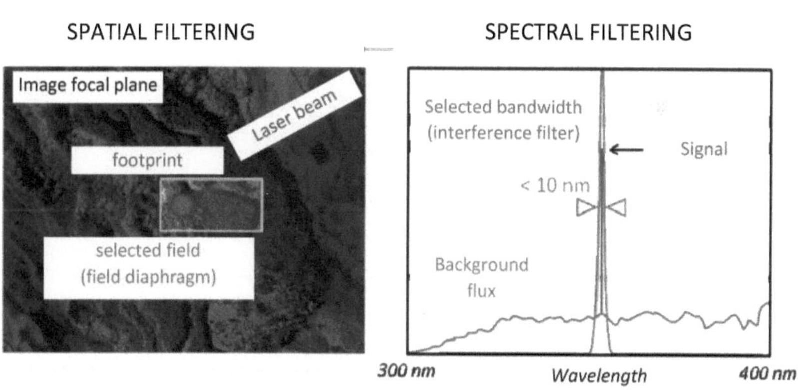

Figure 5.7. *Spatial a) and spectral b) filtering of the incoming flux in the focal and pupil plane, respectively. For a color version of this figure, see www.iste.co.uk/baghdadi/1.zip*

5.3.4.3. *Detection*

The conversion of the optical flux signal into measurable electric voltage is performed by a photodetector, which transforms each photon into a photoelectron (with a probability fixed by its quantum efficiency) discharged in a load resistor. The fast and highly sensitive detectors used in LiDAR, whether avalanche photodiodes (near IR) or photomultipliers (visible and UV), provide a large number of photoelectrons (between 10 and 10,000) for each incident photon via an internal gain phenomenon. Nevertheless, these two types of detectors have very different characteristics in terms of sensitivity, noise or electrical capacity (the latter restricting temporal resolution). These various phenomena have a significant impact on the link budget, in particular as a function of the wavelength selected for the LiDAR. The electronic detection configurations and the spectral ranges of various LiDAR detectors are illustrated in Figure 5.8. Table 5.5 presents the characteristics of these various internal gain photodetectors and more conventional photodiodes. They can be used in analog and photon counting mode.

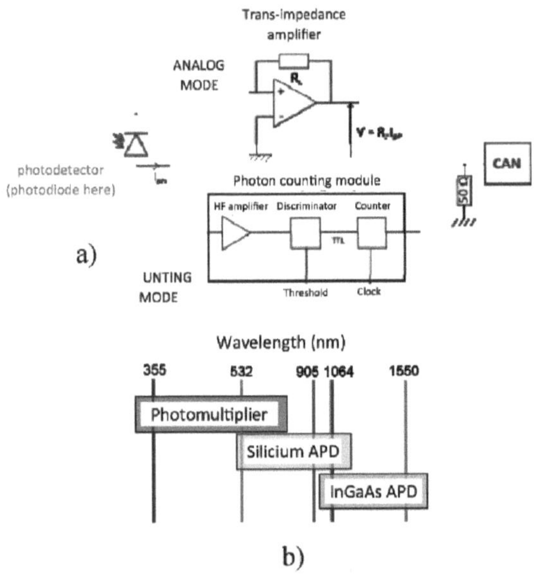

Figure 5.8. *a) Configuration of a photodetector and digitization of the photocurrent I_{ph} amplified by the impedance Rc; b) spectral range of different types of detectors. APD: avalanche photodiode, CAN: analog-to-digital converter, TTL: digital signal 0–5 V. For a color version of this figure, see www.iste.co.uk/baghdadi/1.zip*

Analog detection requires sufficient flux or amplification after the detector. Voltage is acquired at the terminals of a load resistor R_C through which a photocurrent i_{ph} passes, yielding $V = R_C\, i_{ph}$,, a value which is required to exceed a few millivolts in order for the voltage not to be drowned out by the quantization noise of the digital converter. When the load resistance of the converter (usually 50 Ω when adapting the impedance values to coaxial cables) is insufficient, voltage can be taken at a chosen resistor value due to a trans-impedance amplifier (see Figure 5.8(a)). However, the stray capacitance C_p of the detector, which sets the bandwidth of the detection and therefore its time resolution, must also be taken into consideration. The cutoff frequency $f_C = 1/(2\pi R_C C_p)$ has to be above 150 MHz for a range resolution around 1 m in the case of LiDAR systems for forestry applications. The maximum load resistance is therefore comprised between 300 and 1,500 Ω. When this restriction and the characteristics of the detectors provided in Table 5.5 are taken into account, it is possible to calculate the noise equivalent flux of the photodetector and the amplification and digitizer electronic systems. First, it may be noted that conventional photodiodes are affected by the electronic noise to a greater extent due to their low sensitivity. This is not the case of photomultipliers, which are capable of detecting fluxes expressed in pW. Nevertheless, these photodetectors only cover the visible or near UV domain, with the exception of a few oversized and costly models with active cooling, designed specifically for spaceborne LiDARs. Although avalanche photodiodes are preferred in the case of infrared wavelengths, they are affected by internal noise values which can be 100 times higher than in the case of photomultipliers.

Photon counting detection is facilitated by detectors with high internal gain and sufficient bandwidth. A fast electronic circuit counts incident photons as electrical pulses exceeding a predetermined threshold after amplification; it delivers a discrete signal for each time gate, which remains unaffected by the detector or electronic noise. Avalanche photodiodes in the near IR thus achieve the same levels of performance in the photon counting mode as the photomultipliers in the UV medium. The major disadvantage of photon counting is limited dynamics: it becomes saturated when a signal exceeds a few photons per time gate. It is therefore difficult to implement in case of strong background flux due to solar irradiance. This saturation can be compensated by dividing the flux between multiple or sectorized detectors, a method often used by spaceborne LiDARs. Note that photon counting is also possible after the digitization of the analog signal, given a sufficiently sensitive detector (photomultiplier) and a high sampling rate.

Type	Spectral range (μm)	Dimensions (mm)	Quantum efficiency	Sensitivity (A/W)	Capacitance (pF)	NEP (fW/Hz$^{1/2}$)	N.E.F (W) det	N.E.F (W) elec
Si Photodiode	0.2–1.1	1	60%	0.1–0.8	6	150	~10^{-9}	~10^{-7}
InGaAs Photodiode	0.9–1.7	0.5	60–75%	0.7–0.95	5	5	~10^{-11}	~10^{-8}
PM NUV-VIS	0.3–0.6	8	15–40%	10^4–10^5	1	0.5	~10^{-12}	~10^{-13}
Si-APD VIS-NIR	0.5–1.0	0.2	60–75%	60	1.5	2	~10^{-11}	~10^{-10}
Si-APD VIS-NIR	1.0–1.2	1.5	40%	40	3	100	~10^{-9}	~10^{-9}
InGaAs-APD NIR	1.0–1.6	0.2	80%	10	1.5	30	~10^{-10}	~10^{-9}

Table 5.5. *Typical characteristics of commercially available detectors and order of magnitude of optical fluxes equivalent to the detection noise (provided by the NEP, Noise Equivalent Power of the detector) and electronic noise (amplification and digitization), considering a LiDAR system for forestry applications (cutoff frequency 150 MHz and adjusted load resistor). These noises are avoided in the photon counting mode. The noise equivalent flux (N.E.F) is provided both for detection (det) and amplification electronic systems (elec)*

This presentation does not include state-of-the-art or actively cooled photodetectors, such as those designed specifically for spaceborne LiDARs. One should refer to the research conducted by NASA [KRA 09] in order to obtain an overview of the characteristics of these detectors, which occasionally exceed the performances indicated in Table 5.5.

5.4. LiDAR equation of the profiler

The case of a profiler or full waveform LiDAR system is considered below. This is a more complex instrument than the echo systems, which simply detect the position of one or several structures without completely inverting the LiDAR signal for the density of the target medium. As in the section 5.3, the major focus will be on forest applications which constitute a relevant example in this context.

5.4.1. *Interaction with scattering medium and vertical resolution*

5.4.1.1. *Backscattering*

A laser pulse emitted by the LiDAR system propagates through the target medium, within which it experiences scattering and absorption. The former interaction is characterized by the total volume backscattering coefficient β_t which represents the contribution of air molecules (β_m), airborne particles (β_p) and surface diffusers, for example in the forest canopy (β_f). The volume backscattering coefficient of a medium is proportionate to the concentration of the number of diffusers in this medium and to the effective average backscattering cross-section of the diffusers. Therefore, this parameter contains information on the morphological and chemical properties of the diffusers. This parameter is closely linked to the probability that a photon emitted by the laser will be backscattered toward the telescope.

5.4.1.2. *Attenuation*

Since the laser pulse and backscattered radiation propagate in the same scattering and absorbent medium, they are inevitably attenuated along their propagation. The extinction coefficient (α_e) characterizes this attenuation, which is also proportionate to the number concentrations of diffusers. The extinction coefficient merges the scattering and absorption contributions of each type of diffuser.

5.4.1.3. *Depth of interaction*

These interactions are characterized by a certain depth because the duration of the laser pulse does not make it possible to punctually locate the backscattering altitude. A pulse having a duration of Δt_0 is emitted at an instant t_0, as illustrated in the time/altitude diagram in Figure 5.9 in which a look angle of 0° has been considered in order to facilitate this reasoning. If the speed of light is denoted by c, the average interaction altitude of the laser beam with the scattering mediums is calculated using the following relationship:

$$z = \frac{c \cdot (t_r - t_0)}{2} \qquad [5.2]$$

when considering a LiDAR signal received at an instant of reception t_r. The pulse propagates in the scattering medium in order to reach an altitude comprised between $z-\Delta z_0/2$ and $z+\Delta z_0/2$, where $\Delta z_0 = c\Delta t_0/2$ is the resulting vertical resolution of the LiDAR system due to the laser pulse duration. Furthermore, the signal is not recorded instantaneously by the electronic receiving system. An acquisition "gate" characterized by a duration Δt_r, which corresponds, for example, to the aperture of a counting gate or to the sampling period of the LiDAR signal must be considered. Similarly to the duration of the laser pulse, this gate is associated with a vertical resolution Δz_r. The vertical resolution Δz of a LiDAR system is therefore represented by the sum of these two contributions (Figure 5.9):

$$\Delta z = \Delta z_0 + \underbrace{\Delta z_r}_{\frac{c\,\Delta t_r}{2}} \qquad [5.3]$$

In general, $\Delta z_0 << \Delta z_r$.

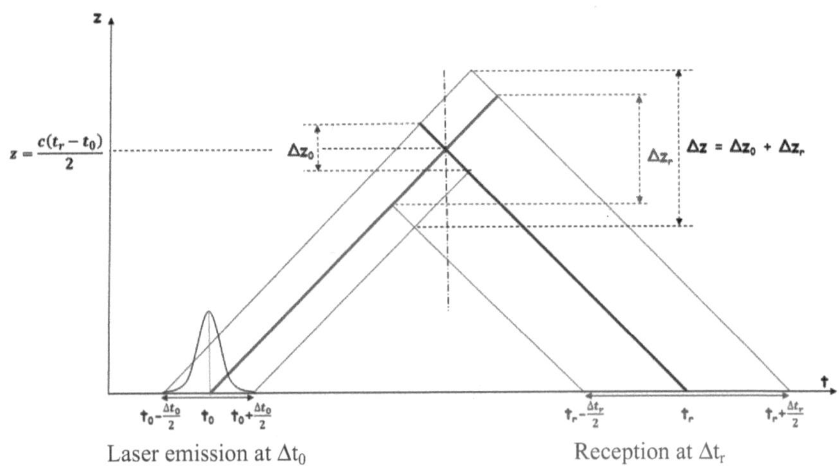

Figure 5.9. *Altitude/time illustration of the vertical resolution of a full waveform LiDAR system, assuming a vertical emission. The emitted pulse has a duration of Δt_0, whereas the reception time gate has a duration of Δt_r. For a color version of this figure, see www.iste.co.uk/baghdadi/1.zip*

5.4.2. *Backscattered power*

A complete demonstration of the LiDAR equation, which may be found in summary books such as Measures [MEA 84], will not be presented in this section. The scattering medium in which the laser beam propagates is considered as a filter in this context. This filter is associated with an impulse response h (expressed in s^{-1}) which involves the backscattering (β_j) and extinction (α_j) coefficients of the components (denoted by j) of the target scattering medium:

$$h(\lambda,t) = \frac{c}{2} \cdot \frac{\cos(\theta)^2 \cdot A \cdot T_{emi}(\lambda) \cdot T_{rec}(\lambda) \cdot F(z(t))}{(z-z_0)^2} \cdot \sum_j \beta_j(\lambda, z(t)) \cdot \exp\left[-\frac{2}{\cos(\theta)} \cdot \sum_j \tau_j(\lambda, z(t))\right] \quad [5.4]$$

where:

– z is the altitude;

– z_0 is the altitude at which the laser pulse is emitted;

– A is the effective surface of the sensor and $\frac{A}{(z-z_0)^2}$ corresponds to the solid angle from which the sensor is seen from the diffuser;

– F is the geometric factor of the LiDAR (see section 5.4.5);

– θ is the look angle;

– T_{emi} and T_{rec} are the transmittances of the emitting and receiving optics, respectively;

– τ_j is the optical thickness of the scattering medium j at the altitude z: $\tau_j(\lambda, z(t)) = \int_{z_0}^{z(t)} \alpha_j(\lambda, z(t')) \cdot dz(t')$.

The backscattered power at the level of the detector is thus provided by the convolution product (denoted by ⊗) between the power waveform of the laser pulse P_e and the impulse response of the scattering medium:

$$P_d(\lambda, z(t)) = P_e(\lambda, t) \otimes h(\lambda, t) \quad [5.5]$$

This convolution can be interpreted graphically, as shown in Figure 5.10. In the case of emission at instant $t = 0$ and reception at instant t, the signal received at instant t', thus for an altitude value z', is associated with emitted power $P_e(t-t')$. A standard level of backscattered power $dP_d(\lambda, z(t)) = h(\lambda, t')P_e(\lambda, t-t')dt'$ which returns to the LiDAR system may thus be identified. When considering the time values represented by t', it is necessary to calculate the sum of each basic contribution in order to obtain the backscattered power P_d:

$$P_d(\lambda,t) = \int_{t-\frac{\Delta t_0}{2}}^{t+\frac{\Delta t_0}{2}} h(\lambda,t')\, P_e(\lambda,t-t')\, dt' \qquad [5.6]$$

This may be regarded as the definition of the convolution product since it is possible to consider the energy emitted outside the time interval Δt_0 as negligible.

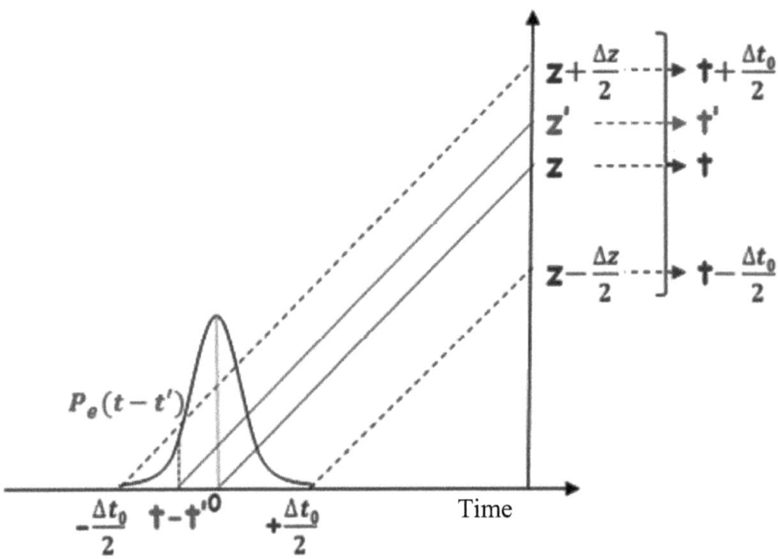

Figure 5.10. *Diagram illustrating the convolution product between the laser pulse and the impulse response of the tested scattering medium. For a color version of this figure, see www.iste.co.uk/baghdadi/1.zip*

5.4.3. *Case of a short-duration laser emission*

If the duration of the laser pulse is short compared to the acquisition time gate, the emitted power can be expressed as a function of the emitted energy E_e as:

$$P_e(\lambda,t) = E_e(\lambda,t) \cdot \delta(t-t_0) \qquad [5.7]$$

with $\delta(t-t_0)$ (expressed in s^{-1}) representing the Dirac distribution centered on t_0. The equation of the backscattered power is therefore expressed in a simplified manner as:

$$P_d(\lambda,t) = \frac{c}{2} \cdot \frac{\cos(\theta)^2 \cdot A \cdot T_{emi}(\lambda) \cdot T_{rec}(\lambda) \cdot F(z(t))}{(z(t)-z_0)^2} \cdot E_e(\lambda,t_0) \cdot \sum_j \beta_j(\lambda,z(t)) \cdot \exp\left[-\frac{2}{\cos(\theta)} \cdot \sum_j \tau_j(\lambda,z(t))\right] \quad [5.8]$$

This equation may be expressed differently depending on the envisaged detection type. The cases of analog detection and photon counting modes are discussed below.

5.4.3.1. *Photon counting detection*

This type of detection makes it possible to count the number of incident photons detected by the photodetector. Photoelectrons are recorded by the pulse counter in time gates with a width of Δt_r and the result is attributed to the instant t_r.

The counting system can also be associated with an impulse response filter h_c (expressed in J^{-1}) defined as:

$$\hbar_c(\lambda,t) = \begin{cases} \frac{\lambda \cdot \eta(\lambda)}{\hbar \cdot c} \; if \; |z| \leq z_r + \frac{\Delta z_r}{2} \\ 0 \; if \; |z| > z_r + \frac{\Delta z_r}{2} \end{cases} \quad [5.9]$$

where η is the quantum efficiency of the photodetector and $\hbar$ represents the Planck constant ($\hbar \sim 6.626\ 10^{-34}$ J·s).

The number of photons N_d detected during Δt_r can therefore be expressed on the basis of successive convolution products of the laser power emitted by the impulse responses of the scattering medium and the sensor:

$$N_d(\lambda,z(t)) = P_e(\lambda,t) \otimes h(\lambda,t) \otimes h_c(\lambda,t) \quad [5.10]$$

As indicated above, the detection by means of photon counting is limited if the backscattered flow is at a high level. A saturation phenomenon may therefore be observed since several coincident photons are only capable of producing a single pulse. In such a case, a desaturation is required, but this operation is limited in its application to a few hundred photons per second.

When the signal backscattered by the scattering layers is strong, it is preferable to use the analog detection mode.

5.4.3.2. *Analog detection*

In the case of analog detection, the average current generated by the detector is recorded over a sample period T_r. Similarly to Δt_r, this sample period is adapted to the desired vertical resolution of the LiDAR. It is possible to reconstruct the actual current evolution in time provided that the maximum frequency of the power spectrum of the latter does not exceed $\frac{1}{2T_r}$ (Nyquist–Shannon sampling theorem).

In the same way as for the photon counting mode, the electronic acquisition system in the analog mode is characterized by impulse response h_a which takes into account the influence exerted at the level of the detector, the transient analyzer and the detector preamplifier. This impulse response expressed in A/W can be initially formulated as [MEA 84]:

$$h_a(\lambda, t) = \begin{cases} \frac{G(\lambda)\cdot G_A}{T_r} \cdot exp\left(-\frac{t}{T_r}\right) \ if\ t \geq 0 \\ 0\ if\ t < 0 \end{cases} \quad [5.11]$$

G_A is the gain of the preamplifier (non-unitary). The detector is associated with a sensitivity G, generally expressed in A/W, which is proportionate to the quantum efficiency:

$$G(\lambda) = \frac{e\cdot\lambda\cdot\eta(\lambda)\cdot G_i(\lambda)}{\hbar\cdot c} \quad [5.12]$$

with e representing the elementary charge of the electron (e ~1.6 10^{-19} C) and G_i representing the intrinsic gain of the detector characterizing its internal amplification.

The sampled current I_d can therefore also be expressed as the successive convolution of the power emitted by the impulse responses of the scattering medium and the sampling system for a time sample nT_r ($n = 0, 1, 2, \ldots$):

$$I_d(\lambda, z(nT_r)) = P_e(\lambda, nT_r) \otimes h(\lambda, nT_r) \otimes h_a(\lambda, nT_r) \quad [5.13]$$

This current is transmitted to a load resistor R_c, and a voltage expressed as $U_d = R_c I_d$ is recorded. The surface parameters to be measured are extracted from the signal represented by this voltage.

5.4.3.3. *General form of the LiDAR equation*

Assuming that the duration of the laser pulse is short compared to the detection integration time Δt_r and the sample period T_r, and that the backscattered power is constant throughout each temporal sampling stage, the LiDAR signal S may be expressed using the following generic form:

$$S(\lambda, z(t)) = K(\lambda) \cdot \frac{F(z(t))}{(z(t)-z_0)^2} \cdot \sum_j \beta_j(\lambda, z(t)) \cdot \exp\left[-\frac{2}{\cos(\theta)} \cdot \sum_j \tau_j(\lambda, z(t))\right] + F_c(\lambda) \quad [5.14]$$

This form is deduced directly from equations [5.10] or [5.13], with K representing the system constant which in the case of the counting or analog modes may be expressed as:

$$K(\lambda) = \begin{cases} \Delta z \cdot \cos(\theta) \cdot \dfrac{E_e(\lambda, t_0) \cdot \lambda \cdot \eta(\lambda)}{\hbar c} \cdot A \cdot T_{emi}(\lambda) \cdot T_{rec}(\lambda) \\ \text{photon counting} \\ \dfrac{c}{2} \cdot cos(\theta)^2 \cdot E_e(\lambda, t_0) \cdot G(\lambda) \cdot G_A \cdot A \cdot T_{emi}(\lambda) \cdot T_{rec}(\lambda) \cdot R_c \\ \text{analog} \end{cases} \quad [5.15]$$

F_c is the background flux used to characterize the natural light supplementing the backscattered laser radiation and which must be taken into account during daytime measurements. This natural light is related to the scattering solar radiation following its atmosphere and surface interaction. Naturally, the units of signal S are different in the photon counting and analog mode. The signal is expressed as the number of photoelectrons per time interval Δt_r in the photon counting mode, and in Volts (V) in the analog mode.

In the particular case of LiDAR measurements of the forest structure, the canopy diffusers (branches, leaves and trunk) have much higher backscattering and extinction coefficients than those which may be found in the atmosphere. The attenuation at the level of the atmospheric medium expressed by its transmission Γ_{atm} can thus be separated from that of the vegetation canopy Γ_f. The equation of the LiDAR signal S within the forest canopy can therefore be formulated as follows:

$$S(\lambda, z(t)) = K(\lambda) \cdot \frac{F(z(t))}{(z(t)-z_0)^2} \cdot \beta_f(\lambda, z(t)) \cdot \Gamma_{atm}(\lambda) \cdot \underbrace{\exp\left[-\frac{1}{cos(\theta)} \cdot \tau_f(\lambda, z(t))\right]}_{\Gamma_f(\lambda)} + F_c(\lambda) \quad [5.16]$$

where index f is used to characterize the forest environment. The atmospheric transmittance due to air molecules ($j = m$), aerosols ($j = a$) and clouds ($j = n$) is indicated by:

$$\Gamma_{atm}(\lambda) = \exp\left[-\frac{2}{cos(\theta)} \cdot \left(\tau_m(\lambda, z(t)) + \tau_a(\lambda, z(t)) + \tau_n(\lambda, z(t))\right)\right] \quad [5.17]$$

The optical thickness of the forest leads to attenuation only in the forward direction of propagation of the laser radiation and not on the backward path. This is due to the significant dimensions of the diffusers in the forest environment in comparison to those present in the atmospheric medium. If a photon does not encounter a diffuser during its forward propagation, this will also apply when it is backscattered toward the detector.

In the forest LiDAR equation, it may be necessary to take into consideration the effects of multiple scattering. Multiple scattering artificially increases the number of backscattered photons once the laser beam has penetrated the forest environment. As in the case of clouds, a multiple scattering coefficient ξ related to transmission is introduced [SHA 15]:

$$\Gamma_f(\lambda) = \exp\left[-\frac{\xi(\lambda)}{cos(\theta)} \cdot \tau_f(\lambda, z(t))\right] \quad [5.18]$$

5.4.4. *Background flux*

One of the main noise sources in the LiDAR measurement is represented by the luminance of the sky background flux L_{FC}. In the case of a small receiving aperture, the latter contributes to the power received according to the following relationship:

$$P_{FC}(\lambda) \approx A \cdot \pi \cdot \left(\frac{\theta}{2}\right)^2 \cdot \Delta\lambda \cdot L_{FC}(\lambda) \quad [5.19]$$

The components of the LiDAR system capable of limiting the effect of background flux are represented by the bandwidth $\Delta\lambda$ of the reception interference filter, the surface A of the collecting instrument and the solid field of view $\pi\left(\frac{\theta}{2}\right)^2$ of the instrument. Figure 5.11 presents a diagram illustrating the solid field of view (FOV) of the LiDAR system reception.

In the same way as for the useful LiDAR signal, it is shown that the background signal in the case of the photon counting and analog modes is provided by the following relationships:

$$F_c(\lambda) = \begin{cases} \dfrac{2 \cdot \eta(\lambda) \cdot T_{emi}(\lambda) \cdot T_{rec}(\lambda) \cdot A \cdot \pi \cdot \left(\frac{\theta}{2}\right)^2 \cdot \lambda \cdot \Delta\lambda \cdot \Delta z}{\cos\theta \cdot \hbar \cdot c^2} \cdot L_{FC}(\lambda) & \text{Photon counting} \\ G(\lambda) \cdot G_A \cdot A \cdot \pi \cdot \left(\frac{\theta}{2}\right)^2 \cdot \Delta\lambda \cdot T_{emi}(\lambda) \cdot T_{rec}(\lambda) \cdot R_c \cdot L_{FC}(\lambda) & \text{Analog} \end{cases} \quad [5.20]$$

The evaluation of the background flux can be carried out independently for each LiDAR profile. In fact, this interference remains constant over the LiDAR profile. Therefore, this may be calculated, after the LiDAR signal is acquired, at a high altitude where the laser contribution is negligible in the case of vertical measurements or below ground level in the case of airborne or satellite measurements.

5.4.5. *Overlap function*

The overlap function F (also referred to as the geometric factor of the LiDAR system) can be modeled knowing the optical elements within the LiDAR system. It represents the effective amount of backscattered energy which reaches the detector. The laser beam divergence, the reception field of view and the distance between the emission and reception optical axes in particular have to be known precisely. Figure 5.12 illustrates the overlap between the emission and reception fields. The laser beam penetrates the reception cone in a gradual manner. In the proximity of the instrument, the overlap is practically nil, and it reaches 1 at sufficient distance. Depending on the characteristics of LiDAR systems, the overlap function can be very different. Most large systems are characterized by a total overlap only at several hundred meters distance, and a correction is required in order to extend the detection limit in the near field.

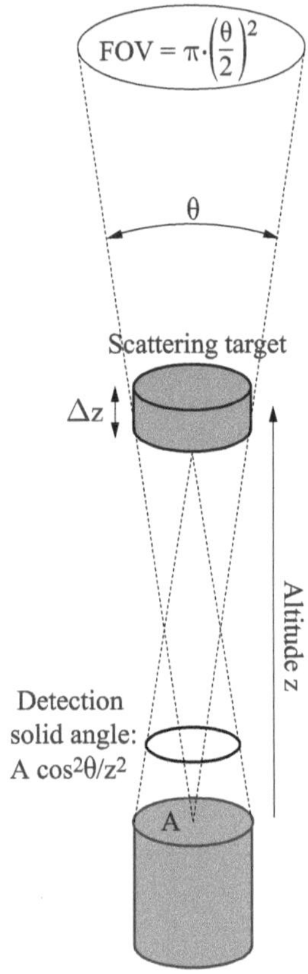

Figure 5.11. *Diagram of the field of view (or FoV) of LiDAR system reception*

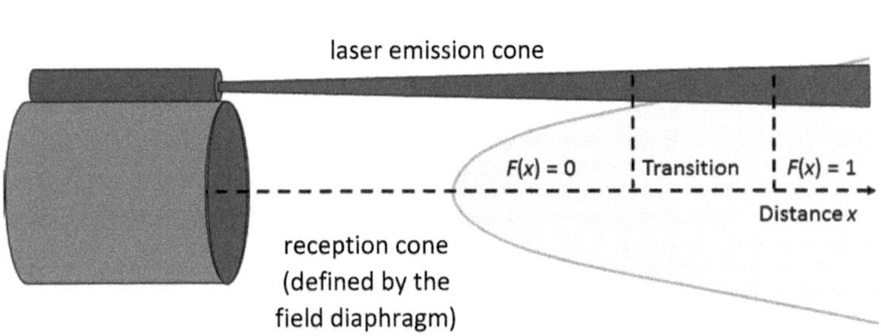

Figure 5.12. *Diagram illustrating the overlap between the laser beam and the FoV of the collecting system*

The overlap function can be determined by modeling the optical system of the LiDAR [MEA 84] or experimentally by using horizontal shots on the basis that the scattering medium is homogeneous along the line of sight (over a sufficient distance from the LiDAR). Considerable differences may be identified between these two approaches. These are mainly linked to a poor understanding of the optical characteristics of the LiDAR components and to inaccuracies in alignment. The experimental method is not only easy to implement, but it is also the most suitable option for conducting the evaluation of the overlap function. In this case, the LiDAR equation is formulated with respect to a horizontal distance x of the laser:

$$\left(S(\lambda,x)-F_c(\lambda)\right)\cdot x^2 = K(\lambda)\cdot F(x)\cdot \Gamma_a(\lambda)\cdot \underbrace{\textstyle\sum_j \beta_j(\lambda,x)}_{\beta_{total}(\lambda,x)}\cdot \exp\left[-2\cdot \underbrace{\textstyle\sum_j \tau_j(\lambda,x)}_{\int_0^x \alpha_{total}(\lambda,x\prime)\cdot dx\prime}\right] \quad [5.21]$$

where β_{total} and α_{total} are the total backscattering and extinction coefficients of the medium in which the laser beam propagates. In general, the calibration is carried out in the atmospheric medium. By considering the logarithm of the previous expression, the following equation may be obtained:

$$\log\left(\left(S(\lambda,x)-F_c(\lambda)\right)\cdot x^2\right) = \log\left(F(x)\right)\underbrace{-2\cdot \alpha_{total}(\lambda)\cdot x + Constant}_{Line\ y} \quad [5.22]$$

When located sufficiently far from the emission $F(x) = 1$, the previous equation becomes the equation of a line of slope $-2\alpha_{total}(\lambda)$. A line y can therefore be adjusted by means of regression when function $F(x) = 1$ in order to deduce it in the area where the overlap is not complete. The expression of the overlap function of the LiDAR system is therefore formulated as:

$$F(x) = \exp\left[\log\left(\left(S(\lambda,x)-F_c(\lambda)\right)\cdot x^2\right) - y\right] \quad [5.23]$$

An example is provided in Figure 5.13(a) where the regression line is drawn in red on the measured horizontal LiDAR profile represented in black. Conventional geometric factors for a full waveform LiDAR system are also represented in Figure 5.13(b).

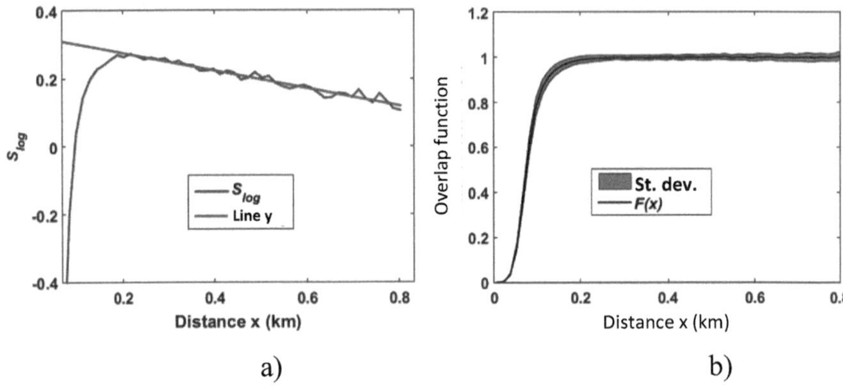

Figure 5.13. *a) Example of a horizontal LiDAR signal (in blue) and the regression line used in order to calculate its overlap factor (in red); b) example of an overlap factor corresponding to a full waveform LiDAR system. The standard deviation is represented in gray. For a color version of this figure, see www.iste.co.uk/baghdadi/1.zip*

5.5. Sources of uncertainty in the LiDAR measurement

In the case of an onboard LiDAR measurement, uncertainties are derived either from the LiDAR instrumental system itself, or from the measurement environment which may not always be well known. Therefore, intrinsic noises at the level of the acquisition chain have to be distinguished from the ones derived from the measurement environment where the background signals (atmosphere, surface solar reflection, etc.) may be encountered.

5.5.1. *Instrumental noise*

5.5.1.1. *Detection noise*

The case of a Poisson noise distribution is introduced in the remainder of this section. Five main sources of noise may be distinguished at the level of the recorded signal:

– the shot noise, also referred to as quantum noise, photon noise or Schottky noise;

– the dark current noise;

– the Nyquist–Johnson noise;

– the amplification noise;

– the quantization noise.

These five sources exert an influence over the analog signal. The photon counting signal is only affected by the shot noise and quantization noise. The expression of each of these noises will be analyzed separately. For additional details, please refer to [MEA 84].

5.5.1.1.1. Shot noise

This type of noise is linked to the fluctuations in the flow of electrons due to the random arrival of photons at the level of the detector. Its standard deviation in the analog detection mode σ_G may be formulated as:

$$\sigma_G(\lambda, \mathrm{z(t)}) = \sqrt{\frac{G_A \cdot \hbar \cdot \frac{c^2}{\lambda} G(\lambda) \cdot \zeta \cdot S(\lambda, z(t))}{2 \cdot p \cdot \left(\Delta z / \cos(\theta) \right) \cdot \eta}} \quad [5.24]$$

where the excess gain factor of the gain detector is defined by ζ. In the case of a PMT, this factor decreases concomitantly with the applied voltage and it is limited to approximately 1.1 and 1.2 when very high gains are considered. This noise, such as illustrated in the subsequent examples, takes into account the average number of LiDAR shots p.

These photon noise calculations (shot noise) are only carried out in the case of a PMT or photovoltaic photodetector. In the case of a photoconductor, the photoelectron generation process is a recombination of the electron–hole pairs. An additional factor 2 must therefore be applied.

In the case of photon counting detection, shot noise may be calculated directly as:

$$\sigma_G(\lambda, \mathrm{z(t)}) = \sqrt{S(\lambda, z(t))} \quad [5.25]$$

5.5.1.1.2. Dark current noise

Dark current noise is constituted by two components:

– the residual current derived from the operating temperature of the detector different from zero, which creates electron/hole pairs, even in the absence of an optical signal (this phenomenon is susceptible to gain);

– the existence of surface defaults at the level of the photodetector, or in the case of a PMT, of a direct leakage current of the photocathode, which generates a “stray” dark current often referred to as *current leakage* (this current is not susceptible to gain, but its significance increases as a function of the size of the photodetector/photocathode surface).

The standard deviation σ_O of these two dark currents is linked to the NEP (expressed in $WHz^{-1/2}$) by means of the following relationship:

$$\sigma_O(\lambda) = NEP \cdot G_A \cdot G(\lambda) \cdot \sqrt{\frac{c}{4 \cdot p \cdot \left(\Delta z / cos(\theta)\right)}} \qquad [5.26]$$

5.5.1.1.3. Nyquist–Johnson noise

Thermal noise, also referred to as the Nyquist–Johnson noise, is linked to the fluctuation in the flow of electrons generated by the thermal agitation of the carriers (Brownian motion) of electric charges across the load resistor as a function of its temperature T (expressed in Kelvin). Its standard deviation σ_{NJ} may be calculated as:

$$\sigma_{NJ} = G_A \cdot \sqrt{\frac{c \cdot k_B \cdot T}{R_C \cdot p \cdot \left(\Delta z / cos(\theta)\right)}} \qquad [5.27]$$

with k_B representing the Boltzmann’s constant.

5.5.1.1.4. Amplification noise

The amplification noise can be characterized by an equivalent noise power NEP_A as in the case of the detector. Its standard deviation σ_A may therefore be calculated as follows:

$$\sigma_A = \frac{NEP_A \cdot \sqrt{\frac{c}{4 \cdot p \cdot \left(\Delta z / cos(\theta)\right)}}}{R_C} \qquad [5.28]$$

5.5.1.1.5. Quantization noise

This type of noise is associated with the analog-to-digital converter during the digitization process of the LiDAR signal. It is affected both by the analog detection and the photon counting detection. Its standard deviation σ_N may be calculated using the following relationship:

$$\sigma_N(\lambda, \mathrm{z(t)}) = \frac{2}{2\sqrt{3}} \cdot \frac{\Delta S(\lambda, z(t))}{2^n \mathrm{b} - 1} \qquad [5.29]$$

where ΔS represents the dynamics of the measured signal and n_b represents the number of bits of the digitizer.

5.5.1.2. *Signal-to-noise ratio*

The signal-to-noise ratio (denoted by SNR) represents the ratio of the useful signal to the total noise component. Assuming independent noises, the SNR may be calculated using the following relationship:

$$SNR = \frac{S(\lambda, z(t)) - F_c(\lambda)}{\sqrt{(\sigma_G(\lambda, z(t)))^2 + (\sigma_O(\lambda))^2 + (\sigma_{NJ})^2 + (\sigma_A)^2 + (\sigma_N(\lambda, z(t)))^2}} \qquad [5.30]$$

In the case of daytime detection, the shot noise is often dominant. In the case of analog detection, shot and dark current noises are generally dominant, as opposed to the other types, when a digitization of at least 10 bits is considered.

The advantage provided by using a detector with gain (including a preamplifier) is to suppress the influence of thermal, amplification and leakage current noises, in particular when the signal is weak. *A contrario*, in the case of a strong signal, it is not required to use an amplified sensor, since the SNR is reduced by the excess noise factor associated with the random gain fluctuation of the detector. In the case of photon counting detectors, gains are very important as they can be used in order to eliminate amplification, thermal and leakage current noises. The excess noise factor remains close to 1 for photomultipliers. In general, only the shot noise is taken into consideration.

Some authors define the SNR on the basis of a differential measurement with respect to a reference measurement in the absence of a useful optical signal. It is otherwise not possible to differentiate the signal component from the noise component in order to calculate the SNR. In this case, the variances of the background noise, dark current noise and Nyquist–Johnson noise are to be doubled in order to calculate the SNR, given the fact that the sum between the useful and stray signal is therefore taken into account.

As soon as the signal to be measured becomes weak, the analog detection, as in the case of gain detectors, reaches its limits and the photon counting mode becomes advantageous. The number of photons in the integration time gate can generate a peak voltage on the basis of the gain. This peak voltage is recorded as a detection event (a count) once a certain threshold has been exceeded. Two modes may be identified: linear and Geiger. In the linear mode (LM), the dead time of the detector is considered to be sufficiently short with respect to the laser pulse in order to count the events by means of a peak voltage discriminator upon exiting the detector and the amplifier assembly. The known gain makes it possible to link the generated signal to the incident photon flow. This is a typical case for photomultipliers. In the Geiger mode (GM), the "dead time" period of the detector (the time required for the evacuation of all charges generated by the avalanche) is significantly longer. It is generally possible to count an event for a laser pulse. The substantially high, yet poorly known, gain makes it possible to carry out a detection in the presence of very low flux, without being able to quantify them in each case. Silicon (SPAD) or InGaAs avalanche photodiodes operating in the Geiger mode (when a higher voltage than the breakdown voltage is applied) are characterized by dead times of the order of 100 ns and 100 µs in each case. This long period results in a more difficult implementation, and only topographic systems are built on the basis of this method [SHR 12].

5.5.1.3. *Impact on the LiDAR distance measurement*

5.5.1.3.1. Impact of the SNR on the LiDAR distance measurement

The precision of ranging measurements is conditioned by the capacity to measure the same relative position on the emitted and received pulses in order to deduce the time of flight interval associated with the propagation of the laser beam. This is limited not only by the detector capacity, but also by the pulse length, the pulse reproducibility and the jitter of the laser. It is sought to obtain a laser pulse with the shortest rising edge possible, below a ns in order to optimize the precision of topographic measurements. For a detector, the rise times (and bandwidth) are essentially conditioned by the characteristics of the load resistor. A compromise between the sensitivity to weak signals, as seen in the expression of the SNR, and bandwidth must therefore be identified. It is also necessary to limit all timing jitters at the level of the detection triggers with respect to the laser emission, as well as to ensure that the time indexing precision is accurate and that temperature fluctuations at the level of the instrument are controlled.

In the case of analog detection, the (optical) SNR beyond defining the maximal measurable range can also affect the precision of distance measurements. If the detection of a peak in the return signal used to perform ranging is then analyzed by means of a threshold or *Constant Fraction Discriminator* (CFD) method [THI 04], the measurement precision σ_z over a distance z with respect to this peak detection may be expressed as a function of the response time τ of the detector [WEH 99]:

$$\sigma_Z = \frac{c}{k\,SNR}\tau \quad [5.31]$$

where k = 2 for a method of the thresholding type and $2\sqrt{2}$ in the case of a CFD method.

In general, other factors pertaining to instrumental uncertainties to be determined by means of an adequate calibration can be added to this predominant factor. For greater clarity, refer to [MOR 02] and [RES 06].

5.5.1.3.2. Further uncertainties in distance measurement

A first category of uncertainty is related to atmospheric refraction. If the line of sight direction is θ, deviation δ can be formulated according to [CHA 96]:

$$\delta = cs\, t \tan\theta \quad [5.32]$$

with:

$$cst = 10^{-6}\left[\left(\frac{2410\,H}{H^2-6H+250}\right) - \left(\frac{2410\,H}{\hbar^2-6h+250}\right)\left(\frac{\hbar}{H}\right)\right] \quad [5.33]$$

where *h* (in km) represents the altitude above sea level of the measured area and *H* represents the LiDAR altitude above sea level, as shown in Figure 5.14.

In the case of a LiDAR located at an altitude *H* of 2 km scanning an area at an altitude h of 0.2 km, the *cst* parameter has a value of approximately 2 10^{-5}, namely, a displacement of 20 µrad in the case of a 45° viewing angle, which corresponds to one-tenth of the usual divergences of a mapping LiDAR. The errors occurring in terms of ranging and positioning of the measuring point with respect to laser sighting are estimated to be 5 and 8 cm, respectively. Commercial LiDAR instruments often incorporate a customized correction model and are capable of correcting the measurement.

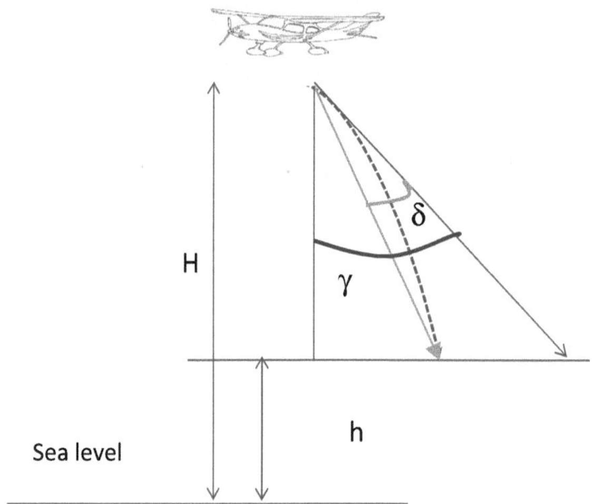

Figure 5.14. *Parameters used when the atmospheric refraction is taken into account. For a color version of this figure, see www.iste.co.uk/baghdadi/1.zip*

The time of flight measurement principle requires that the atmospheric refractive index, denoted by n, be known. This index may be determined as a function of temperature and water vapor content. If n_{lidar} represents the value used by the LiDAR and n_{atm} represents the true atmospheric value (particular models such as Ciddor's model [CID 96]), the bias pertaining to the measurement of distance z is expressed as:

$$\sigma_z = z\frac{n_{LiDAR} - n_{atm}}{n_{atm}} \qquad [5.34]$$

It is not uncommon for systems to incorporate precise predictive models in order to reduce this bias.

5.5.2. *Georeferencing uncertainties*

In the case of an onboard 3D LiDAR, the laser scanning mechanism is designed to move continuously in a three-dimensional space. As a result, in order to produce 3D georeferenced data, the global position of the system and the orientation (position) of the laser scanning system must be known at all times. This method for estimating the position and altitude of a remote sensing platform is commonly referred to as direct georeferencing. For currently available scanning systems (ALS), georeferencing is effected on the basis of a combination between the Global Navigation Satellite System (GNSS) and an inertial unit Inertial Measurement Unit (IMU). The differential GNSS is capable of providing improved aircraft positioning at decimeter level and of refreshing information at a rate of 1 and 50 Hz. The GNSS alone cannot estimate the attitude of the aircraft, and as a result GNSS measurements are combined with the measurements carried out using an IMU. The latter is constituted by three accelerometers and orthogonal gyroscopes, capable of providing the estimations of 6 degrees of freedom of movement with a refresh rate comprised between 100 and 2,000 Hz.

By using the diagram illustrated in Figure 5.15, the coordinates of target point P at ground level considered with respect to the global point of reference can be expressed as:

$$R_P^{global} = X_{nav}^{global} + M_{nav}^{global}\{M_{lidar}^{nav}\,V_{lidar} + V_{trans}\} \qquad [5.35]$$

with X_{nav}^{global} representing the position vector with respect to a global point of reference of the GPS/IMU unit at an instant t, $M_{nav}^{global} = f(r, p, y)$ represents the rotation matrix at an instant *t* with respect to the local point of reference of the carrier toward the global point of reference as a function of roll angle (*r*), pitch (*p*) and yaw (*y*); M_{LiDAR}^{nav} is the rotation matrix of the LiDAR point of reference with respect to the local point of reference of the carrier, V_{LiDAR} is the target vector of point *P* with respect to the LiDAR point of reference and V_{trans} is the translation vector of the LiDAR point of reference with respect to the point of reference of the carrier.

Vector V_{lidar} can be expressed as:

$$V_{LiDAR} = M_{LiDAR}^{int}(\theta(t) + \Delta\theta)\begin{bmatrix} 0 \\ 0 \\ \rho + \Delta\rho \end{bmatrix} \qquad [5.36]$$

with M_{LiDAR}^{int} representing the internal LiDAR rotation matrix with respect to the LiDAR point of reference as a function of the look angle of scanner θ at an instant *t*, and Δθ the associated error, ρ representing the distance measured by the scanner and Δρ representing the associated error. The errors in terms of look direction are related to the precision of the optical system/galvanometer encoder assembly as well as to the mechanical effects resulting from look direction misalignment during rotation. It may be inferred that the associated error Δθ is a linear function of look angle

$\Delta\theta = a\,\theta(t)$ with a representing a constant. The maximum levels of precision estimated by manufacturers are of approximately 0.001°.

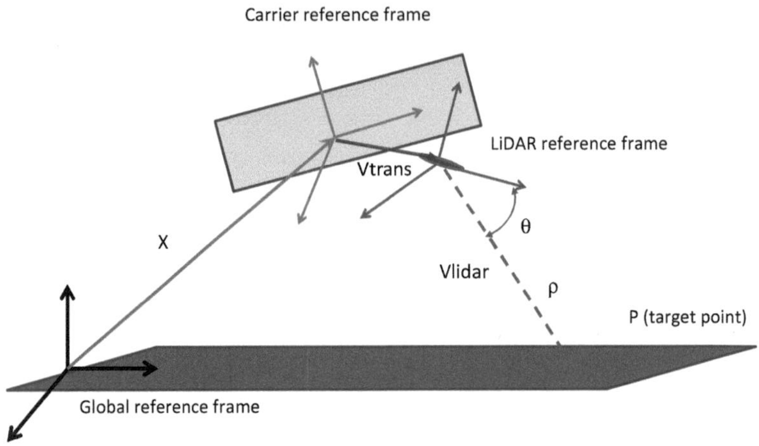

Figure 5.15. *Diagram for the calculation of georeferencing uncertainties. For a color version of this figure, see www.iste.co.uk/baghdadi/1.zip*

The errors in distance measurement $\Delta\rho$ have been discussed previously. A few orders of magnitude relating to the errors arising from referencing by means of commercial systems are indicated below: 2 cm–250 m for the LMS-Q1560 by Riegl, 2 cm for the AX80 by Trimble (without range mentioned). These values degrade as a function of distance (SNR) as seen above.

The errors in terms of the GPS position are dependent on the operating conditions (number of satellites ensuring signal coverage) and on the operating mode. At present, differential systems (DGPS) can ensure a resolution of from 5 to 15 cm, while the latter is of the order of meters in the absence of a differential correction. The effects related to the tropospheric and ionospheric propagation can also modify measurement precision and can lead to an error in the range of 5–30 cm.

The absolute errors of the angles delivered by the inertial unit are preferably situated (with the exception of the military inertial unit) at approximately 0.0025–0.0005° with respect to the roll/pitch and 0.005–0.01° with respect to the yaw. They do not have the tendency to affect the relative points precision, thus facilitating a postprocessing adjustment of data with a view to the modeling thereof. Nevertheless, they are to be taken into consideration for spaceborne applications. The main default of inertial units is represented by their temporal drift. While the accuracy of the IMU degrades in time, the GNSS, on the other hand, is capable of providing a stable measurement with reduced drift. As a result, an integrated GNSS/IMU operates on the basis of long-term GNSS stability in order to estimate and eliminate biases at the level of the IMU measurements. This integration often employs a Kalman filter in order to provide an optimum estimation of the position and orientation of the aircraft using a high data rate. Flight trajectory estimates are provided to the IMU data clock (100–2,000 Hz for the most efficient units). This data rate can be substantially lower in comparison to the laser pulse repetition rate which is generally between 100 and 400 kHz for multi echo systems. In this case, the trajectory must be interpolated in order to determine the laser position and orientation with each shot. Taking into account the substantial lower values of specific frequency related to the movement dynamics of the carriers used at present (< 10 Hz), a linear or *spline* interpolation is generally deemed as sufficient. This operation mode is associated with a reduction in the number of errors related to bias or degradation due to the temporal synchronization of all instruments required to operate using the same clock.

The major measurement uncertainty will depend on a very accurate knowledge of the laser scanner alignment with respect to the local/navigation point of reference. The effects derived from an alignment bias at the level of the pitch, roll and yaw are illustrated in Figure 5.16 (from left to right). If the theoretical direction is represented in blue, the bias effect (in red) generates a point (in black) characterized by a poor elevation and/or lateral position.

The arithmetic error ε_z may be represented as a function of the pointing angle θ or the uncertainty or poor understanding of pointing ε_θ, as well as of the terrain slope γ by relationship (Figure 5.17):

$$\varepsilon_z = z_0 \cdot \epsilon_\theta \cdot tan(\gamma + \theta) \qquad [5.37]$$

If no corrections are applied, in the case of a flat surface ($\gamma = 0°$) for an aircraft flying at 1,000 m Above Ground Level (AGL), a small error of 0.1°, for a slope characterized by a 30° incidence angle, leads to an error $\varepsilon_z \sim 1$ m. For spaceborne applications, these errors must be limited as much as possible. For the same slope, an error of 17 μrad (~0.001°) in the case of a satellite of the Pleiades type in an orbit with an altitude of 500 km and a look direction of 0.7° generates an error $\varepsilon_z \sim 0.1$ m. Ideally, M_{LiDAR}^{nav} is equal to the identity matrix. In reality, the LiDAR and carrier points of reference must be aligned at best and the residual matrix $\Delta M_{LiDAR}^{nav} = M_{LiDAR}^{nav} - I$ must be determined using a calibration method. For this purpose, it is necessary to complete various overflights over areas including known points of interest (navigation mark, landmark, etc.) in order to determine this residual matrix by means of data postprocessing. For a short overview of calibration methods, additional references may be found, for example, in the works of Skaloud and Litchi [SKA 06], Kumari *et al.* [KUM 11] and Habid *et al.* [HAB 10].

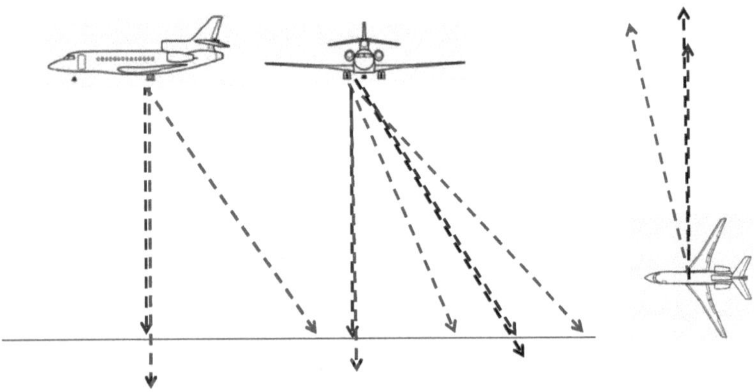

Figure 5.16. *Error due to an incorrect LiDAR alignment with respect to the inertial unit. For a color version of this figure, see www.iste.co.uk/baghdadi/1.zip*

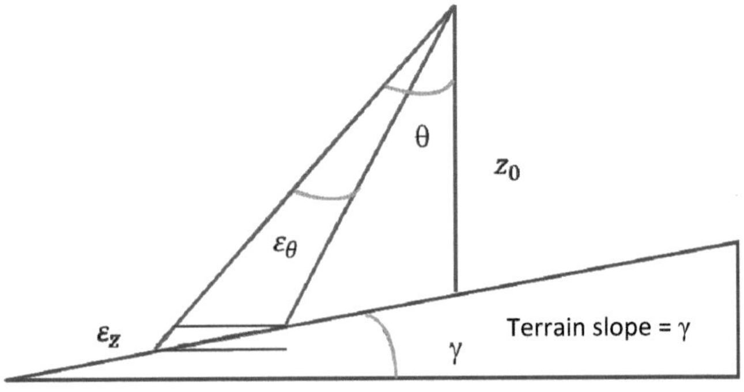

Figure 5.17. *Altimetric error ε_z*

5.6. Key points

Active LiDAR remote sensing has proven to be an efficient tool in the study of terrestrial or planetary surfaces for several decades. It may be implemented both at surface level and by means of satellites or probes. The implementation in the operational field and the analysis of LiDAR data require an understanding of the technology of the instrument used for the restitution of the sought parameters. The latter can be very diverse and conditioned by the type of LiDAR systems and/or scientific applications. It must incorporate a flawless understanding of the sources of uncertainty which may derive from the instrument and from the vector, such as the spatiotemporal variability of the studied medium. In addition to these error sources, it is also necessary to take into account the geopositioning of the instrument which can exert a significant influence on the analytical results. This is particularly true for airborne or spaceborne measurements conducted in the presence of sloped terrains.

5.7. Bibliography

[AHM 83] AHMED F.A., "A laser scanning plotter for measurement of archaeologic objects", *Journal of Field Archaeology*, vol. 10, no. 2, pp. 236–238, 1983.

[AXE 10] AXELSSON A., "Rapid topographic and bathymetric reconnaissance using airborne LiDAR", *Proceedings of SPIE 7835*, *International Society for Optics and Photonics*, p. 783503, 2010.

[BAI 12] BAILLY J.S., KINZEL P.J., ALLOUIS T. *et al.*, "Airborne LiDAR methods applied to riverine environments", in CARBONNEAU P., PIÉGAY H. (eds), *Fluvial Remote Sensing for Science and Management*, Wiley-Blackwell, p. 458, September, 2012.

[BAL 99] BALTSAVIAS E.P., "Airborne laser scanning: existing systems and firms and other resources", *ISPRS Journal of Photogrammetry and Remote sensing*, vol. 54, no. 2, pp. 164–198, 1999.

[BER 06] BERTHIER S., CHAZETTE P., COUVERT P. *et al.*, "Desert dust aerosol columnar properties over ocean and continental Africa from Lidar in-Space Technology Experiment (LITE) and Meteosat synergy", *Journal of Geophysical Research*, vol. 111, p. D21202, 2006.

[BRA 01] BRAUN A., CHENG K.C., CSATHO B. *et al.*, "ICESat laser altimetry in the Great Lakes", *Proceedings of the 60th Annual Meeting of the Institute of Navigation*, pp. 409–416, 2001.

[BRO 06] BROCK J.C., WRIGHT C.W., KUFFNER I.B. *et al.*, "Airborne lidar sensing of massive stony coral colonies on patch reefs in the northern Florida reef tract", *Remote Sensing of Environment*, vol. 104, no. 1, pp. 31–42, 2006.

[BRU 01] BRUNEAU D., QUAGLIA P., FLAMANT C. *et al.*, "Airborne lidar LEANDRE II for water-vapor profiling in the troposphere. I. System description", *Applied Optics*, vol. 40, pp. 3450–3461, 2001.

[CAI 99] CAIRO F., DI DONFRANCESCO G., ADRIANI A. *et al.*, "Comparison of various linear depolarization parameters measured by lidar", *Applied Optics*, vol. 38, no. 21, pp. 4425–4432, 1999.

[CHA 95] CHAZETTE P., DAVID C., LEFRÈRE J. *et al.*, "Comparative lidar study of the optical, geometrical, and dynamical properties of stratospheric post-volcanic aerosols, following the eruption of El Chichon and Mount Pinatubo", *J. Geophys. Res.*, vol. 100, no. D11, pp. 23195–23207, 1995.

[CHA 03] CHAZETTE P., "The monsoon aerosol extinction properties at Goa during INDOEX as measured with lidar", *Journal of Geophysical Research*, vol. 108(D6), p. 4187, 2003.

[CHA 08] CHANIN M.L., *L'Ecole de l'Espace, Le service d'Aéronomie 1958-2008: histoire et science*, CNRS Editions, Paris, 2008.

[CHA 12] CHAZETTE P., DABAS A., SANAK J. *et al.*, "French airborne lidar measurements for Eyjafjallajökull ash plume survey", *Atmospheric Chemistry And Physics*, vol. 12, pp. 7059–7072, 2012.

[CHA 14] CHAZETTE P., MARNAS F., TOTEMS J., "The mobile Water vapor Aerosol Raman LIdar and its implication in the framework of the HyMeX and ChArMEx programs: application to a dust transport process", *Atmospheric Measurement Techniques*, vol. 7, pp. 1629–1647, 2014.

[CHU 98] CHURNSIDE J.H., TATARSKII V.V., WILSON J.J., "Oceanographic lidar attenuation coefficients and signal fluctuations measured from a ship in the Southern California Bight", *Applied Optics*, vol. 37, pp. 3105–3112, 1998.

[CHU 09] CHURNSIDE J.H., DONAGHAY P.L., "Thin scattering layers observed by airborne lidar", *ICES Journal of Marine Science*, vol. 66, pp. 778–789, 2009.

[CID 96] CIDDOR P.E., "Refractive index of air: new equations for the visible and near infrared", *Applied Optics*, vol. 35, pp. 1566–1573, 1996.

[ESP 14] ESPOSITO S., MURA M., FALLAVOLLITA P. *et al.*, "Performance evaluation of lightweight LiDAR for UAV applications", *Geoscience and Remote Sensing Symposium (IGARSS)*, IEEE International, pp. 792–795, 13–18 July 2014.

[FLA 05] FLAMANT P.H., "Atmospheric and meteorological LiDAR: from pioneers to space applications", *C. R. Physique*, vol. 6, no. 8, pp. 864–875, 2005.

[FOW 11] FOWLER A., KADATSKIY V., "Accuracy and error assessment of terrestrial, mobile and airborne lidar", *Proceedings of American Society of Photogrammetry and Remote Sensing Conference (ASPRP)*, Milwaukee, USA, 1–5 May 2011.

[FRY 97] FRY E.S., EMERY Y., QUAN X. *et al.*, "Accuracy limitations on Brillouin lidar measurements of temperature and sound speed in the ocean", *Applied Optics*, vol. 36, no. 27, pp. 6887–6894, 1997.

[GUE 78] GUENTHER G.C., GOODMAN L.R., "Laser applications for near-shore nautical charting", *Proc. SPIE Ocean Optics V*, vol. 160, pp. 174–183, 1978.

[HAB 10] HABIB A., BANG K., KERSTING A. *et al.*, "Alternative methodologies for LiDAR system calibration", *Remote Sensing*, vol. 2, no. 3, pp. 874–907, 2010.

[HOG 83] HOGE F.E., SWIFT R.N., "Experimental feasibility of the airborne measurement of absolute oil fluorescence spectral conversion efficiency", *Applied Optics*, vol. 22, no. 1, pp. 37–47, 1983.

[KRA 09] KRAINAK M.A., SUN X., YANG G. *et al.*, "Photon detectors with large dynamic range and at near-infrared wavelength for direct detection space lidars", *Proc. SPIE*, vol. 7320, p. 732005, 2009.

[KRE 98] KREKOV G.M., KREKOVA M.M., SHAMANAEV V.S., "Laser sensing of a subsurface oceanic layer. II. Polarization characteristics of signals", *Applied Optics*, vol. 37, no. 9, pp. 1596–1601, 1998.

[KUM 11] KUMARI P., CARTER W., SHRESTHA R., "Adjustment of systematic errors in ALS data through surface matching", *Advances in Space Research*, vol. 47, pp. 1851–1864, 2011.

[LEO 79] LEONARD D.A., CAPUTO B., HOGE F.E., "Remote sensing of subsurface water temperature by Raman scattering", *Applied Optics*, vol. 18, no. 11, pp. 1732–1745, 1979.

[LOG 03] LOGNOLI D., CECCHI G., MOCHI I. *et al.*, "Fluorescence lidar imaging of the cathedral and baptistery of Parma", *Applied Physics B*, vol. 76, no. 4, pp. 457–465, 2003.

[MEA 84] MEASURES R.M., *Laser Remote Sensing: Fundamentals and Applications*, John Wiley, Krieger Publishing Company, Malabar, USA, 1984.

[MEG 85] MEGIE G., ANCELLET G., PELON J., "Lidar measurements of ozone vertical profiles", *Applied Optics*, vol. 24, no. 21, pp. 3454–3463, 1985.

[MEG 89] MÉGIE G., *Ozone, l'équilibre rompu*, CNRS Editions, Paris, 1989.

[MOR 02] MORIN K.W., Calibration of Airborne Laser Scanners, UCGE Reports Number 20179, 2002.

[NEL 88] NELSON R., KRABILL W., TONELLI J., "Estimating forest biomass and volume using airborne laser data", *Remote Sensing of Environment*, vol. 24, no. 2, pp. 247–267, 1988.

[RES 06] RESHETYUK Y., Investigation and calibration of pulsed time-of-flight terrestrial laser scanners, PhD thesis, Royal Institute of Technology (KTH), Department of Transport and Economics Division of Geodesy, Stockholm, Sweden, 2006.

[ROY 11] ROYER P., CHAZETTE P., LARDIER M. *et al.*, "Aerosol content survey by mini-N2-Raman lidar: application to local and long-range transport aerosols", *Atmospheric Environment*, vol. 45, no. 39, pp. 7487–7495, 2011.

[SHA 14] SHANG X., CHAZETTE P., "Interest of a full waveform flown UV lidar to derive forest vertical structures and aboveground carbon", *Forests*, vol. 5, pp. 1454–1480, 2014.

[SHA 15] SHANG X., CHAZETTE P., "End-to-end simulation for a forest-dedicated full waveform lidar onboard a satellite initialized from airborne ultraviolet lidar experiments", *Remote Sensing*, vol. 7, pp. 5222–5255, 2015.

[SHR 12] SHRESTHA K.Y., CARTER W.E., SLATTON K.C. *et al.*, "Shallow bathymetric mapping via multistop single photoelectron sensitivity laser ranging", *IEEE Transactions on Geoscience and Remote Sensing*, vol. 50, no. 11, pp. 4771–4790, 2012.

[SKA 06] SKALOUD J., LICHTI D., "Rigorous approach to bore-sight self-calibration in airborne laser scanning", *ISPRS Journal of Photogrammetry & Remote Sensing*, vol. 61, pp. 47–59, 2006.

[SMU 62] SMULLIN L., FIOCCO G., "Optical echoes from the Moon", *Nature*, vol. 194, p. 1267, 1962.

[SOL 77] SOLODUKHIN V., ZUKOV A., MAZUGIN I., "Possibilities of laser aerial photography for forest profiling", *Lesnoe Khozyaisto (Forest Management)*, vol. 10, pp. 53–58, 1977.

[SVA 95] SVANBERG S., "Fluorescence lidar monitoring of vegetation status", *Physica Scripta*, vol. T58, pp. 79–85, 1995.

[THI 04] THIEL K.H., WEHR A., "Performance capabilities of laser scanners – an overview and measurement principle analysis", *Proceedings of the ISPRS Working Group VIII/2 Laser- Scanners for Forest and Landscape Assessment*, Fribourg, Germany, pp. 14–18, 3–6 October 2004.

[WEH 99] WEHR A., LOHR U., "Airborne laser scanning – an introduction and overview", *ISPRS Journal of Photogrammetry & Remote Sensing*, vol. 54, pp. 68–82, 1999.

[WEI 05] WEITKAMP C., *Lidar Range-Resolved Optical Remote Sensing of the Atmosphere*, Springer, New York, 2005.

[YOU 03] YOU S., HU J., NEUMANN U. *et al.*, "Urban site modeling from lidar", *Computational Science and Its Applications*-ICCSA 2003, Springer Berlin Heidelberg, pp. 579–588, 2003.

[YOU 06] YOUNG A.P., ASHFORD S.A., "Application of airborne LIDAR for seacliff volumetric change and beach-sediment budget contributions", *Journal of Coastal Research*, vol. 22, no. 2, pp. 307–318, 2006.

[ZWA 02] ZWALLY H.J., SCHUTZ B., ABDALATI W. *et al.*, "ICESat's laser measurements of polar ice, atmosphere, ocean, and land", *Journal of Geodynamics*, vol. 34, no. 3, pp. 405–445, 2002.

6

Airborne LiDAR Data Processing

6.1. Introduction

This chapter introduces the main data analysis methods associated with topographic and bathymetric airborne LiDAR systems. Data delivered by these sensors can be of two types: the majority of commercial systems deliver three-dimensional (3D) point clouds (systems referred to as *multi-echo*), whereas a limited number directly provides the whole laser signal backscattered by the Earth surface (systems referred to as full-waveform (FW)).

The developments in the fields of semiconductor electronics and photonics over the past 50 years have facilitated the construction of reliable and accurate laser systems characterized by a very high spatial resolution. The main types of systems used for metrological or environmental purposes are represented by rangefinders; these systems measure the round-trip time of the emitted photons and make it possible to estimate the distances between the source and the recognized objects. This chapter focuses on 3D scanning systems which facilitate a very high spatial resolution as a result of a low-divergence laser beam emission. These types of systems differentiate themselves from the systems based on phase measurements, triangulation, two-dimensional (2D) sensor array flash imaging and 3D focal plane imaging. Additional details on LiDAR technology are provided in Chapter 5, Volume 1.

Chapter written by Clément MALLET, Nesrine CHEHATA and Jean-Stéphane BAILLY.

A LiDAR system is generally constituted by at least the laser source, the synchronized system designed to detect the emitted waveform and the hybrid georeferencing system which makes it possible to locate the devise and measure its attitude. These systems are either spaceborne, airborne (aircraft, helicopters unmanned aerial vehicles (UAVs)) or terrestrial (static or carried on board a mobile mapping vehicle, see Chapter 2, Volume 5). The lasers which emit electromagnetic waves in a single direction are referred to as profilers. The measurement is spatialized by means of a vector displacement (satellite, airborne and terrestrial), often in conjunction with a beam deflection mechanism (oscillating mirror, prism and rotating mirrors or rotating polygon), as is the case of the airborne systems or mobile mapping vehicles.

The main focus of this chapter is the processing of topographic and bathymetric LiDAR data with an emphasis on airborne sensors. The types of systems seeking to estimate the morphology of the terrain and to extract its constituent objects emit waves in the infrared domain (usually 1,064 or 1,550 nm, representing the most suitable atmospheric windows). Although they facilitate a good discrimination between land and water areas, these wavelengths are not adapted to water, riverine or coastal areas [SME 13]. Nevertheless, in the case of shorter wavelengths (532 nm), the laser pulses penetrate the water column and make it possible for depth-limited river and seabed measurements to be recorded. The system referred to as bathymetric LiDAR benefits in particular from wavelength coupling in the infrared and green domains in order to measure the depth of the water column (Chapter 5, Volume 5).

The latest generation of airborne LiDAR systems is multispectral and facilitates the simultaneous acquisition of multiple wavelengths (3-10). As a result, some commercial systems are hereafter characterized by dual topographic and bathymetric applications [COL 10].

6.2. LiDAR Waveform analysis

6.2.1. *Technology*

Conventional LiDAR systems (referred to as *multi-echo*) provide a geometric representation of scenes in the form of 3D point clouds: this represents the most common data type. However, a certain level of processing is required to extract a number of products, such as Digital Terrain Models, 3D city models or forest variables. These are presented in more detail in section 6.3. Full-Waveform (FW) systems for were used for the first time almost 30 years ago for applications in bathymetry, and have become very popular over the past decades with the development of commercial topographic systems and their implementation in the riverine and coastal environments. These sensors make it possible to record backscattered signals from each laser pulse emitted. These signals are referred to as "waveforms". Therefore, such a sequence represents the progression of the laser pulse interacting with the reflecting surfaces. The shape of these waveforms varies according to the objects reached: this variation is due both to their geometric and spectral properties (Figure 6.1). As a result, FW data provide additional information with respect to conventional multi-echo data. Instead of poorly structured 3D point clouds, FW systems provide a succession of one-dimensional (1D) profiles representing 3D scenes, thus providing additional physical knowledge in relation to the objects reached by the laser beam, as well as a finer spatial discrimination between the objects or surfaces located in close proximity. Recording these waveforms facilitates their *a posteriori* analysis of the acquisition phase and therefore the development of more advanced processing methods than those implemented at the level of the onboard sensor software [PFE 15]. These processing methods are described in more detail in the following section.

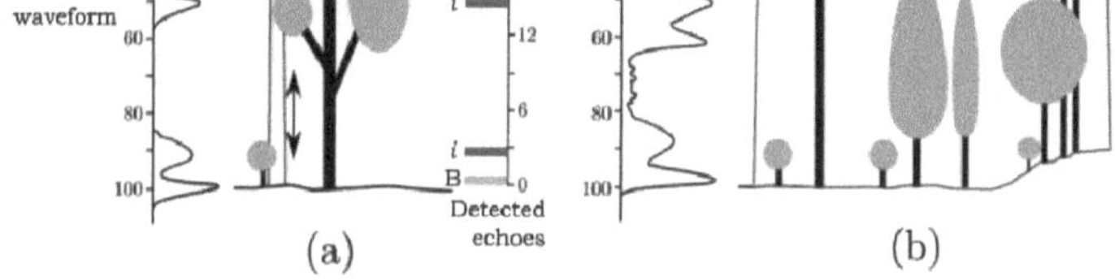

Figure 6.1. *Two Different waveforms as a function of the dimensions of the emitted laser beam: a) airborne sensor: each object reached specifically contributes to the shape of the waveform. A multi-echo sensor will only detect a limited number of echoes (i.e. altitudes): the first echo (A) the final echo (B) and the intermediate echoes (i); b) satellite or large-footprint sensor: the waveform represents the integration according to altitude of the existing objects, with a significant influence from the terrain in sloping areas*

The first systems were designed for bathymetric applications, and then under the direction of NASA, experimental topographic systems, often spaceborne, were introduced (SLA-02 or ICESat/GLAS). These systems demonstrated the relevance of recording full waveforms in order to characterize areas of tropical vegetation. Then in 2004, the small-footprint commercial airborne system (laser beam footprint below 1 m), LiteMapper 5600, designed without a specific application, was introduced for the first time [MAL 09]. Various studies have been conducted so far on this type of data, both in forested and urban areas. This number has increased significantly over the past 5 years owing to the development of commercial systems with this digitization capacity (Optech Titan, RIEGL LMS-Q680i, Leica ALS-80). The benefits of this type of data have been demonstrated within the context of a more accurate determination of the position of reached objects [JUT 06], a fine separation of objects in close proximity [LIN 09], land-cover mapping [MAL 10] or monitoring continental and river waters [ABD 13].

A waveform represents the laser energy backscattered by the Earth surface and its constituent elements as a function of time. This profile is sampled approximately every nanosecond. Waveform shapes vary according to the geometric and spectral properties of the illuminated objects and their arrangement within the laser beam. As a result, these are conditioned by the footprint of this beam. Small, medium or large footprints may therefore be distinguished (Figure 6.1). The first category corresponds to commercial topographic systems (footprint <1 m) illuminating a limited number of objects and making it possible to collect returns which are very well defined in space. The second category of footprints relates to high-power bathymetric lasers. The third category (10–70 m, in particular the ICESat/GLAS sensor) is used to obtain an average signal at the level of local altimetric profiles corresponding to the vertical distribution of the objects included in these large footprints [MAL 10]. This makes it possible to simultaneously highlight ground areas and estimate the terrain elevation [ZWA 02], as well as the height of vegetation (on the basis of the information derived therefrom, such as tree height and biomass [DUN 10, BAG 14]). Nevertheless, the latter are not adapted to the altimetry of continental waters if the targets of interest are not sufficiently large with respect to the footprint [BAG 11].

The processing methods of LiDAR waveforms, regardless of the sensor and wavelength of interest, may be divided into two main categories [PAR 11]:

– decomposition methods aimed at identifying the main or a maximum number of modes present in the waveforms [WAG 06]. These modes – referred to as *echoes* – represent objects or groups of objects located in spatial proximity. Their detection makes it possible to extract a standard distance measurement therefrom, and subsequently their geographical position, in order to generate 3D point clouds such as those obtained by means of standard multi-echo systems;

– global methods which preserve and analyze the waveform on the whole without seeking to decompose it into a sum of specific contributions. The information is therefore carried by the entire waveform, which is considered as an altimetric profile out of which metrics can be extracted in a direct manner: this is particularly useful in the estimation of forest parameters (waveform = canopy height profile) [HAN 15]. This analysis can also be explored further in conjunction with the analysis conducted on spatially neighboring waveforms. It is therefore possible to recognize patterns in the existing spatiotemporal data cube [JUT 05] or to invert radiative transfer and geometric optics models in order to identify the variables of interest [KOE 06].

The LiDAR equation which expresses the power received *Pr* as a function of the power transmitted *Pt* may be formulated as follows:

$$Pr(t) = \sum_{i=1}^{n} \frac{D^2}{4\pi\lambda^2 R_i^4} Pt(t) \star (\eta \star \sigma_i)(t) \qquad [6.1]$$

where $\star$ corresponds to the convolution product, n represents the number of the targets reached by the laser beam, D represents the aperture diameter of the LiDAR (constant), λ represents the wavelength (constant), R represents the distance between the sensor and the target i, η represents the atmospheric influence (assumed to be constant) and σ represents the backscattering cross-section of the target i (by analogy with the radar).

The following is in addition provided: $\eta = exp(-2Ra)$, where a represents the atmospheric attenuation coefficient assumed to be constant and $\sigma = \pi\rho\beta^2R^2$, with ρ representing the reflectance of the target and β the divergence of the LiDAR beam (constant).

6.2.2. *Waveform deconvolution*

A recorded LiDAR waveform represents the convolution of the waveform emitted by the backscattering cross-section of the scene, constituted in practice by the backscattering cross-sections of all objects or groups of reached objects. Deconvolution methods are implemented in order to eliminate the influence of the transmitted waveform, as well as to detect echoes. The main obstacle is represented by the fact that is an ill-posed problem if no *a priori* constraint is introduced to constrain the problem. It is for this reason that Fourier-quotient methods fail to produce satisfactory results.

The convolution of the transmitted wave $Pt(t)$ and the surface response $S(t)$ providing the received wave form $Pr(t)$ may be expressed in the Fourier frequency domain as follows (f representing the frequency):

$$S(f) = \frac{Pr(f)}{Pt(f)} \qquad [6.2]$$

In order to calculate $S(f)$, the $Pr(f)$ and $Pt(f)$ functions have to be measured. Emitted and received LiDAR waveforms are available, however, these are associated with a noise term "*B*" being added to the signal. As a function of the latter, the estimation of S can be impaired by severe numerical errors.

Four main approaches have been developed specifically for LiDAR waveforms, which vary according to the implemented method, the desired level of location accuracy, and also according to the selected type of modeling applied to the transmitted wave form (Gaussian hypothesis used in most cases) and the background noise. The objective is often to reduce the background noise without impairing the estimation of the surface response.

By using a Fourier domain deconvolution as a basis and hypothesizing an emitted wave of the Gaussian type, Jutzi and Stilla [JUT 06] estimate the surface response by means of a Wiener linear filter. This facilitates a resistance to the background noise which can be present in the LiDAR waveforms generated by experimental or poorly calibrated systems. This estimated response is used as a parametric function by the standard wave form decomposition methods detailed in section 6.2.2.

The Wiener filter minimizes the mean squared error between the surface response *S(f)*, which is not corrupted by the noise, and the estimated response $\acute{S}(f)$. A solution can be formulated if the Wiener filter is represented by a function of the following type:

$$W(f) = \frac{|Pr(f)^2|}{|Pr(f)^2| + |B(f)^2|} \quad [6.3]$$

| . | corresponds to the module for each contribution. $B(f)=N(f)/S(f)$ can be estimated on the basis of the background noise (*S* is the mean power spectral density of the noise and *N* is the density of the incoming signal), *Pr(f)* is known, it is only required to estimate $|Pr(f)^2|$ in order to build such a filter.

Still on the basis of a Gaussian hypothesis, Parrish *et al.* [PAR 09] formulate this issue as a Bayesian estimation problem or the *a posteriori* maximum is obtained iteratively by means of the expectation–maximization method. This approach includes a parameter which makes it possible to settle the compromise between the number of accurate detections and false alarms.

Nevertheless, the emitted waveforms and the backscattering targets do not feature Gaussian characteristics in each case, for which reason approaches excluding an analytic function have been proposed: the emitted and received waveforms together with the backscattering cross-sections are modeled by piecewise linear functions in [WAN 09] and uniform B-splines in [RON 11], the parameters of which are estimated by means of a least squares system.

6.2.3. *Unsupervised detection and decomposition methods*

Advanced echo detection methods focus on the extraction of intensity local maxima, whereas decomposition methods seek in addition to represent a waveform as the sum of the contributions which can often be modeled by means of an analytic function (Figure 6.2). The parameters of the latter are therefore associated with the features which can be calculated by means of local 3D point cloud analysis for subsequent filtering and classification purposes (see section 6.3). Individual echo detection and modeling are primarily examined in the case of airborne topographic and bathymetric LiDARs, where the local amplitude maximum refers to the distance between the sensor and a well-defined target.

In addition to the deconvolution methods mentioned above, detection methods are primarily based on a local analysis of the first and second derivatives of the signal, and often conducted in an iterative manner in order to eliminate false detections due to noise and partially overlapping echoes [LIN 10]. The most commonly used techniques together with their advantages and disadvantages are presented in [WAG 04]. Additionally, a number of authors focus on the detection of low-amplitude echoes which are difficult to detect in a single waveform (excessive noise). The adopted principle often consists of jointly analyzing the waveforms emitted before or after by the LiDAR as follows: the spatio-temporal aggregation of the amplitude profiles thus highlights, by means of amplitude stacking, the response of a surface partially occluded by an object placed in the foreground, non-visible at the level of a single waveform [STI 07, ZHU 14]: in practice, this is the case of the ground located beneath the forest canopy.

Decomposition methods seek to fit a parametric or empirical model at the level of each one of the n echoes detected in the waveform y:

$$y_i = \sum_{k=1}^{n} \phi_k(x_i) + b \qquad [6.4]$$

where ϕ_k is the model selected, $\{x_i\}_{i=1..N}$ are the sequence of N points sampling the waveform in a uniform manner and b is the background noise.

Priority is often given to the Gaussian distribution by hypothesizing that the emitted waveform itself is of the Gaussian type (in practice, this is slightly asymmetrical to the right): the convolution with the backscattering cross-sections (namely, the return waveform of aerial and satellite topographic systems) can also be modeled in this manner [HOF 00]. Such a model has a dual advantage: in addition to its simplicity, its parameters can be easily interpreted and estimated by using the least squares or expectation–maximization methods [MAL 09]. Its expression is:

$$\phi_k(x) = A_k exp\left(\frac{-(x-\mu_k)^2}{2w_k^2}\right)^2 \qquad [6.5]$$

μ_k is the echo position, w_k is the echo half-width and A_k represents its amplitude (Figure 6.2).

This model is sufficient for the majority of topographic, satellite or airborne sensors, except in the case of a significant atmospheric influence and an urban environment where asymmetries are recorded. These asymmetries are due either to the geometry of the objects (surface slopes, grounds or building roofs accentuate the asymmetry already present in the emitted waveform), to their spectral properties, or to the position of the objects within the laser beam. More complex parametric models have been proposed [MAL 10], aiming to improve the adjustment of waveforms but without a significant contribution to subsequent land-cover classification.

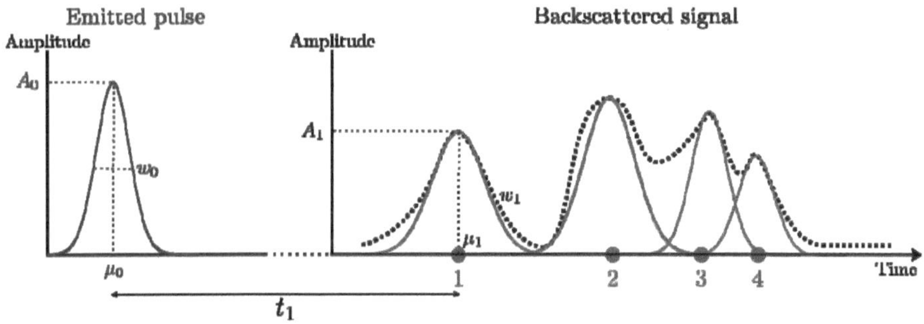

Figure 6.2. *Parametric decomposition and modeling of a LiDAR waveform. The main echoes are initially detected (four in this case). The timing difference between the emitted wave and one of these echoes makes it possible to identify the distance between the target and the space/aircraft and to generate a 3D point. Modeling (Gaussian in this case) makes it possible to adjust each detected echo and to obtain additional morphologic parameters (amplitude A and half-width w in this case)*

In the case of waveforms generated and recorded by bathymetric sensors, the contribution of the water column is not symmetrical: priority is therefore given to exponential, or quadrilateral functions [LES 07, BOU 14].

Terrestrial sensors are characterized by very high reverse power dynamics and the responses of the systems indicate nonlinear scale behaviors. This accentuates the initial asymmetry of the generated peak and results in non-Gaussian echoes. Thus, empirical models are preferred [HAR 15]. The main disadvantage is that the method has to be calibrated by monitoring targets characterized by different standard reflectance values over a number of indicated distances.

6.2.4. *Supervised methods*

Another family of methods for processing LiDAR waveforms encountered in the literature is based on statistical modeling. In order to calibrate a statistical model, it is therefore necessary to provide learning samples (where the term supervised is derived from), for which pairs of "waveforms-value of the variable" to retrieve are given. In most cases, only one variable of the target is retrieved in LiDAR waveforms from these methods (for example, canopy height in the case of near-infrared topographic systems, or water depth in the case of green bathymetric systems). This variable is deduced from relative positions in the waveforms, instead of from absolute positions. Furthermore, this confines the applicability of these methods.

Following the example of the functional analysis models [RAM 06] or curve regression [COR 03], this method is based on three main stages:

– a selection stage of a useful section of waveform with constant length;

– a reduction stage of the dimension of the useful section of the waveform to a number of synthetic variables;

– a regression or classification stage of the variable of interest by means of the synthetic variables constructed during phase 2 [PEE 07, COL 08, ALL 10, ZHO 15].

Indeed, a straightforward regression or classification between the variable of interest of the target (tree height, water depth, etc.) and the whole waveform, in which case each (instant) sample of the waveform is a predictor variable, cannot be envisaged (excessive dimensionality and predictor variables strongly correlated).

In order to illustrate these methods and individual stages, these will be applied in the case of a small set of 18 theoretical waveforms including 512 length samples. These waveforms are generated on the basis of a simple combination of three Gaussians characterized by different noise parameters, following the example of the waveform represented in Figure 6.2. These 18 waveforms simulate the categories of data acquired at the level of a forest canopy in a mountainous region, according to a sampling frequency of 5 ns at the level of the detectors. In the case of each individual waveform, it is sought to illustrate the method which makes it possible to identify the distance (canopy height) between the central positions of two of the most eccentric Gaussians.

6.2.4.1. *Extraction of the useful section of waveforms*

The useful section of a waveform is the part which indicates the interaction of the laser beam with the target of interest in time (for example, the forest canopy). It is generally extracted on the basis of the signal-to-noise ratio criteria (when the intensity is n times higher than the standard deviation of the background noise). The 18 initial simulated waveforms, the threshold (gray horizontal line) selected in order to record the initiation of the signal (in this case, the threshold is equal to 4 times the standard deviation of the background noise) and the initiation of the corresponding useful part of the signal (red vertical line on the left-hand side) can be observed in Figure 6.3. Similarly, the end of the useful signal can be extracted as the final value, or the intensity is above the threshold (vertical dotted line in gray). However, in order to illustrate the methods from stage 2, this can also be arbitrarily set to a fixed number of samples (in this case 64) with respect to the initiation of the signal, covering the most delayed end of signal, and having a dyadic length (number of samples), i.e. calculated as a power of 2. The end of the "useful" signal is represented by the vertical red line on the right-hand side in Figure 6.3.

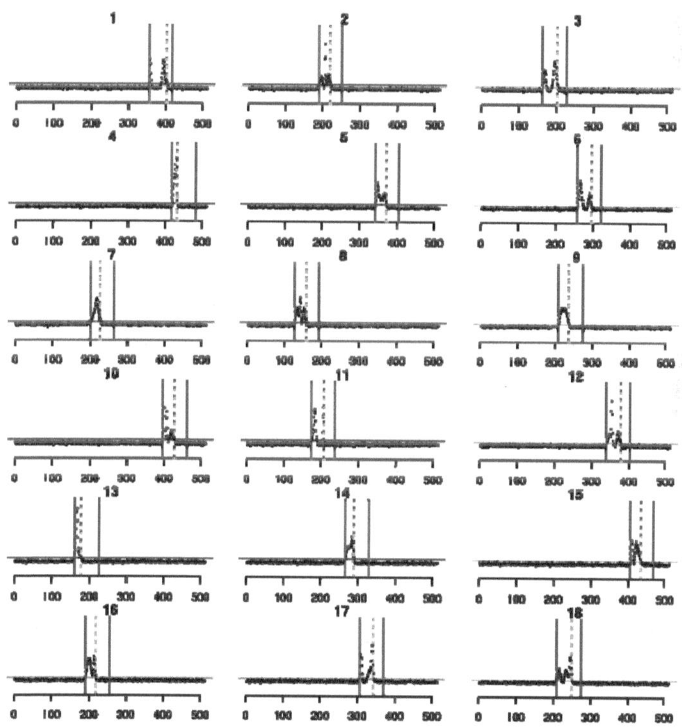

Figure 6.3. *Example of a basis of 18 waveforms having a length of 512 and an extracted useful part of 64 in length (between two red vertical lines). The x-axis represents time and the y-axis represents waveform intensity. For a color version of this figure, see www.iste.co.uk/baghdadi/1.zip*

At the end of this stage, a conventional individual-character data table is obtained (18 lines × 64 columns) where each line represents the "useful" part of the waveform and each column is a (temporal) sample of the waveform (V1–V64) commencing upon initiation of the useful signal.

6.2.4.2. *Reduction of waveform dimensions*

Numerous methods have been applied in the literature for the purpose of reducing the dimensions of LiDAR waveforms: methods based on "expert" metrics, methods by means of a Principal Component Analysis (PCA) or "spectral" methods (wavelets and Fourier). There is nothing to prevent the implementation of other conventional functional analysis methods (e.g., approximation by means of splines functions); however, this has not yet been used.

In the case of specialized metrics pertaining to a waveform, the *waveform extent* used at the level of the forest canopy [LEF 05] and characterized by the duration between the initiation and the end of the part of the signal above a noise threshold can be cited as an example (for example, 4 times the standard deviation of the background noise). However, these last two methods include the advantage of proposing non-correlated synthetic variables which subsequently allow for increasingly stable regressions.

The PCA is a conventional method for reducing the dimensions of multivariate data. In this case, each time sample of the waveform is considered to be a variable (V1–V64, Figure 6.4). The objective of the PCA is to reconstruct novel variables (the main components or PC) which are linear combinations of the former variables (V), constructed in a hierarchical manner so as to maintain the maximum information (variance) of the data table (Figure 6.5). In this case, the PCA must be carried out by means of covariance and not by means of correlation (without standardization of variables V1–V64). In Figure 6.5, it may be noted that at the level of the base of 18 waveforms having a length of 64 (Figure 6.4), the first two components (two novel variables) contain 54% of the waveform information and make it possible to discriminate the studied characteristic (canopy height) as shown in Figure 6.5. In this figure, the novel variable PC1 (axis of abscissae) appears to be well correlated to the canopy height. This axis appears to oppose the two mode waveforms (PC1 > 0) to mono mode waveforms (PC1 < 0), as shown by the waveforms taken as an example (in gray) for each of the four angular sectors in the PC1-PC2 plane. Instead, the novel variable PC2 (axis of ordinates) appears to be related to the intensity of the peaks at the level of the waveforms or conversely to their width.

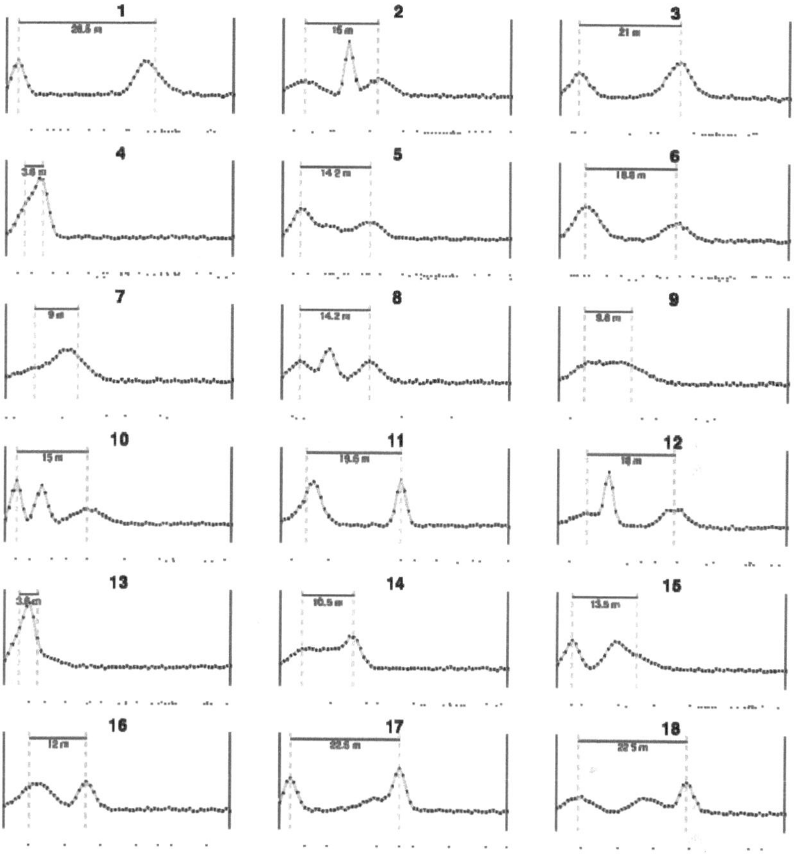

Figure 6.4. *Example of a base of 18 extracted waveforms having a length of 64 (useful part from V1 to V6 64, respectively, beginning and end of the useful signal, represented by the red vertical lines) and studied characteristics (distance between extreme peaks expressed in m, in green). The x-axis represents time and the y-axis represents waveform intensity*

For the purpose of water depth regression at the level of the bathymetric LiDAR waveforms, Allouis *et al.* [ALL 10] have extracted the bathymetric waveforms from the Shoals 1000-T system. Waveforms were characterized by a fixed length of 51 samples by means of a simple extraction at the level of the original waveforms (the initiation of the signal being located at the same instant in each case, owing to a constant distance between the LiDAR platform and the water surface which indicates the initiation of the signal). At the end of the PCA, Allouis *et al.* [ALL 10] retained only the first two components (PC1 and PC2), containing 99% of the former information (Figure 6.4). In this case, a transition was made from 51 variables to two variables.

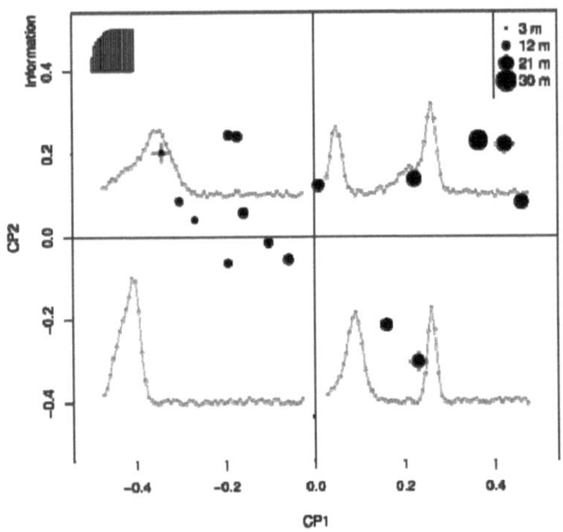

Figure 6.5. *Projection of the 18 waveforms (round black points) on the first factorial plane generated by the PCA (PC1 and PC2) with a point size relative to the studied characteristic (tree height). A representative waveform for each angular sector (the most distant waveform from the reference origin per angular sector) whose corresponding point is marked with a gray cross is represented by a gray solid line. The evolution of information as a function of the number of PC is represented by the bar diagram in the top left-hand side*

Similarly, for the purpose of canopy height regression, Fayad *et al.* [FAY 14] have selected ICESat/GLAS waveforms at the level of a useful part characterized by a fixed length of 470 samples (every ns), with the first sample where the intensity is 4.5 times higher than the standard deviation of the background noise being considered as the initiation of the "forest" signal. After the PCA, Fayad *et al.* [FAY 14] retained 51 components with respect to the 470 original variables, retaining 94% of the initial information. A transition was therefore made from a number of 470 variables to 51 variables.

The decomposition of a time signal *z(t)* (a waveform in this case) into wavelets can also be used in order to compress or filter LiDAR waveforms [FAN 04, LAK 10], in particular when these waveforms are characterized by a dyadic length (length is calculated as a power of 2) and as in the case of the PCA, when these are organized into a table according to their useful part. The principle of a decomposition into wavelets is to successively approach the waveform using a mixture of functions, linear combinations of a set of orthogonal wavelet functions (father and mother), according to resolutions (*j*) which increase by a factor of 2. The signal approximation equation is expressed as:

$$z(t) \approx \sum S_{J,k}\, \Phi_{J,k}(t) + \sum\sum d_{j,k}\, \psi_{j,k}(t) \qquad [6.6]$$

In this equation, *S* represents the fitting coefficients of the father function ϕ, whose integral is equal to 1, and *d* represents the adjustment coefficients of the mother function ψ, whose integral is equal to 0. The index *j* represents the resolution level ranging from the finest resolution ($j = 1$) to the coarsest resolution ($j = J$), *k* representing the location index at the level of the time signal, ranging from 1 (at the initiation of the signal) to n_{j-1} (location at the end of the signal when the resolution *j* is considered). The coarsest resolution generally corresponds to $J = n - 2$.

In order to accurately understand the behavior of this decomposition, Figure 6.6 represents, for example, the decomposition at the level of the Haar wavelet set of the fifth waveform in the base of the 18 theoretical waveforms (Figure 6.3), as well as the values of the *S* and *d* coefficients obtained (bar diagrams on the right-hand side). The decomposed part of the signal *z(t)* corresponding to different resolutions ranging from $j = 1$ (resolution = 2), to $J = 4$ (resolution = 16) is represented by the red line in the diagrams located on the left-hand side of Figure 6.6. These red lines simply correspond to the product between the wavelet coefficients (values recorded in the bar diagrams on the right-hand side) for each location and each scale and the father (coefficients *S*) or mother (coefficients *d*) functions. In Figure 6.7, it may be observed how this decomposition can be used as a signal simplification filter by rebuilding the signal (red curve) by means of inverse transformation and successive integration of increasingly fine resolution levels. In this figure, it may be observed that stopping at the second level of resolution (resolution of 8) appears to be sufficient to represent the location of the peaks.

The wavelet coefficients S or d (father and mother) can therefore be synthetic variables which resume the useful waveform for a certain resolution. In each case, the mother coefficients characterized by a resolution 8 (eight coefficients) are the most informative in terms of peak positioning in Figure 6.6. A regression of the variable of interest (tree height) on the basis of these eight coefficients would in this case facilitate a transition from a number of 64 variables to eight variables.

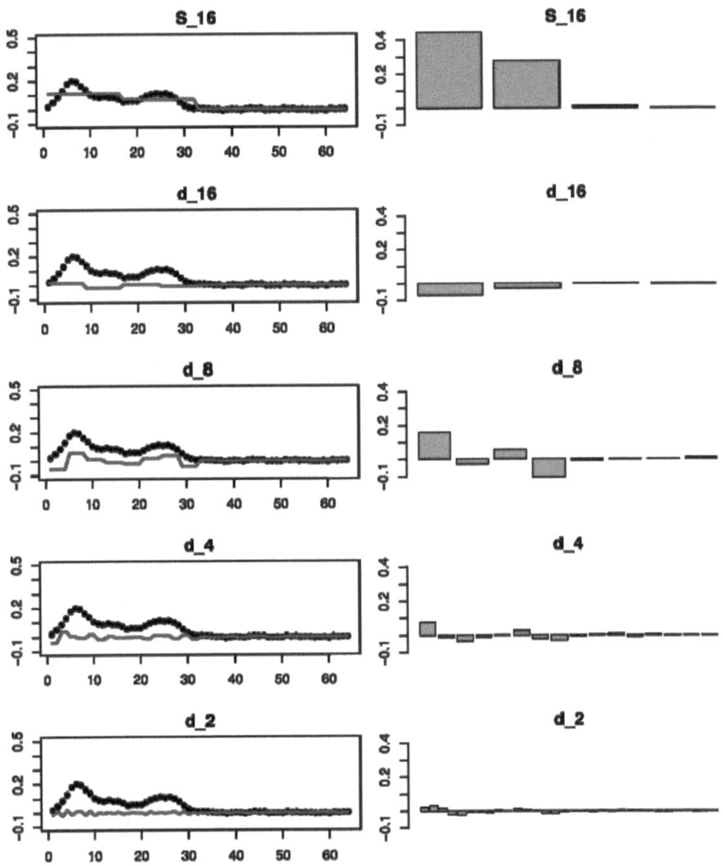

Figure 6.6. *Example of the wavelet decomposition of the theoretical waveform N°5 (black curves on the left-hand side) (Figure 6.3) at different levels (red curves on the left-hand side). The axis of abscissae represents time and the axis of ordinates represents waveform intensity in the figures on the left-hand side. The values of the wavelet coefficients are represented in the bar diagrams on the right-hand side*

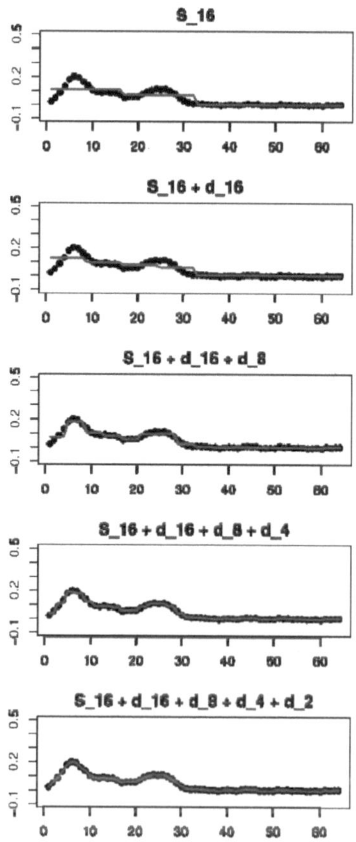

Figure 6.7. *Use of the wavelet coefficients of the theoretical waveform No. 5 (Figure 6.3) (black curves) in order to produce various waveform simplification filters (red curves). The x-axis represents time and the y-axis represents waveform intensity*

To this day, two studies have used such an approach in order to process LiDAR waveforms [WAN 13, POU 15]. Given the fact that the wavelet coefficients are associated with a scale and location within the signal, the selection of the wavelet coefficients is supervised. Pourrahmati *et al.* [POU 15] have, for example, used the Haar wavelet decomposition of the ICESat/GLAS waveforms in order to estimate the volume and height of the trees in the forests of Iran on the basis of wavelet coefficients: the results of these approaches have also proven to be less efficient compared to more direct approaches or approaches based on expert metrics. More anecdotally, but by the same token, the Fourier spectral decomposition of LiDAR waveforms has also been used as a classification basis of forest species by [VAU 11].

6.2.4.3. *Statistical models*

The most common statistical models, when the target variable of interest is quantitative, are either linear models (multiple linear regression) or threshold regression models: regression trees [BRE 84] and random forests [BRE 01], this latter type of model being often more efficient in the predictive term. In the case of multiple linear models, the selection of the most suitable model is often effected by means of a step-by-step model selection procedure which maximizes the Akaike criterion. In the case of qualitative target variables [COL 08], the statistical models encountered are represented by the classification trees in the case of supervised classifications [BRE 84], or K-means in the case of non-supervised classifications [COL 08].

Allouis *et al.* [ALL 10] have, for example, used a regression tree (CART) using as predictors the two first components generated by a PCA of the HawkEye II bathymetric LiDAR waveforms in order to predict water depths below 1.5 m. However, Fayad *et al.* [FAY 14] have built multiple linear regression models in order to estimate tree height on the basis of IceSAT-GLAS waveforms, based on 41 main components (containing 94% of the baseline information) and expert metrics (the *Waveform extent*).

At first sight, a statistical regression approach in the case of LiDAR waveforms may be considered less desirable owing to the necessity of having a comprehensive terrain data base associated with the waveforms and representative of their diversity. However, this approach can represent one of the efficient thresholds when other methods (for example, peak detection) are not applied: this is, for example, the case when detecting very low depths at the level of the bathymetric LiDAR waveforms when two peaks cannot be distinguished [ALL 10]. In this case, it is necessary to consider these approaches based on LiDAR signal processing as terrain data interpolation methods [WAN 13]. When the target variable of the medium is qualitative (for example, benthic floor types), these approaches are more adapted [COL 08], especially when no physical model is provided making it possible to associate a waveform characteristic (for example, intensity of the final peak) with the type of the studied target.

6.2.5. *Inversion methods of LiDAR equations*

One of the limitations of the statistical approaches is represented by the fact that these are often of the monovariable type. A more physical approach based on the inversion of the equations of radiative transfer models (RTM) has the advantage of being capable of estimating several target variables of interest at the same time. This allows the unknown parameters of the LiDAR equations to be estimated. If the data sets do not make it possible to sufficiently constrain the estimation, other issues can arise, such as non-identifiability or equifinality problems, namely, obtaining several parameter sets as a solution to the problem (for example, solutions of the Pareto front). In the case of these inverse approaches, it is sought to optimize one or several equation parameters (which correspond to one or several studied variables of the target medium) such that the waveforms simulated by the equations correspond most closely to the data sets (recorded waveforms).

In methodological terms, the inversion methods are therefore parameter optimization methods which seek to minimize a criterion (for example, the sum of the squares of the deviations between the observed waveform and the waveform simulated by the parameter radiative transfer model) for an indicated parameter value. These methods include the set of monodimensional, deterministic (golden ratio, simple gradient, etc.), stochastic (simulated annealing and genetic algorithms) or multi-dimensional (multi-dimensional Newton, Levenberg–Marquardt, etc.) optimization methods (algorithms). Although this is a conventional approach in the case of atmospheric [KLE 85, CAR 86] or oceanographic [CHU 14] LiDAR waveforms, it is less often encountered in the case of LiDAR signals used in the observation of variables at the level of terrestrial targets, where the number of unknown variables and parameters of the target interacting with the LiDAR signal is often high and where the validated number of waveform models is low. According to an inversion method applied to a simplified radiative transfer model, Koetz *et al.* [KOE 06] have, for example, succeeded in deducing multiple forest variables at the same time (rate of ground vegetation, LAI, maximum canopy height and vertical crown extensions) on the basis of the large-footprint airborne LiDAR data collected in Switzerland.

6.3. 3D point cloud processing

The analysis of the 3D point cloud consists in characterizing each 3D LiDAR point by a set of features which will then make it possible to segment the point cloud into homogeneous regions, to recognize predefined shapes or to classify each 3D point or each 3D region and to assign it a predefined label in a given nomenclature.

6.3.1. *LiDAR Feature extraction*

Each 3D point is characterized by a certain number of features, mostly calculated in 3D, by analyzing the distribution of its neighboring 3D points having the same characteristics. The neighbors are obtained either by means of a "nearest neighbor" request or by means of inclusion in a neighborhood having a predefined form centered on the point of interest: the most common types are cylindrical (without constraints on the altitude neighboring points, it becomes a request in 2D, useful for the calculation of the features based on histograms or altitude differences) or even spherical, if it is only sought to recover all the points within a predefined distance. The form can also be selected after the distribution analysis of the 3D points located in a cylindrical neighborhood by default. A Principal Component Analysis makes it possible to determine if the neighborhood is predominantly spherical, planar or linear. The dimension of the neighborhood is the only parameter to be determined and it is subject to the density of the 3D points and the selected scale of analysis.

The fixed neighborhoods are easier to set up but they are not adapted to different contexts and when a mixture of objects is involved. Thus the derived geometric features may not be optimal. However, adaptive neighborhoods [FIL 05, DEM 11] take into account the distribution of 3D points, thus ensuring the reliability of the 3D neighborhoods (in the sense of a predefined geometric criterion such as the planarity or the orientation of the normals) and an improved calculation of the 3D features [UNN 10]. However, these require an extended calculation time.

In practice and in the particular case of a fusion with optical imagery, the features of the 3D points are often interpolated to a regular 2D grid in order to facilitate the management of large volumes of data and the high dimensionality (number of features) of the 3D points. Multi-echo LiDAR features (namely derived from 3D points) can be classified into four categories. They can be based on:

– altitudes, heights;

– multiple backscattering (echoes);

– local 3D geometry;

– LiDAR intensity.

In the case of FW LiDARs, specific features derived from the waveforms can also be included.

6.3.1.1. *Height-based features*

These features appear to be the most discriminating in the urban environment [GUO 11, ME 14] because they make it possible to separate high-rise, low-rise and ground elements (for example, in order to separate vegetation and buildings). Thus, the difference in altitude between the point of interest and the lowest point in its neighborhood (ground/off-ground discrimination), the difference between the first and the last echo of the point of interest (ground/roofs), the local slope (ground/roofs) or the local altitude variance, which returns high values in the case of vegetation, are taken into consideration.

In addition, other features can be calculated on the basis of altitude histograms in a cylindrical neighborhood. Statistical features can be derived such as the mean, median and the different quartiles. They make it possible to filter the ground and off-ground points, to detect the different strata in a forest [HER 10] or even to estimate the forest biophysical variables [VAN 08, TON 11].

6.3.1.2. *Echo-based features*

The simplest features consist in using the first and last echoes of the same waveform or in a given neighborhood. The first echoes correspond to the first objects encountered: roofs, top of canopy or even uncovered ground. The last echoes correspond to the last objects having backscattered the LiDAR signal. In the case of roofs and uncovered ground (such as roads), a single echo per LiDAR pulse is generally obtained. In contrast, in vegetated environments, the number of echoes per LiDAR pulse will depend on foliage density and the type of vegetation. As a result, it becomes a useful variable for the classification of tree species [KOR 10]. In the case of very dense vegetation, the LiDAR signal cannot penetrate the canopy and as a consequence, the last echo does not necessarily correspond to the ground. It is important to take into account this characteristic when filtering ground/off-ground points by using the number of the echo (see Chapter 7).

Other features are based on the total number of echoes and the position of the current echo in the waveform, such as:

– the total number of echoes of the waveform; which returns high values in the case of vegetation and generally a double-echo for building facades in the urban environment;

– the normalized number of echoes; obtained by dividing the number of the current echo by the total number of echoes in the waveform. This feature returns high values in the case of vegetation.

6.3.1.3. *3D geometry-based features*

A variance-covariance matrix is calculated at the level of the 3D points of a spherical neighborhood (whose dimensions can be set for the whole scene of interest with the knowledge of mean point density). The obtained eigenvalues $\lambda_1 > \lambda_2 > \lambda_3$ are used as features. In addition, they can be combined in order to generate other features used to describe the local spatial distribution of the 3D points in a small neighborhood and which are associated with the dimensionality of the neighborhood: 3D (spherical), 2D (planar) or 1D (linear) [GRO 06]. The best-known examples are:

$$\begin{aligned} &\text{Anisotropy} = A_\lambda = \frac{\lambda_1 - \lambda_3}{\lambda_1} \\ &\text{Sphericity} = S_\lambda = \frac{\lambda_3}{\lambda_1} \\ &\text{Planarity} = P_\lambda = \frac{\lambda_2 - \lambda_3}{\lambda_1} \\ &\text{Linearity} = L_\lambda = \frac{\lambda_1 - \lambda_2}{\lambda_1} \end{aligned} \qquad [6.7]$$

λ_3 returns low values in the case of planar objects (ground, building roof tops, see Figure 6.8) and high values in the case of non-planar objects (vegetation, building edges, ridges, etc.). In addition, if only λ_1 is high, then reference is made to a linear distribution of points for example (power lines).

The anisotropy, sphericity, planarity and linearity features are classified between 0 and 1. Planarity provides high values in the case of planar objects [DEM 11]. However, sphericity returns high values in the case of 3D neighborhoods whose spatial distribution of points is isotropic (Figure 6.8).

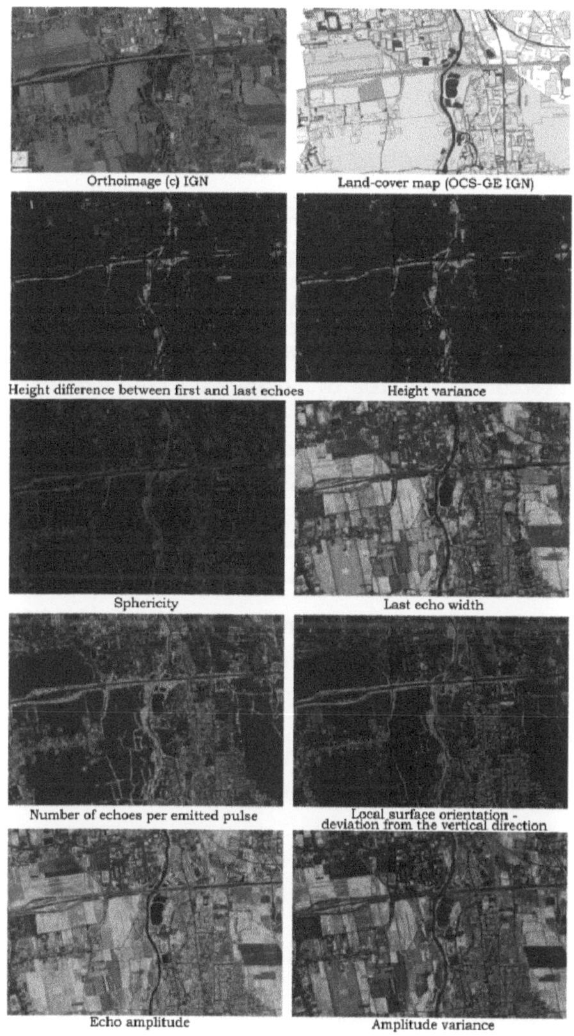

Figure 6.8. *Examples of LiDAR features in a peri-urban area (2 km × 2.5 km) top view. Each image pixel is calculated by using all the 3D points included in a cylindrical neighborhood of 5 m around the pixel of interest. The minimum values are represented in blue, and the highest values are represented in red. For a color version of this figure, see www.iste.co.uk/baghdadi/1.zip*

Other features making it possible to discriminate planar objects (such as buildings or roads) from vegetation can also be calculated by estimating a local 3D plane in the 3D neighborhood of the current 3D point: surface normal, verticality, 3D plane estimation residues and altitude variance distributions enable discrimination of planar surfaces in general, ground or facades in the case of the most specialized features. They are also commonly used in the classification and segmentation of terrestrial point clouds (Chapter 2, Volume 5).

6.3.1.4. *LiDAR intensity-based features*

In addition to the 3D position of the LiDAR point, the intensity feature is increasingly used to discriminate a certain number of categories of interest (ground types and tree species) as well as to perform accurate adjustments between 3D point clouds and optical images [MAS 09, GRE 13]. In the remote sensing community, the term *amplitude* is used to describe the value of the backscattered energy for the detected echo, whereas the intensity corresponds to the energy of the echo at the level of this waveform (namely, to the integration of all amplitude variables corresponding to this echo). A distinction is often not made between these terms, which is rendered even more complicated by the fact that:

– few LiDAR sensors record the full waveforms;

– the manufacturers of standard multi-echo sensors rarely indicate how this amplitude value is obtained (which evolves as a function of the echo detection method integrated in the system itself [WAG 04]).

If the echo has been modeled by means of a Gaussian function, then the intensity I is expressed as a function of the amplitude A and the echo width w:

$$I = \sqrt{2\pi}Aw \quad [6.8]$$

Prior to any implementation, this feature needs to be calibrated and corrected by using internal and external factors of the LiDAR sensor. Atmospheric fluctuations and energy variations during the emission of the laser beam can be attenuated by the calibration process implemented at the level of the surveyed area and LiDAR strip [WAG 08]. The variations in the backscattered energies at the level of the sensor, which are due to the geometry of the reached objects or the distance to the plane, are included in the correction methods [HOF 07]: the latter locally analyzes the neighborhood of each 3D point for the purpose of achieving a correction adapted to each point.

If the targets are Lambertian, then $\rho = \rho_m \cos(\alpha)$, where ρ_m is the reflectance of the material and α is the angle of incidence of the LiDAR beam with the target. Thus, the backscattering cross-section can be expressed as:

$$\sigma = \pi \rho_m cos(\alpha) \beta^2 R^2 \qquad [6.9]$$

β is the divergence of the LiDAR beam.

By returning to the LiDAR equation, and assuming that the LiDAR echoes can be modeled using a Gaussian function, the following may be obtained:

$$\sigma = C_{cal} R^4 A w \text{ with } C_{cal} = \frac{\pi \rho_m \beta^2}{R^2 A_m w_m} \qquad [6.10]$$

A and w represent the estimated amplitude and echo width, whereas A_m and w_m represent the estimated amplitude and the width in the case of an echo of the material m characterized by a known reflectance value ρ_m. In practice, ρ_m is determined using the known spectral databases, and A_m and w_m are obtained by means of the average of a large number of 3D points detected on the calibration surface of interest.

These methods have also been applied within the context of 3D points generated by mobile mapping systems [HAR 15] and thus a main obstacle is yet to be overcome. The main methods developed thus far are based on the hypothesis that the targets have a Lambertian behavior. Nevertheless, significant variations in amplitude as well as specular effects are observed in the same region, including at constant angles of incidence. Although more advanced *bidirectional reflectance distribution function* (BRDF) models have been selected (such as the Phong model [COO 81]), no significant improvements could be recorded [JUT 10]. The types of targets generally used in the literature, such as asphalt roads, are not optimal, due to the variations in backscattered intensity when the angles are constant and reflectance values are low (0.1–0.15), which is detrimental in the case of the sensors characterized by a low signal-to-noise ratio.

6.3.1.5. *Waveform-based features*

As indicated in section 6.2, the extracted features fall into two categories:

– those derived from an echo decomposition of the waveform (Figure 6.2);

– those derived from the analysis of the signal on the whole.

The features extracted from an echo decomposition are morphological: they are either derived from the local analysis of the LiDAR signal around the echo [LAK 10], or directly equal to the parameters of the selected analytic function used to fit the waveform [MAL 11]. The main features are:

– *amplitude*: high amplitudes are recorded for building roofs, gravel and cars. Low values are recorded at road level. The lowest values are recorded at the level of vegetation and are due to a significant heterogeneity of the target and significant attenuation of the signal (Figure 6.8). Variations occurring between tree species have been recorded and are used in the classification of forest environments [REI 08, HEI 11];

– *width*: in the case of small footprint sensors, this feature provides information on the roughness and altimetric scattering of the targets in spatial proximity and which are indistinguishable using the resolution of the sensor [NEU 09]. Higher values are recorded at the level of vegetation. The width of an echo also increases along with the angle of incidence with the reached surface. A low echo width has a high probability of being associated with buildings or even ground level in the urban environment. For example, it may be noted that a positive correlation has been observed between echo width and vegetation height [ROG 15];

– *high-order moments*: skewness and kurtosis [LAK 10, PAR 14, RON 14].

– *asymmetry*: the majority of echoes do not have a symmetrical form, as described in [MAL 10]. This feature has been used for the classification in urban environments, but its relevance in discriminating ground, buildings and vegetation is yet to be demonstrated [MAL 11]. The latter is more relevant in the case of coastal, riverine and forest environments [PAR 14];

– *backscattering cross-section* [WAG 08]: defined as the product of a constant (obtained by means of calibration) with the width and amplitude of an echo (see equation [6.10]). It makes it possible to derive a normalized backscattering cross-section σ^0 (independence of the laser beam with respect to the illuminated area) as well as the backscattering coefficient γ (correction of the angle of incidence).

$$\gamma = \frac{4\sigma}{\pi R^2 \beta^2} \qquad [6.11]$$

Guo *et al.* [GUO 11] and Mallet *et al.* [MAL 11] have demonstrated the advantages of computing the backscattering cross-section for 2D and 3D classification in urban environments, whereas this is demonstrated by Alexander *et al.* [ALE 10] and Fieber *et al.* [FIE 11] in the case of the backscattering coefficient.

A number of studies are also based on the understanding of the full waveform in order to extract the features related to quantiles and the distance between echoes, which are highly relevant for the classification of tree species [YU 14, HOV 16], the extraction of the main vegetation strata [HOF 12], the detailed characterization of the vertical structure of forest plots [CAO 14] and, in bathymetry, to obtain information on suspended matter (turbidity, nature of material, etc.).

6.3.2. *Filtering and segmentation of 3D point clouds*

The 3D point cloud filtering methods for generating Digital Terrain Models are described in more detail in Chapter 7, Volume 1. In this context, exclusive focus is placed on 3D points at the level of the ground surface, whereas all the other targets are not considered.

However, segmentation techniques seek to partition the 3D point cloud into homogeneous regions. The homogeneity criteria differ in terms of the final objective and can be applied to any feature derived from the 3D LiDAR data. The criterion can be associated with altitude, for example when forest strata are detected, to planarity, for example when roads or sloping roofs are detected, or even sphericity when tree crowns are extracted.

Two objectives are possible:

– oversegmenting the point cloud, i.e. partitioning it into small segments to simplify the subsequent shape and object detection stages, or for the purpose of its semantization. This step is often prioritized in the processing chains, because LiDAR data can consist of large volumes. Standard data analysis and image processing techniques are therefore adopted: k-means, k-NN (k- Nearest Neighbors), nearest neighbors, region growing, watershed, mean-shift by using the 3D geometry of the point cloud and in many cases a planarity/curvature and/or intensity criterion;

– directly detecting the objects of interest: tree crowns in forest environments, building roofs, cars, power lines, etc. The analysis scale is the same as that of the objects of interest (Figure 6.9). In the case of a fine discrimination, this step can also be followed by a classification step (see section 6.3.4).

The segmentation of 3D point clouds is essentially used in urban environments to detect urban objects for the purpose of classification: (the production of land-cover maps, urban 3D modeling) or even in the forest environment for the identification of individual tree crowns (ITC, see Figure 6.9). In forested environments, LiDAR data characterized by a low density (~5–10 pts/m^2) make it possible to estimate certain forest variables, to detect the different strata of a forest [JAS 10, FER 12] and to delimitate tree crowns [STR 15]. The segmentation can be carried out in 2D when using the LiDAR features projected in 2D, a Digital Surface Model (DSM) in the urban environment or a Canopy Height Model (CHM) in forested environment. Thus, all conventional image segmentation methods such as region growing, k-means, mean-shift, graph cuts, based on the intra-region spectral homogeneity criteria or sharp gradients between region boundaries, can be identified.

Figure 6.9. *An example 3D point clouds segmentation: automatic extraction of tree crowns in the tropical environment. All 3D points contained in the same region are displayed using a random unique color. The mean-shift method is described in more detail in [FER 12] and in this section. For a color version of this figure, see www.iste.co.uk/baghdadi/1.zip*

The majority of these methods have been adapted to the segmentation of 3D point clouds.

A region growing algorithm using seeds as a basis and to which a neighborhood expansion criterion is applied may be identified among the most common segmentation methods of 3D point clouds The seeds can be initialized using local altitude minima or maxima. A homogeneity hypothesis of the studied 3D region is specified. For example, a threshold imposed on the mean height value or the variance of the criterion used (altitude, density, planarity, sphericity, etc.) is commonly used. The homogeneity hypothesis is verified in an iterative manner by adding new points until the convergence is obtained. The disadvantage of such a parametric approach is that it cannot be adapted to the different types of objects encountered in a complex scene.

Other segmentation methods are based on the partitioning of the point cloud (*clustering*), which corresponds to a non-supervised classification technique. In practice, the k-means algorithm is commonly used. It makes it possible to automatically partition the 3D points into homogeneous groups by affecting the 3D points to the nearest group/cluster in an iterative manner. The principle is to minimize the inter-cluster similarity and to maximize the intra-cluster similarity. The user must define the number of clusters and specify the used distance. The Euclidean distance is most commonly used. The inconvenience of this method is represented by its dependence on the initialization of the cluster centroids. The k-means can be applied in a hierarchical order so as to initially filter the ground/off-ground points and subsequently the different objects above the ground [CHE 08].

Other approaches make it possible to carry out a non-parametric segmentation of 3D point clouds, such as the mean-shift algorithm (in the absence of a hypothesis on the distribution of 3D points in the feature space). This has been initially proposed by Fukunaga and Hostetler [FUK 75] and subsequently adapted to image segmentation [COM 02]. It has then been applied to 3D point cloud segmentation in order to detect power lines, urban objects [MEL 07] or even forest strata [FER 12].

The mean-shift algorithm is a non-parametric analysis method of multi-modal spaces for the delimitation of partitions (or clusters) of any shape. The mean-shift method is used to group the points by associating them with a local mode estimated, namely, to a density peak in the local histogram. This method explores the probability density function of the feature or features considered in a local neighborhood using both spatial and feature information. It is based on the density estimation technique at the level of each kernel, which makes it possible to estimate the probability density function of a random variable x which is distributed in a d dimensional space R^d. Each point x_i contributes according to its distance to the center of the local considered neighborhood. The least complex estimation of the density function $\acute{p}(x)$, while considering a discontinuous set of samples, is to smooth this data set, for example, by convolving it with a kernel of fixed dimensions. Once $\acute{p}(x)$ has been estimated, optimization techniques, such as the gradient descent, make it possible to identify these local maxima.

$$\acute{p}_{h,k}(x) = \frac{1}{nh^d} \sum_{i=1}^{n} K\left(\frac{x - x_i}{h}\right) \qquad [6.12]$$

With n representing the number of samples of the random variable, K the kernel function and h the width of the kernel which influences the bias and variance of the density estimator: a low value of h will result in a low bias but in a high variance (the estimations will be correct on average, but these will vary according to each sample). A high value of h will reduce the variance of the estimator, but will in turn increase the bias. In the case of remote sensing, the width of the kernel is directly related to the dimensions of the objects of interest, given the fact that the spatial information is integrated in the segmentation process: a highly reduced width lead, for example, to an oversegmentation of the data.

The most commonly used kernel is the Gaussian kernel:

$$K(u) = \frac{exp\left(\frac{-1}{2}\|u\|^2\right)}{\left(\sqrt{2\pi}h\right)^d} \qquad [6.13]$$

It is known to provide accurate results despite its very slow convergence with respect to the modes, as opposed to another popular kernel, Epanechnikov, which converges in a finite number of steps: $K_E(u) = max(0,1 - u)$. It is assumed that the kernel function $K(u)$ is radially symmetric, and is only dependent on the norm of u. A profile kernel function $K_{pr}(u), u \geq 0$ may be identified such as $K(u) = AK_{pr}(\|u\|^2)$ in which A is a normalization constant which guarantees that $K(u)$ integrates to 1.

The mean-shift algorithm will determine the local maxima (modes) of $\acute{p}_k(x)$ which correspond to the zero values of gradient $\nabla\acute{p}_k(x) = 0$. In the case of large dimensional spaces, conducting an evaluation $\acute{p}_k(x)$ at the level of the entire search space becomes highly complex from a computational perspective. Thus, the algorithm employs the multiple gradient descent techniques.

By considering a new kernel $\theta_{pr(.)} = K'_{pr}(.)$ such as $G(u) = BG_{pr}(\|u\|^2)$, the following may be obtained:

$$\nabla\acute{p}_k(x) = \frac{2A}{nh^{d+2}}\acute{p}_{h,\theta}(x)ms(x) \qquad [6.14]$$

with:

$$ms(x) = \frac{\sum_{i=1}^{n} x_i\theta_{pr}(\|(x-x_i)/h\|^2)}{\sum_{i=1}^{n} \theta_{pr}(\|(x-x_i)/h\|^2)} - x \qquad [6.15]$$

The mean-shift vector corresponds to the difference between the mean values of the distances of samples x_i to the center x by using kernel θ and the input vector x. This mean-shift vector can be expressed in the following manner:

$$ms(x) = \frac{Bh^2}{2A}\frac{\nabla\acute{p}_k(x)}{\acute{p}_\theta(x)} \qquad [6.16]$$

As a result, the mean-shift vector follows the direction of the maximum gradient of the density function. A recursive shift along the mean-shift vector makes it possible to achieve a local maximum:

$$x(t+1) \leftarrow x(t) + ms(x) \qquad [6.17]$$

with t corresponding to the number of iterations.

Once the mode location has been determined, the region (or cluster) associated with this mode is delimited as a function of the local structure of the feature space.

In the case of 3D point clouds, by considering the planimetric (x,y) and altimetric (z) dimensions, the kernel can be separated into two kernels having two parameters of different widths, which can correspond to their footprint and the height/thickness of the objects $h_{xy}$and h_z, respectively.

This method has been successfully implemented in [FER 12] to delimit forest strata (horizontal segmentation) and subsequently tree crowns (vertical segmentation per stratum).

The advantage of this non-parametric method is that it does not require for the number of studied clusters to be defined *a priori.*

6.3.3. *Shape recognition: extraction of planes, spheres and cylinders*

The recognition of parametric shapes in a 3D LiDAR topographic point cloud is possible due to the low level of noise. Opaque surfaces (ground and buildings) are thus very well-defined geometrically [OUD 10]. The higher the density of the points, the easier it is to recognize complex shapes (roof types, building superstructures, cars, etc.); however, the complexity of the calculation increases along with the global arrangement procedure of the detected shapes. Specific methods are subsequently developed in the case of point clouds generated by terrestrial sensors and make it possible to automatically analyze large volumes of data for the purpose of object detection and surface reconstruction [MAT 14, OES 15].

The most widely implemented methods are RANSAC (*RANdom SAmple Consensus*) and 3D Hough transform. As in the case of the segmentation procedures presented in section 6.3.2, these methods are often based on the information provided on the intensity or local orientation of the point cloud (normals, see Figure 6.10). RANSAC is an iterative estimation method of the parameters of a mathematical model on the basis of a data set comprising artifacts (outliers). These outliers can never be adjusted to the model, regardless of the set of parameters, as opposed to the remaining data. The 3D Hough transform is particularly used to detect planes. It is used to identify the set of points which statistically represent the best planes, not in the sense of the least squares theory, but rather the planes containing the highest possible number of points. In addition to its increased calculation time, its main disadvantages are represented by its complex automatic parameterization and the fact that the set of selected points can often comprise several nearly coplanar but spatially distinct planes (ground, sloping roofs, etc.).

Figure 6.10. *Two examples of 3D point cloud segmentation. Left: orthoimage of an urban area. Segmentations by region growing according to altitude (center) and orientation of the common normals (right) criterion. One color per segment. For a color version of this figure, see www.iste.co.uk/baghdadi/1.zip*

These methods are often used for the reconstruction of building roofs: thus, it is preferable to initially proceed with a classification step to sharply reduce the calculation time [TOS 10] by proposing a focusing mask. In a more optimal manner, in the case of non-noisy or filtered 3D point clouds of the outliers, the planes can be extracted using a simple region growing at the level of surface normals, restricted by a given mean-square error [LAF 12]. Non-planar shapes (spheres, cylinders and cones) are more difficult to estimate because most of the time, the LiDAR point clouds cover only one portion of the total shape (some peculiar scanning mechanisms, such as circular beams, make it possible to describe the shapes in a more accurate manner). The least-squares methods, such as the Levenberg–Marquardt optimization technique, converge given the fact that the parameterization and the first-order approximation of the Euclidean distance proposed in [MAR 01] are used.

6.3.4. *Classification, synergy with other data sources*

Airborne LiDAR data are widely used for land-cover classification (applications using terrestrial and mobile data sets are described in Chapter 2, Volume 5). The data collected by topographic LiDAR sensors essentially provide information regarding the height, the 3D structure of the objects and the intensity measured in the near infrared domain. As described in section 6.3.1, different feature families can be derived. Full waveform LiDAR makes it possible to supplement these features by morphological features describing each echo.

The classification of LiDAR data is a preliminary step for large number of applications, as shown in Figure 6.11. Applications will essentially depend on the density of the LiDAR point clouds, as well as the analysis scale. The per-3D point or per-object labeling techniques used are the same as those employed in image processing, according to which the neighborhood and/or the additional elements are taken into account: Support Vector Machines Random Forests, Markov and Conditional Random Fields, etc. [WEI 15]

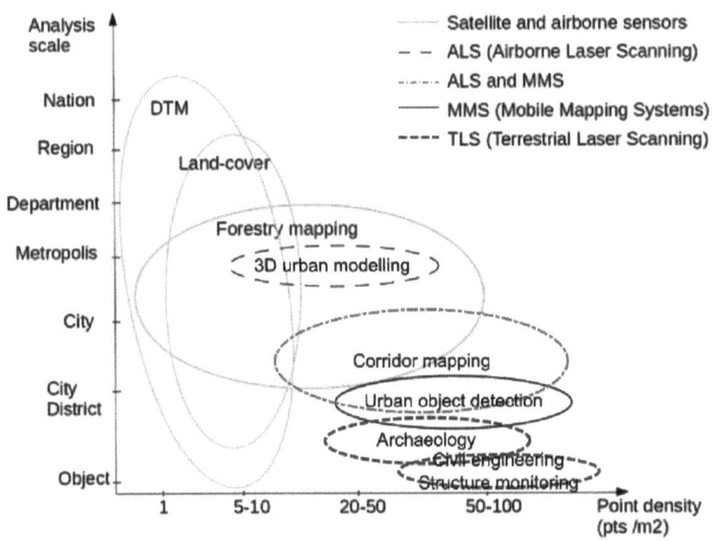

Figure 6.11. *Applications of LiDAR data as a function of point density and analysis scale. ALS: topographic or bathymetric airborne LiDAR, TLS: fixed topographic terrestrial LiDAR, MMS: LiDAR carried on board a mobile mapping vehicle. For a color version of this figure, see www.iste.co.uk/baghdadi/1.zip*

It can be observed that a high point density is required in the case of metrology, civil engineering, infrastructure or land slide monitoring applications, in archeology and for the 3D modeling of cities. A low density makes it possible to produce land cover maps with fairly coarse nomenclatures and large-scale Digital Terrain Model (DTM).

Forest mapping will encompass a wide range of point densities. In fact, low densities make it possible to produce fine DTM below vegetation canopies or to estimate forest variables such as canopy height for inventory purposes [MCR 07]. However, high point densities are required for the delimitation of trees, the detection of strata, the detection of dead trees or even the discrimination of tree species.

Figure 6.11 also shows the various platforms capable of carrying on board a mobile, airborne or satellite LiDAR sensor or a terrestrial LiDAR sensor. These platforms lead to different point densities and analysis scales, and are therefore intended for different applications.

The classification of LiDAR data is often used in the urban environment where ground and off-ground objects are easy to discriminate due to their average and variance of altitudes. Similarly, it is easy to discriminate vegetation due to the high number of echoes in such areas. In the urban environment, when the density of LiDAR points is low (1–3 pts /m^2), the nomenclature remains generalized and limited to standard topographic objects (buildings, roads, high vegetation, low vegetation, water, etc.) with 4–7 classes [YAN 15]. However, the LiDAR data do not allow separation of objects with the same altitude but made of different materials such as roads or rooftops. Thus, the information related to intensity values becomes relevant provided that it is corrected and accurately calibrated [NIE 14]. When the point density increases (>5 pts/m^2), new categories corresponding to the objects of small dimensions can be identified: hedgerows, cars, etc. [BOR 15], but they also require more advanced classification methods as opposed to the standard 3D point classification methods, such as support vector machines or random forests: graphic models (*Markov Random Field* and *Conditional Random Fields*) represent solutions making it possible to take into account the contextual information extracted from 3D data sets.

In the forest environment, the low-density LiDAR data (<5–10 pts/m^2) do not make it possible to classify vegetation species; instead they allow for quantitative analyses to be carried out on forest variables following the segmentation of the point cloud. These densities are nevertheless adapted to forest statistical inventories. In the case of higher point densities (>10 pts/m^2), the structure of trees is defined and it allows the classification of tree species. The accuracy of the classification is improved by 10% in the case of a very high point density (>40 pts/m^2) [LI 13].

In the coastal environment, it is possible to use LiDAR data exclusively for the classification of water/arid areas and coastline extraction (Chapter 6, Volume 5). An operational method has been proposed and validated on a large scale in [SME 13], using point density-based features.

Practically, the different study areas are more complex, and if the interest is focused on more refined nomenclatures, it is necessary to combine the LiDAR data with other optical (multispectral and hyperspectral) or radar data sources. In fact, LiDAR data are characterized by a substantial complementarity to multispectral or hyperspectral optical data. Optical images provide a high spatial resolution, a dense 2D information in a regular grid. The measured values of spectral reflectances cover a large wavelength range, from visible to short-wavelength infrared, by passing through the near and mid-infrared. Hyperspectral imaging provides a more refined spectral reflectance than multispectral imaging, as a result of the hundreds of spectral bands (each spectral band corresponding to a wavelength).

Although LiDAR provides information on the 3D structure and the spatial arrangement of objects, it also provides irregular 2D sampling and intensity measurements in the near-infrared or medium green, respectively, in the case of a topographic or bathymetric LiDAR. In addition, the LiDAR has the capacity to penetrate the vegetation until reaching the ground, thus returning information on vegetated regions (for example, roads underneath forest canopies) which cannot be accessed only by means of optical images.

The fusion of multi-source data requires data adjustment. LiDAR data are obtained by means of direct georeferencing; the positional precision is lower than 0.5 m. The georeferencing of optical images is more complex. This requires the presence of ground control points and can result in severe residual errors in the case of high reliefs, for example. Knowing the external orientation parameters of the optical sensors (available in the case of a direct georeferencing hybrid system constituted by a GPS and an inertial unit), two methods may be employed in order to adjust LiDAR and optical data:

– rasterization of LiDAR points on the image grid. This method consists of reprojecting LiDAR points in 2D and interpolating them in order to obtain a 2D altitude or LiDAR intensity image. This represents the most commonly used method because it facilitates the application of all 2D image processing techniques. However, this operation leads to a loss of information and precision due to the interpolation techniques;

– use of 3D LiDAR points by assigning them the corresponding spectral features derived from the optical image. The direct terrain to image transformation makes it possible to connect each LiDAR point to the corresponding pixel in the image. This technique makes it possible to benefit from the 3D structure of LiDAR data; however, only the first echoes will have image features, as these will become visible in the image.

Once the data sets have been adjusted, the fusion of optical and LiDAR data can occur at three different levels:

– at data level, which requires a physical modeling and a calibration of the signal (atmospheric, radiometric corrections and directional effects). However, it is not always clear how this fusion may be carried out and this aspect is not widely addressed in the literature;

– at feature level by combining multi-source features derived from each type of data. In the case of LiDAR data, the features described in section 6.2 can be calculated. However, in particular, vegetation, brightness and water spectral indices as well as textural features can be extracted on the basis of optical data. Feature selection methods for the purpose of classification then make it possible to retain the best features for each class. Chehata *et al.* [CHE 09] and Guo *et al.* [GUO 11] have indicated the relevance of multispectral and full waveform LiDAR data fusion for urban classification, respectively in 2D and 3D. Two fusion methods were tested: (1) the fusion is carried out by assigning the corresponding multispectral features to each 3D LiDAR point, and (2) the LiDAR points have been projected and interpolated in 2D. The studies conducted in the bathymetric context have not demonstrated the relevance of such fusion;

– at the level of the final products, which are obtained separately by using Bayesian fusion techniques [DUD 73] or the Dempster–Shafer theory

[POU 11] in the case of each data source. A methodological panorama is presented in [ZHA 10].

In practice, LiDAR data represent the type of data which are most often fused with hyperspectral data. The hyperspectral data provide a very fine spectral reflectance, measured by wavelength, and thus make it possible to finely discriminate between the categories having similar 3D structures but different spectral characteristics.

Initially, the hyperspectral/LiDAR fusion is used in the urban environment in order to discriminate the different types of urban materials. Subsequently, it is used in order to exploit the 3D geometry of buildings and thus correct the measured spectral reflectance values by means of the hyperspectral sensor: in the case of land-cover classification methods, they can be more robust to illumination conditions and shade problems [ZHA 13].

The hyperspectral and LiDAR data are essentially fused in the forest environment for the purpose of estimating forest variables [FIN 13] and discriminating tree species [DAL 12, DAL 14]. A low point density (~4 pts /m^2) makes it possible, once a digital surface model (DSM) has been generated, to segment tree crowns. The hyperspectral information is then used in order to characterize and categorize the obtained regions. An overview of the hyperspectral/LiDAR fusion methods in the case of forest ecosystems is presented in detail in [TOR 14].

The fusion of LiDAR and radar data has rarely been addressed in the literature up to this point and it is mostly based on the estimation and monitoring of the forest resource [HYD 07], and the generation of a DSM in the urban environment [GAM 03] by means of LiDAR and radar DSM product fusion. Other fusion works have been carried out at the feature level for the classification in a natural environment [BAR 15]. The relevance of data fusion has been significant for forestry and agricultural categories.

6.4. Perspectives

The perspectives are primarily related to the technological developments of the LiDAR sensors and to the capacity of free softwares, open sources and commercial solutions to[1], process large volumes of data and extract

1 http://potree.org.

increasingly relevant and specific features for a larger range of applications (detection of power lines, reconstruction of buildings, digital terrain models and extraction of hedgerows) and particular environments (urban, coastal and forest).

The increasingly sampling of the terrestrial surface by means of the airborne sensors facilitates the detection and monitoring of increasingly fine objects: corridors (rails, power lines and dams), hedgerows, cars, dead trees below the forest canopy, etc. This will be accentuated by:

– the installation of LiDAR systems on UAVS in order to facilitate monitoring on an even larger scale (but thus excessively small surfaces);

– airborne surveys covering increasingly wide territories;

– the provision of 3D data on various dates (and on a country scale such as Switzerland or the Netherlands). Subsequently, this must be supplemented by a better structured management of storage and processing of large volumes of data by employing *data mining* or *Big Data* techniques [BOE 15].

Thus, the emergence of commercial multispectral LiDAR sensors supporting two or three wavelengths may be noted. Initially designed for dual application in topography and bathymetry, they could facilitate a better discrimination of land-cover categories (tree species, types of ground and roof materials), as shown in the first studies conducted using experimental hyperspectral LiDAR sensors [HAK 15].

6.5. Key points

– the raw LiDAR data are either represented by time signals referred to as full waveform, or 3D point clouds characterized by a high altimetric quality;

– they are acquired by single-wavelength geospatial and terrestrial sensors, despite the emergence of multispectral sensors;

– LiDAR sensors make it possible to acquire data below the forest canopy, in particular owing to the separability of the LiDAR beam generating multiple backscattering, referred to as echoes, in the case of the same laser pulse;

– processing stages are required in the case of LiDAR waveforms in order to obtain the information related to the distance between the reached targets and the sensor and to extract the 3D point clouds;

– additional information referring to the backscattered energy is often available in addition to the 3D position and number of echoes;

– the majority of remote sensing data analysis techniques (segmentation, pattern recognition and classification) are applied to LiDAR data sets, in three dimensions, by integrating the geometric features related to the reached targets;

– the point density restricts the applications relating to LiDAR data processing: more or less dense sampling of the terrestrial surface does not make it possible to detect nor recognize all its constituent objects.

6.6. Bibliography

[ABD 13] ABDALLAH H., BAILLY J.S., BAGHDADI N. *et al.*, "Potential of space-borne LIDAR sensors for global bathymetry in coastal and inland waters", *IEEE Journal of Selected Topics in Applied Earth Observations and Remote Sensing*, vol. 6, no. 1, pp. 202–216, 2013.

[ALE 10] ALEXANDER C., TANSEY K., KADUK J. *et al.*, "Backscatter coefficient as an attribute for the classification of full-waveform airborne laser scanning data in urban areas", *ISPRS Journal of Photogrammetry and Remote Sensing*, vol. 65, no. 5, pp. 423–432, 2010.

[ALL 10] ALLOUIS T., BAILLY J.-S., PASTOL Y. *et al.*, "Comparison of LiDAR waveform processing methods for very shallow water bathymetry using Raman, near-infrared and green signals", *Earth Surface Processes and Landforms*, vol. 35, no. 6, pp. 640–650, 2010.

[BAG 11] BAGHDADI N., LEMARQUAND N., ABDALLAH H. *et al.*, "The relevant of GLAS/ICESAT elevations data for the monitoring of river networks", *Remote Sensing – Sensors*, vol. 3, pp. 708–720, 2011.

[BAG 14] BAGHDADI N., LE MAIRE G., FAYAD I. *et al.*, "Testing different methods of forest height and aboveground biomass estimations from ICESat/GLAS data in eucalyptus plantations in Brazil", *IEEE Journal of Selected Topics in Applied Earth Observations and Remote sensing*, vol. 7, no. 1, pp. 290–299, 2014.

[BAR 15] BARBANSON C., MALLET C., GRESSIN A. *et al.*, "Fusion of LiDAR and SAR data for land-cover mapping in natural environments", *Proceedings of the IEEE International Geoscience and Remote Sensing Symposium*, Milan, Italy, 26–31 July 2015.

[BOE 15] BOEHM J., LIU K., "NoSQL for storage and retrieval of large LiDAR data collections", *International Archives of Photogrammetry Remote Sensing and Spatial Information Sciences*, vol. XL-3/W3, pp. 577–582, 2015.

[BOR 15] BÖRCS A., BENEDEK C., "Extraction of vehicle groups in airborne LiDAR point clouds with two-level point processes", *IEEE Transactions on Geoscience and Remote Sensing*, vol. 53, no. 3, pp. 1475–1489, 2015.

[BOU 14] BOUHDAOUI A., BAILLY J.S., BAGHDADI N. *et al.*, "Modeling the water bottom geometry effect on peak time shifting in LiDAR bathymetric waveforms", *IEEE Geoscience and Remote Sensing Letters*, vol. 11, no. 7, pp. 1285–1289, 2014.

[BRE 84] BREIMAN L., FRIEDMAN J., STONE C.J. *et al.*, *Classification and Regression Trees*, CRC Press, 1984.

[BRE 01] BREIMAN L., "Random forest", *Machine Learning*, vol. 45, no. 1, pp. 5–32, 2001.

[CAO 14] CAO L., COOPS N.C., HERMOSILLA T. *et al* ., "Using small-footprint discrete and full-waveform airborne LiDAR metrics to estimate total biomass and biomass components in subtropical forests", *Remote Sensing*, vol. 6, no. 8, pp. 7110–7135, 2014.

[CAR 86] CARNUTH W., REITER R., "Cloud extinction profile measurements by LiDAR using Klett's inversion method", *Applied Optics*, vol. 25, no. 17, pp. 2899–2907, 1986.

[CHE 08] CHEHATA N., BRETAR F., "Terrain modeling from LiDAR data: Hierarchical K-means filtering and Markovian regularization", *ICIP*, pp.1900–1903, 2008.

[CHE 09] CHEHATA N., GUO L., MALLET C., "Airborne LiDAR feature selection for urban classification using random forests", *International Archives of Photogrammetry, Remote Sensing and Spatial Information Sciences*, vol. 38, no. Part 3/W8, pp. 207–212, 2009.

[CHU 14] CHURNSIDE J.H., SULLIVAN J.M., TWARDOWSKI M.S., "LiDAR extinction-to-backscatter ratio of the ocean", *Optics Express*, vol. 28, pp. 18698–18706, 2014.

[COL 08] COLLIN A., ARCHAMBAULT P., LONG B., "Mapping the shallow water seabed habitat with the SHOALS", *IEEE Transactions on Geoscience and Remote Sensing*, vol. 46, no. 10, pp. 2947–2955, 2008.

[COL 10] COLLIN A., LONG B., ARCHAMBAULT P., "Salt-marsh characterization, zonation assessment and mapping through a dual-wavelength LiDAR", *Remote Sensing of Environment*, vol. 114, no. 3, pp. 520–530, 2010.

[COM 02] COMANICIU D., MEER P., "Mean shift : a robust approach toward feature space analysis", *IEEE Transactions on Pattern Analysis and Machine Intelligence*, vol. 24, no. 5, pp. 603–619, 2002.

[COO 81] COOK R.L., TORRANCE K.E., "A reflectance model for computer graphics", *Proceedings of SIGGRAPH*, pp. 307–316, 1981.

[COR 03] CORNILLON P.A., SAINT-ANDRE L., BOUVET J.M. *et al.*, "Using B-splines for growth curve classification: applications to selection of eucalypt clones", *Forest Ecology and Management*, vol. 176, no. 1, pp. 75–85, 2003.

[DAL 12] DALPONTE M., BRUZZONE L., GIANELLE D., "Tree species classification in the Southern Alps based on the fusion of very high geometrical resolution multispectral/hyperspectral images and LiDAR data", *Remote Sensing of Environment*, vol. 123, pp. 258–270, 2012.

[DAL 14] DALPONTE M., ØRKA H.O., ENE L.T. *et al.*, "Tree crown delineation and tree species classification in boreal forests using hyperspectral and ALS data", *Remote Sensing of Environment*, vol. 140, pp. 306–317, 2014.

[DEM 11] DEMANTKE J., MALLET C., DAVID N. *et al.*, "Dimensionality-based scale selection in 3D LiDAR point clouds", *International Archives of Photogrammetry, Remote Sensing and Spatial Information Sciences*, vol. 38, no. 5/W12, pp. 97–102, 2011.

[DUD 73] DUDA R., HART P., *Pattern Classification and Scene Analysis*, John Wiley & Sons, pp. 98–105, 1973.

[DUN 10] DUNCANSON L.I., NIEMANN K.O., WULDER M.A., "Estimating forest canopy height and terrain relief from GLAS waveform metrics", *Remote Sensing of Environment*, vol. 114, no. 1, pp. 138–154, 2010.

[FAN 04] FANG H.T., HUANG D.S., "Noise reduction in LiDAR signal based on discrete wavelet transform", *Optics Communications*, vol. 233, no. 1, pp. 67–76, 2004.

[FAY 14] FAYAD I., BAGHDADI N., BAILLY J.S. *et al.*, "Canopy height estimation in French Guiana using LiDAR ICESat/GLAS data", *Remote Sensing*, vol. 6, pp. 11883–11914, 2014.

[FER 12] FERRAZ A., BRETAR F., JACQUEMOUD S. *et al.*, 3-D mapping of a multi-layered Mediterranean forest using ALS data", *Remote Sensing of Environment*, vol. 121, pp. 210–223, 2012.

[FIE 13] FIEBER K.D., DAVENPORT I.J., FERRYMAN J.M. *et al.*, "Analysis of full-waveform LiDAR data for classification of an orange orchard scene", *ISPRS Journal of Photogrammetry and Remote Sensing*, vol. 82, pp. 63–82, 2013.

[FIL 05] FILIN S., PFEIFER N., "Neighborhood systems for airborne laser data", *Photogrammetric Engineering & Remote Sensing*, vol. 71, no. 6, pp. 743–755, 2005.

[FIN 13] FINLEY A.O., BANERJEE S., COOK B.D. *et al.*, "Hierarchical Bayesian spatial models for predicting multiple forest variables using waveform LiDAR, hyperspectral imagery, and large inventory datasets", *International Journal of Applied Earth Observation and Geoinformation,* vol. 22, pp. 147–160, 2013.

[FUK 75] FUKUNAGA K., HOSTETLER L., "The estimation of the gradient of a density function, with applications in pattern recognition", *IEEE Transactions on Information Theory*, vol. 1, pp. 32–40, 1975.

[GAM 03] GAMBA P., DELL'ACQUA F., HOUSHMAND B., "Comparison and fusion of LiDAR and InSAR digital elevation models over urban areas", *International Journal of Remote Sensing*, vol. 24, no. 22, pp. 4289–4300, 2003.

[GRE 13] GRESSIN A., MALLET C., DAVID N. *et al..*, "Towards 3D LiDAR point cloud registration improvement using optimal neighborhood knowledge", *ISPRS Journal of Photogrammetry and Remote Sensing*, vol. 79, pp. 240–251, 2013.

[GRO 06] GROSS H., THOENNESSEN U., "Extraction of lines from laser point clouds", *International Archives of Photogrammetry, Remote Sensing and Spatial Information Sciences*, vol. 36, no. 3, 2010.

[GUO 11] GUO L., CHEHATA N., MALLET C. *et al.*, "Relevance of airborne LiDAR and multispectral image data for urban scene classification using random forests", *ISPRS Journal of Photogrammetry and Remote Sensing*, vol. 66, no. 1, pp. 56–66, 2011.

[HAK 15] HAKALA T., NEVALAINEN O., KAASALAINEN S. *et al.*, "Technical note: multispectral LiDAR time series of pine canopy chlorophyll content", *Biogeosciences,* vol. 12, pp. 1629–1634, 2015.

[HAN 15] HANCOCK S., ARMSTON J., LI Z. *et al.*, "Waveform LiDAR over vegetation: An evaluation of inversion methods for estimating return energy", *Remote Sensing of Environment*, vol. 164, pp. 208–224, 2015.

[HAR 15] HARTZELL P.J., GLENNIE C.L., FINNEGAN D., "Empirical waveform decomposition and radiometric calibration of a terrestrial full-waveform laser scanner", *IEEE Transactions on Geoscience and Remote Sensing,* vol. 53, no. 1, pp. 162–172, 2015.

[HEI 11] HEINZEL J., KOCH B., "Exploring full-waveform LiDAR parameters for tree species classification", *International Journal of Applied Earth Observation and Geoinformation,* vol. 13, no. 1, pp. 152–160, 2011.

[HER 10] HERMOSILLA T., ALMONACID J., FERNÁNDEZ-SARRÍA A. *et al.*, "Combining features extracted from imagery and LiDAR data for object-oriented classification of forest areas", *International Archives of Photogrammetry, Remote Sensing and Spatial Information Sciences*, vol. 38, no. 4/C7, 2010.

[HOF 00] HOFTON M.A., MINSTER J.B., BLAIR J.B., "Decomposition of laser altimeter waveforms", *IEEE Transactions on Geoscience and Remote Sensing*, vol. 38, no. 4, pp. 1989–1996, 2000.

[HOF 07] HÖFLE B., PFEIFER N., "Correction of laser scanning intensity data: data and model-driven approaches", *ISPRS Journal of Photogrammetry and Remote Sensing*, vol. 62, no. 6, pp. 1415–1433, 2007.

[HOF 12] HÖFLE B., HOLLAUS M., HAGENAUER J., "Urban vegetation detection using radiometrically calibrated small-footprint full-waveform airborne LiDAR data", *ISPRS Journal of Photogrammetry and Remote Sensing*, vol. 67, pp. 134–147, 2012.

[HOV 16] HOVI A., KORHONEN L., VAUHKONEN J. *et al.*, "LiDAR waveform features for tree species classification and their sensitivity to tree- and acquisition related parameters", *Remote Sensing of Environment*, vol. 173, pp. 224–237, 2016.

[HYD 07] HYDE P., NELSON R., KIMES D. *et al.*, "Exploring LiDAR-RaDAR synergy – predicting aboveground biomass in a southwestern ponderosa pine forest using LiDAR, SAR and InSAR", *Remote Sensing of Environment*, vol. 106, no. 1, pp. 28–38, 2007.

[JAS 10] JASKERNIAK D., LANE P.N.J., ROBINSON A. *et al.*, "Extracting LiDAR indices to characterize multilayered forest structure using mixture distribution functions", *Remote Sensing of Environment*, vol. 115, pp. 573–585, 2010.

[JUT 06] JUTZI B., STILLA U., "Range determination with waveform recording laser systems using a Wiener filter", *ISPRS Journal of Photogrammetry and Remote Sensing,* vol. 61, no. 2, pp. 95–107, 2006.

[JUT 07] JUTZI B., Analyse der zeitlichen Signalform von rückgestreuten Laserpulsen, PhD thesis, Technical University of Munich, Germany, 2007.

[JUT 10] JUTZI B., GROSS H., "Investigations on surface reflection models for intensity normalization in airborne laser scanning (ALS) data", *Photogrammetric Engineering & Remote Sensing,* vol. 76, no. 9, pp. 1051–1060, 2010.

[KLE 85] KLETT J.D., "LiDAR inversion with variable backscatter/extinction ratios", *Applied Optics*, vol. 24, no. 11, pp. 1638–1643, 1985.

[KOE 06] KOETZ B., MORSDORF F., SUN G. *et al.*, "Inversion of a LiDAR waveform model for forest biophysical parameter estimation", *IEEE Geoscience and Remote Sensing Letters,* vol. 3, no. 1, pp. 49–53, 2006.

[KOR 10] KORPELA I., ØRKA H.O., MALTAMO M. *et al.*, "Tree species classification using airborne LiDAR – effects of stand and tree parameters, downsizing of training set, intensity normalization, and sensor type", *Silva Fennica*, vol. 44, no. 2, pp. 319–339, 2010.

[LAF 12] LAFARGE F., MALLET C., "Creating large-scale city models from 3D-point clouds: a robust approach with hybrid representation", *International Journal of Computer Vision*, vol. 99, no. 1, pp. 69–85, 2012.

[LAK 10] LAKY S., ZALETNYIK P., TOTH C., "Compressing LiDAR waveform data", *Proceedings of International LiDAR Mapping Forum*, Denver, USA, pp. 3–5, 2010.

[LEF 05] LEFSKY M.A., HARDING D.J., KELLER M. *et al.*, "Estimates of forest canopy height and aboveground biomass using ICESat", *Geophysical Research Letters*, vol. 32, p. L22S02, 2005.

[LES 07] LESAIGNOUX A., BAILLY J.S., ALLOUIS T. *et al.*, "Epaisseur d'eau minimale mesurable en rivière sur fronts d'ondes LiDAR simulés", *Revue française de photogrammétrie et de télédétection*, vol. 186, pp. 48–53, 2007.

[LI 13] LI J., HU B., NOLAND T.L., "Classification of tree species based on structural features derived from high density LiDAR data", *Agricultural and Forest Meteorology*, vol. 171–172, no. 15, pp. 104–114, 2013.

[LIN 09] LIN Y.C., Full waveform airborne laser scanning for DTM generation, PhD thesis, Newcastle University, 2009.

[LIN 10] LIN Y.C., MILLS J.P., SMITH-VOYSEY S., "Rigorous pulse detection from full-waveform airborne laser scanning data", *International Journal of Remote Sensing*, vol. 31, no. 5, pp.1303–1324, 2010.

[MAL 09] MALLET C., BRETAR F., "Full-waveform topographic lidar: state-of-the-art", *ISPRS Journal of Photogrammetry and Remote Sensing*, vol. 64, no. 1, pp. 1–16, 2009.

[MAL 10] MALLET C., Analysis of full-waveform LiDAR data for urban area mapping, PhD thesis, Télécom ParisTech, Paris, France, 2010.

[MAL 11] MALLET C., BRETAR F., ROUX M. *et al.*, "Relevance assessment of full-waveform LiDAR data for urban area classification", *ISPRS Journal of Photogrammetry and Remote Sensing*, vol. 66, no. 6, pp. S71–S84, 2011.

[MAR 01] MARSHALL D., LUKACS G., MARTIN R., "Robust segmentation of primitives from range data in the presence of geometric degeneracy", *IEEE Transactions on Pattern Analysis and Machine Intelligence*, vol. 23, no. 3, pp. 304–314, 2001.

[MAS 09] MASTIN A., KEPNER J., FISHER J., "Automatic registration of LIDAR and optical images of urban scenes", *IEEE Conference on Computer Vision and Pattern Recognition*, pp. 2639–2646, 2009.

[MAT 14] MATTAUSCH O., PANOZZO D., MURA C. *et al.*, "Object detection and classification from large-scale cluttered indoor scans", *Computer Graphics Forum*, vol. 33, no. 2, pp. 11–21, 2014.

[MCR 07] MCROBERTS R., TOMPPO E., "Remote sensing support for national forest inventories", *Remote Sensing of Environment*, vol. 110, no. 4, pp. 412–419, 2007.

[MEL 07] MELZER T., "Non-parametric segmentation of ALS point clouds using mean shift", *Journal of Applied Geodesy*, vol. 1, pp. 159–170, 2007.

[NEU 09] NEUENSCHWANDER A.L., MAGRUDER L.A., TYLER M., "Landcover classification of small-footprint, full-waveform LiDAR data", *Journal of Applied Remote Sensing*, vol. 3, no. 1, 2009.

[NIE 14] NIEMEYER J., ROTTENSTEINER F., SOERGEL U., "Contextual classification of LiDAR data and building object detection in urban areas", *ISPRS Journal of Photogrammetry and Remote Sensing*, vol. 87, pp. 152–165, 2014.

[OES 15] OESAU S., LAFARGE F., ALLIEZ P., "Planar shape detection and regularization in tandem", *Computer Graphics Forum*, vol. 35, pp. 203–213, 2015.

[OUD 10] OUDE ELBERINK S., Acquisition of 3D topography: automated 3D road and building reconstruction using airborne laser scanner data and topographic maps, PhD thesis, Delft University, 2010.

[PAR 09] PARRISH C.E., NOWAK R., "Improved approach to LIDAR airport obstruction surveying using full-waveform data", *Journal of Surveying Engineering*, vol. 135, no. 2, pp. 72–82, 2009.

[PAR 11] PARRISH C.E., JEONG I., NOWAK R.D. *et al.*, "Empirical comparison of full-waveform LiDAR algorithms: range extraction and discrimination performance", *Photogrammetric Engineering & Remote Sensing*, vol. 77, no. 8, pp. 825–838, 2011.

[PAR 14] PARRISH C.E., ROGERS J.N., CALDER B.R., "Assessment of waveform features for LiDAR uncertainty modeling in a coastal salt marsh environment", *IEEE Geoscience and Remote Sensing Letters*, vol. 11, no. 2, pp. 569–573, 2014.

[PEE 07] PE'ERI S., PHILPOT W., "Increasing the existence of very shallow-water LiDAR measurements using the red-channel waveforms", *IEEE Transactions on Geoscience and Remote Sensing*, vol. 45, no. 5, pp. 1217–1223, 2007.

[PFE 15] PFEIFER N., MANDLBURGER G., GLIRA P. *et al.*, "Lidar: exploiting the versatility of a measurement principle in photogrammetry", *Proceedings of the Photogrammetric Week*, Stuttgart, Germany, 2015.

[POU 11] POULAIN V., INGLADA J., SPIGAI M. *et al.*, "High-resolution optical and SAR image fusion for building database updating", *IEEE Transactions on Geoscience and Remote Sensing,* vol. 49, no. 8, pp. 2900–2910, 2011.

[POU 15] POURRAHMATI M.R., BAGHDADI N., DARVISHSEFAT A.A. *et al.*, "Capability of GLAS/ICESat data to estimate forest canopy height and volume in mountainous forests of Iran", *IEEE Journal of Selected Topics in Applied Earth Observations and Remote Sensing*, vol. 8, pp. 5246–5261, 2015.

[RAM 06] RAMSAY J.O., *Functional Data Analysis*, John Wiley & Sons, 2006.

[REI 08] REITBERGER J., KRZYSTEK P., STILLA U., "Analysis of full waveform LIDAR data for the classification of deciduous and coniferous trees", *International Journal of Remote Sensing*, vol. 29, no. 5, pp. 1407–1431, 2008.

[ROG 15] ROGERS J.N., PARRISH C. E., WARD L.G. *et al.*, "Evaluation of field-measured vertical obscuration and full waveform LiDAR to assess salt marsh vegetation biophysical parameters", *Remote Sensing of Environment*, vol. 156, pp. 264–275, 2015.

[RON 11] RONCAT A., BERGAUER G., PFEIFER N., "B-spline deconvolution for differential target cross-section determination in full-waveform laser scanning data", *ISPRS Journal of Photogrammetry and Remote Sensing*, vol. 66, no. 4, pp. 418–428, 2011.

[RON 14] RONCAT A., BRIESE C., JANSA J. *et al.*, "Radiometrically calibrated features of full-waveform LiDAR point clouds based on statistical moments", *IEEE Geoscience and Remote Sensing Letters*, vol. 11, no. 2, pp. 549–553, 2014.

[SME 13] SMEECKAERT J., MALLET C., DAVID N. *et al.*, "Large-scale classification of water areas using airborne topographic LiDAR data", *Remote Sensing of Environment*, vol. 138, pp. 134–148, 2013.

[STI 07] STILLA U., YAO W., JUTZI B., "Detection of weak laser pulses by full-waveform Stacking", *International Archives of Photogrammetry, Remote Sensing and Spatial Information Sciences*, vol. 36, pp. 25–30, 2007.

[STR 15] STRÎMBU V.F., STRÎMBU B.M., "A graph-based segmentation algorithm for tree crown extraction using airborne LiDAR data", *ISPRS Journal of Photogrammetry and Remote Sensing*, vol. 104, pp. 30–43, 2015.

[TON 11] TONOLLI S., DALPONTE M., VESCOVO L. *et al.*, "Mapping and modeling forest tree volume using forest inventory and airborne laser scanning", *European Journal of Forest Research*, vol. 130, no. 4, pp. 569–577, 2011.

[TOR 14] TORABZADEH H., MORSDORF F., SCHAEPMAN M.E., "Fusion of imaging spectroscopy and airborne laser scanning data for characterization of forest ecosystems – a review", *ISPRS Journal of Photogrammetry and Remote Sensing*, vol. 97, pp. 25–35, 2014.

[TOS 10] TOSHEV A., MORDOHAI P., TASKAR B, "Detecting and parsing architecture at city scale from range data", *IEEE Conference on Computer Vision and Pattern Recognition*, pp. 398–405, 2010.

[UNN 10] UNNIKRISHNAN R., LALONDE J.F., VANDAPEL N. *et al.* "Scale selection for geometric fitting in noisy point clouds", *International Journal on Computational Geometry and Applications*, vol. 20, no. 5, pp. 543–575, 2010.

[VAN 08] VAN AARDT J.A.N., WYNNE R.H., SCRIVANI J.A., "Lidar-based mapping of forest volume and biomass by taxonomic group using structurally homogenous segments", *Photogrammetric Engineering & Remote Sensing*, vol. 74, no. 8, pp. 1033–1044, 2008.

[VAU 11] VAUGHN A., MOSKAL L.M., TURNBLOM E.C., "Fourier transformation of waveform LiDAR for species recognition", *Remote Sensing Letters,* vol. 2, no. 4, pp. 347–356, 2011.

[WAG 04] WAGNER W., ULLRICH A., MELZER T. *et al.*, "From single-pulse to full-waveform airborne laser scanners: potential and practical challenges", *International Archives of Photogrammetry, Remote Sensing and Spatial Information Sciences,* vol. 35, no. B3, pp. 201–206, 2004.

[WAG 06] WAGNER W., ULLRICH A., DUCIC V. *et al.*, "Gaussian decomposition and calibration of a novel small-footprint full-waveform digitising airborne laser scanner", *ISPRS Journal of Photogrammetry and Remote Sensing*, vol. 60, no. 2, pp. 100–112, 2006.

[WAG 08] WAGNER W., HYYPPA J., ULLRICH A. *et al.*, "Radiometric calibration of full-waveform small-footprint airborne laser scanners", *International Archives of Photogrammetry, Remote Sensing and Spatial Information Sciences*, vol. 37, no. B1, pp. 163–168, 2008.

[WAN 09] WANG Y., ZHANG J., RONCAT A. *et al.*, "Regularizing method for the determination of the backscatter cross section in LiDAR data", *Journal of the Optical Society of America* A, vol. 26, no. 5, pp. 1071–1079, 2009.

[WAN 13] WANG C., TANG F., LI L. *et al.*, "Wavelet analysis for ICESat/GLAS waveform decomposition and its application in average tree height estimation", *IEEE Geoscience and Remote Sensing Letters*, vol. 10, no. 1, pp. 115–119, 2013.

[WEI 15] WEINMANN M., JUTZI B., HINZ S. *et al.*, "Semantic point cloud interpretation based on optimal neighborhoods, relevant features and efficient classifiers", *ISPRS Journal of Photogrammetry and Remote Sensing*, vol. 105, pp. 286–304, 2015.

[YAN 15] YAN W.Y., SHAKER A., EL-ASHMAWY N., "Urban land cover classification using airborne LiDAR data: a review", *Remote Sensing of Environment*, vol. 158, pp. 295–310, 2015.

[YU 14] YU X., LITKEY P., HYYPPÄ J. *et al.*, "Assessment of low density full-waveform airborne laser scanning for individual tree detection and tree species classification", *Forests*, vol. 5, no. 5, pp. 1011–1031, 2014.

[ZHA 10] ZHANG J., "Multi-source remote sensing data fusion: status and trends", *International Journal of Image and Data Fusion*, vol. 1, no. 1, pp. 5–24, 2010.

[ZHA 13] ZHANG Q., PAUCA V.P., PLEMMONS R.J. *et al.*, "Detecting objects under shadows by fusion of hyperspectral and LiDAR data: a physical model approach", *Proceedings of 5th Workshop Hyperspectral Image Signal Process: Evol. Remote Sens.*, pp. 1–4, 2013.

[ZHO 15] ZHOU Y., QIU F., AL-DOSARI A.A. *et al.*, "ICESat waveform-based land-cover classification using a curve matching approach", *International Journal of Remote Sensing*, vol. 36, no. 1, pp. 36–60, 2015.

[ZHU 14] ZHUANG W., MOUNTRAKIS G., "An accurate and computationally efficient algorithm for ground peak identification in large footprint waveform LiDAR data", *ISPRS Journal of Photogrammetry and Remote Sensing*, vol. 95, pp. 81–92, 2014.

[ZWA 02] ZWALLY H., SCHUTZ B., ABDALATI W. *et al.*, "ICESat's laser measurements of polar ice, atmosphere, ocean, and land", *Journal of Geodynamics*, vol. 34, no. 3, pp. 405–445, 2002.

7

Digital Terrain Models Derived from Airborne LiDAR Data

7.1. Introduction

This chapter focuses on the automatic generation of Digital Terrain Models (DTMs) on the basis of the three-dimensional (3D) point clouds generated by airborne topographic LiDAR sensors. A DTM is a mathematical representation (model) of the ground surface, most often in the form of a regular grid, in which a unique elevation value is assigned to each pixel. Thus, objects, including permanent off-ground components such as vegetation and buildings, can be detected. When the latter are included, reference is made to the digital surface model (DSM), used to analyze the ground surface visible from the sky (i.e. the surface which can be obtained natively in standard optical imagery).

The difference between DSM and DTM makes it possible to extract a product which provides a direct indication of the height of the objects with respect to the ground. In the LiDAR community, this is often referred to as the Canopy Height Model (CHM), as it is primarily used for forestry applications (Figure 7.1). It is more generally known as the digital elevation model (DEM).

LiDAR remains the preferred tool for the generation of DTM due to its precision in height measurement (< 15 cm) and the capacity of the laser beam to penetrate the forest canopy and subsequently to provide ground-

Chapter written by Clément MALLET and Nicolas DAVID.

based measurements below the canopy. LiDAR acquisitions for DTM generation are prioritized under "leaf-off" conditions, namely in winter, in order to minimize the presence of backscattering objects between the sensor and the ground (Figure 7.2).

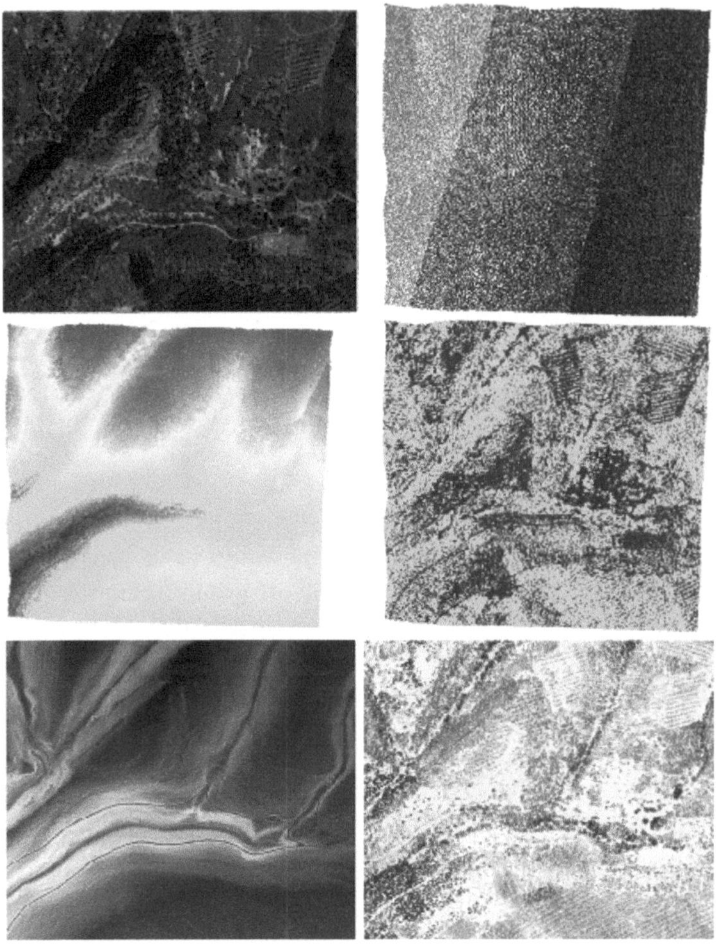

Figure 7.1. *Airborne LiDAR data processing for the generation of DTM. Top left: orthoimage of the area of interest (1 km x 1 km © IGN, Lat: 44.176° – Long: 5.322°). Top right: raw 3D LiDAR data colored according to the flight line (1 point/m²). Middle left: raw 3D data colored according to altitude (blue: 1180 m → red: 1340 m). Middle right: filtered 3D data (brown: ground - green: offground. Bottom left: Generated DEM (yellow → brown: 1180 m --> 1325 m) associated with the slope map. A very high spatial resolution of the product and the presence of roads in this area can be noted. Bottom right: Canopy Height Model (yellow: 0 → dark green: 15 m). For a color version of this figure, see www.iste.co.uk/baghdadi/1.zip*

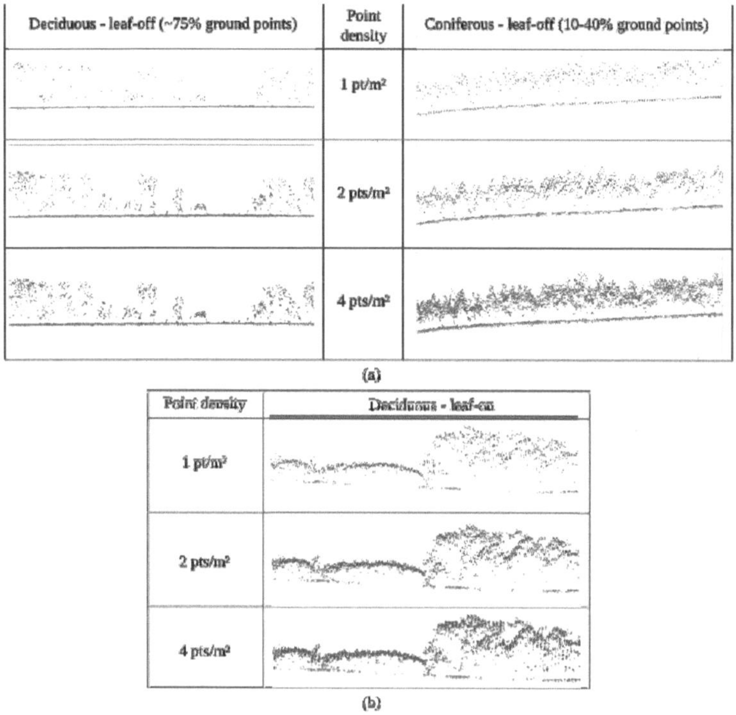

Figure 7.2. *Evolution of the LiDAR point distribution at ground level as a function of the emitted point density, season and type of obstacles (in this case two types of forests). 3D point clouds are colored with respect to the class obtained with an automatic filtering step (orange = ground; green = offground). Increasing the point density allows for a more accurate ground characterization when the vegetation is not too dense a), as opposed to when the vegetation is dense b). For a color version of this figure, see www.iste.co.uk/baghdadi/1.zip*

Commercial solutions have been identified in order to produce DTMs on a large scale and in an operational manner, and numerous countries have adopted this remote sensing technique in order to obtain a national or regional altimetric coverage characterized by high altimetric quality and spatial resolution: France, in particular within the framework of the Litto3D® program, Switzerland, the Netherlands, United Kingdom, Finland, Sweden, certain Spanish provinces and American states, etc. The applications serve many purposes and can be generally classified into two categories:

– applications which require reliable height information, for the reconstruction of terrain morphology or the detection of the high frequencies at the level of the relief (hydrology, archaeology, coastal areas – dunes, channels, erosion – or the estimation of the volume of mines);

– applications whose topographic information serves as the basis for advanced environmental measurements and modeling (extraction of forestry parameters, 3D reconstructions of urban areas or the detection of structures, such as bridges).

The generation of DTM from 3D LiDAR point clouds is conventionally divided into two steps (Figures 7.1 and 7.3):

– the separation of the points located off the ground from the ones located at ground level. Only the latter set is further retained, which explains why this step is often referred to as *filtering*;

– the interpolation of ground points in a regular grid in order to generate the final surface. This surface can be created by integrating additional external constraints such as break lines or existing topographic geodatabases.

A certain number of methods combine these two steps in order to obtain an iterative process in which the ground surface evolves as a function of the position of the 3D points in close proximity. Other methods initially prioritize the interpolation of LiDAR data in the form of DSMs in order to adopt standard image processing methods for off-ground object filtering. These are described hereby in conjunction with simple filtering methods.

The majority of existing methods can in addition be implemented in the case of 3D point clouds generated from other data sources such as photogrammetry and image matching [WAN 14, REE 15] or SAR tomography [SCH 15].

7.2. Filtering of 3D point clouds

7.2.1. *Introduction*

Filtering consists of separating the 3D points located at ground level from the ones which are located off the ground (Figure 7.2). In this case, it is assumed that outliers, which could be generated by the LiDAR systems and which sometimes appear below the ground surface, have been eliminated in advance (isolated altitude points below the cluster of 3D points characterizing the ground surface). It is also hypothesized that LiDAR data generated from multiple consecutive overlapping acquisitions (referred to as strips) have been registered. This makes it possible to obtain a denser terrain sampling, thus reducing the number of occluded areas, given the fact that multiple points of view are available for the same portion of the observed terrain.

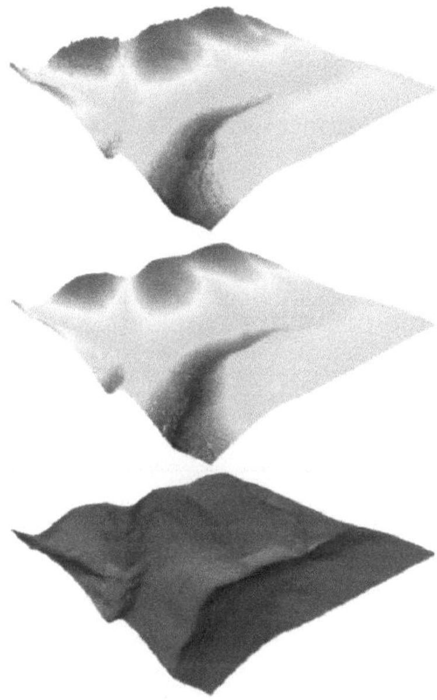

Figure 7.3. *Standard processing chain for DTM generation on the basis of 3D LiDAR point clouds. Top: raw cloud, colored according to altitude (blue: 1180 m → red: 1340 m). Middle: point cloud including only ground 3D points, colored according to altitude (blue: 1180 m → red: 1310 m). Bottom: DTM generated on the basis of ground points using the progressive triangulation method (section 7.2.3). The visualized area is the area presented in detail in Figure 7.1. For a color version of this figure, see www.iste.co.uk/baghdadi/1.zip*

The literature of the past 15 years has seen the publication of a great number of filtering methods, a proportion of which has been synthesized and analyzed in [MEN 10]. These are based on shared and realistic hypotheses, namely, that in a given local neighborhood, the ground corresponds to the lowest altitudes, the slope is limited and the ground surface is continuous, without abrupt changes of curvature. The main difficulties identified are therefore represented by:

– terrains characterized by sharp discontinuities (bridges, cliffs and buildings of various dimensions);

– low vegetation;

– the lack of ground-level LiDAR points over large areas (dense vegetation or very large buildings).

Despite the penetration capacity of airborne LiDAR at the level of the forest canopy, there is no guarantee that the last LiDAR echoes (namely, the final target to have backscattered a sufficient quantity of emitted energy) correspond to the ground points: certain surfaces are in fact opaque (buildings), in the same manner as vegetation structures can be dense and prevent a deep penetration of the laser beams (see Figure 7.2), in particular in the case of high incidence angles (>20°). In the case of very dense forests, such as tropical forests, the proportion of LiDAR points reaching the ground can amount to 1%, and in many cases, increasing the density of the emitted points does not result in an improved ground characterization [ASN13, LEI 15].

The filtering of 3D points can be regarded as a conventional classification or segmentation method for 3D data, and standard techniques pertaining to this field can be implemented [SIT 05, BAR 10]. Each 3D point is characterized by a certain number of geometric features related to its neighborhood (See Chapter 6, Mallet *et al.*). A supervised or unsupervised data analysis method is adopted in order to label it in the *ground* category, or to cluster it with the neighboring 3D points sharing the same characteristics. Supervised methods use learning samples collected by a user, in order to characterize the categories of interest in advance (defined via a nomenclature) and to predict, in the case of each new 3D point, the label to which it belongs. Unsupervised methods simply cluster the most similar 3D points by using a set of characteristics (referred to as descriptors) calculated for each point. The advantage of these techniques is represented by their simplicity: filtering is regarded as a substep of a land-cover classification step. The main disadvantage is represented by the fact that the analysis and decision-making process are often local, whereas the methods dedicated specifically to the generation of DTMs make it possible to take account of more comprehensive set of criteria (slope break, morphological continuity notions, etc.) at larger scale. Certain segmentation methods prioritize an implementation in the "object" space, namely, in the geographical space, primarily based on planarity and maximum curvature criteria [TOV 05].

In the following we discuss, three methods developed specifically for the purpose of filtering: morphological methods, adaptive densification and surface filters.

7.2.2. *Morphological filtering*

Morphological methods are based on image-based filters, operating in two dimensions in a regular grid: these are based on the analysis of DSMs mainly by means of mathematical morphology. The disadvantage of image-based methods is constituted by the loss in geometric precision. The discriminative information is in this case represented by the altitude of point clouds often including an *a priori* slope gradient. The methods vary as a function of the operators and structural elements selected. On the one hand, standard opening filters which make it possible to determine the minimum and maximum values can be replaced by other rank operators (the pixels included in the structural element are classified as a function of the ascending altitude: 0, 50 and 100% indicate the minimum, average and maximum values, respectively), for example of a *dual-rank* type (morphological opening or closing according to a given rank [ZHA 03]). On the other hand, the dimensions of the surroundings are critical as well, and the most effective methods are hierarchical: the dimensions increase with each filtering stage up to a maximum value, which is in most cases related to the dimensions of the largest elements off ground level (buildings, see [PIN 13]).

In order to avoid leveling-off the topography, a simple strategy consists of conducting a before/after surface analysis after each filtering stage and subsequently reincorporating as ground points the sections whose elevation shift is "low"; reference is often made to an *a priori* height threshold applied to buildings and vegetation, which can be learned statistically, provided that, for example, a classification process has been conducted in advance [ZHA 03, MON 14]. Geodetic operations can be implemented so as not to incorporate the *a priori* knowledge pertaining to the categories of interest and the structure thereof: these act as connected filters seeking to preserve the "objects" [BEN 03] and define an optimal neighborhood in the case of each pixel [FAU 07].

7.2.3. *Progressive densification*

A second category of methods is based on a different strategy: using a small set of ground points as a basis, this classification is characterized by an iterative refinement of the terrain, wherein the proportion of *ground* points used to redefine the ground surface increases according to a set of criteria (Figure 7.4). Reference is made to the principle of iterative/progressive densification. The most popular method is the adaptive *triangular irregular network* (TIN) method [AXE 00]. A triangular network is initiated on the basis of a set of ground points, for example the lowest points located in a large area (100 m) regularly lining the area of interest, or in four corners of the latter [SOH 08]. *Ground* points are added one-by-one in each triangle, as long as certain criteria are taken into account: maximum point density, edge length, slope between the triangle and the neighboring facets and difference in height (Figure 7.4). This simple method is highly parametrizable and flexible, as it is only based on thresholds: it is thus adapted to large-scale operational contexts, such as implemented at the level of the TerraSolid software suite [SOI 15]. Recent methods extending this approach primarily seek to obtain a denser and more robust initial set of points, by integrating the topography hypotheses mentioned above [ZHA 13].

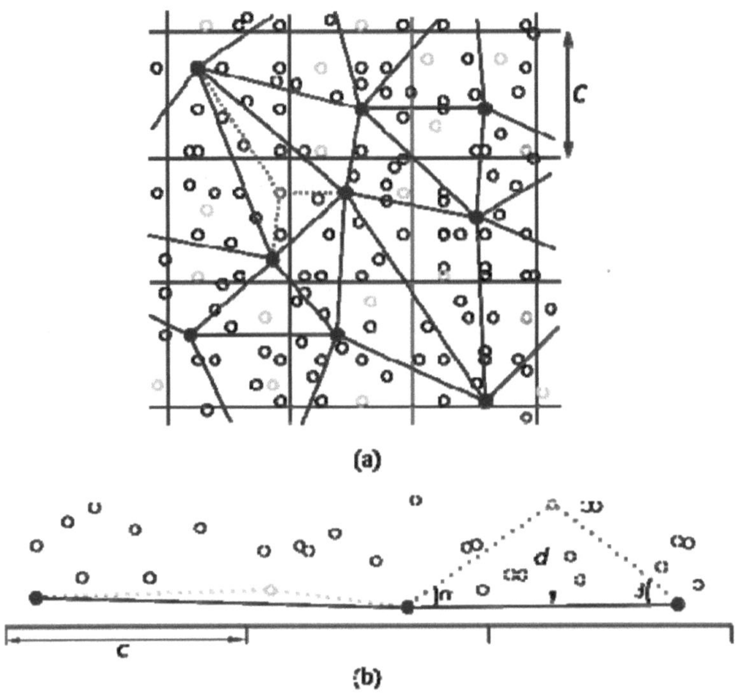

Figure 7.4. *Illustration of the iterative densification principle by means of the TIN approach (see section 7.2.3): a) the initial triangulation (in blue) is generated by selecting, for example, the LiDAR point with the lowest altitude at the level of each c-dimensioned cell in a regular grid (for example, 50 or 100 m). The triangulation can be densified by adding LiDAR points as new vertices fin the network while fully respecting the set criteria (green points); b) these criteria are, for example, the distance d to the TIN or the maximum permissible angles according to the type of landscape. For a color version of this figure, see www.iste.co.uk/baghdadi/1.zip*

7.2.4. *Filtering based on surface adjustment*

The final category of methods, namely, the surface filters, often follows a pattern similar to that of the progressive densification. These methods evaluate the ground surface in an iterative manner on the basis of a set of labeled 3D points but instead of progressively increasing this set of 3D points and without reconsidering the latter, these methods reduce or increase the influence of the points according to their distance to the estimated surface in an iterative manner. In most cases, the process ends when the surface is stable. The various methods developed differentiate themselves according to two aspects: the functional model and the weighting point. The functional model can be represented by Kriging, a least-squares interpolation method [KRA 01] or a thin plate spline method [EVA 07]. An asymmetric weight function is often prioritized in order to assign, for a given iteration, additional weight at the level of the points located below the current surface (as these are most likely ground points), as well as at the level of those located above the surface (off the ground), up to a certain permissible distance h (Figure 7.5). The weight function p can thus be expressed as:

$$p(r)=\left\{\begin{matrix} 1 & \text{if} \quad r>g \\ \dfrac{1}{1+\left(a.\left(r-g()\right)^{b}\right.} & \text{if} \quad g\leq r\leq g+h \\ 0 & \text{if} \quad r>g+h \end{matrix}\right\} \qquad [7.1]$$

where:

– a, b are the parameters which determine the decrease rate of p toward 0;

– g indicates where the decrease is initiated and can be estimated on the basis of the altitude histogram of the LiDAR points (a cluster of 3D points can most likely be present at the level of a locally identifiable ground surface);

– h is used in order to exclude the LiDAR points, as these are located too high above the surface (Figure 7.5).

In practice, the respective values of a, b and h of 2 m^{-1}, 2 and 5 m produce good results in the case of point densities ranging from 1 to 5 points/m².

Certain articles focus on the optimal definition of the local neighborhood in order to conduct the analysis of the height histogram of the LiDAR points [BRE 10, MON 12].

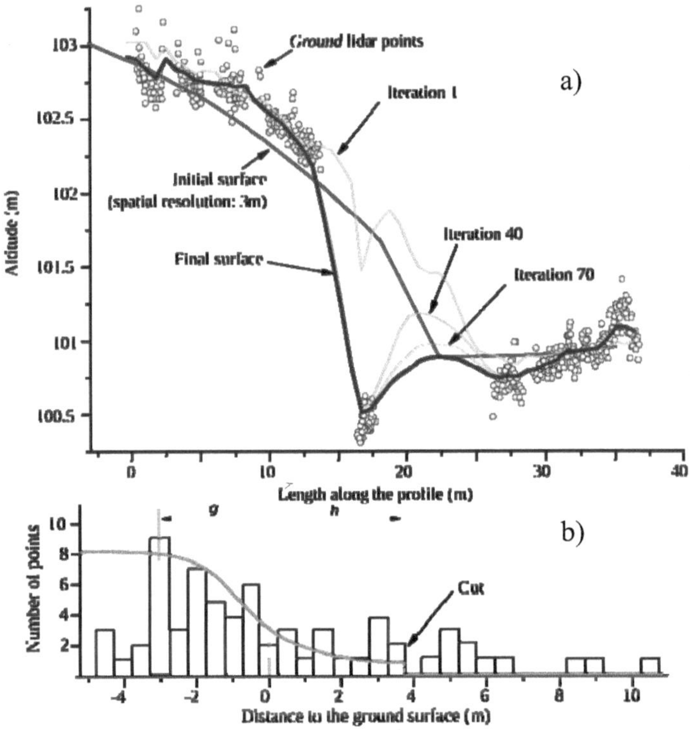

Figure 7.5. *Filtering based on the robust interpolation principle of Kraus and Pfeifer (2001): a) iterative refinement of the ground surface on the basis of the existing LiDAR points; b) asymmetric weighting of these LiDAR points such that the lowest points can be used as ground surface attractors. For a color version of this figure, see www.iste.co.uk/baghdadi/1.zip*

The progressive densification technique will provide good results if a balanced distribution of *ground* and *offground* points exists. Otherwise, a hierarchical method must be adopted by gradually decimating the point cloud starting with the coarsest level [CHE 13]. This also has the advantage of accelerating the process, which is highly relevant for upscaling purposes. Variants are also provided in order to facilitate an adaptation to the specific conditions of the media most susceptible to error, such as mountain forests [KOB 07] or dense urban areas, which make it possible to reduce the influence of above ground points even further.

7.2.5. *Integration of non-geometric information*

These three methods are based on the geometric analysis of the point cloud conducted according to the density, slope or altitude point distribution criteria.

The information pertaining to the backscattered intensity, often available with 3D coordinates, is barely used in order to improve this filtering stage. On the one hand, it is not easy to calibrate it over large surfaces (see Chapter 6), and, on the other hand, there are no methods which can be used in order to correct the influence of objects above the ground (in particular vegetation) on the value of ground measurements [MAG 14]. However, this remains relevant in the case of the terrains which are clear of all vegetation and in which the backscattered intensity is highly discriminative:

– the objects located just above the ground surface, such as low vegetation, hedgerows, barriers, which do not represent high frequencies at the level of the relief [GOE 08]. These must therefore be eliminated in order to avoid the generation of a ground surface comprising artifacts [CHE 09];

– the high frequencies of the relief which may not sufficiently backscatter the LiDAR pulses and are thus not visible in the raw 3D point cloud. This is, for example, the case of water-filled ditches or channels in coastal environments which can appear as flat surfaces in a LiDAR point cloud, but whose low-intensity values make it possible to isolate them with the purpose of conducting a specific processing operation thereafter [BRZ 08].

The advent of full waveform LiDAR sensors [MAL 09] has facilitated the extraction of additional information and in particular the retrieval of a width feature for each LiDAR echo. The latter is estimated as the standard altitude deviation of each echo and provides information on the height distribution of the backscattering targets located in a small area (the one represented by the LiDAR beam footprint). A LiDAR echo is emitted with a particular echo width. Uncovered ground areas will generate a return echo with a similar width, whereas vegetated areas, with high or low-rise objects, will create a set of echoes characterized by a higher width, due to the presence of multiple targets of small dimensions which are almost colocated in altitude (Figure 7.6). The echo width has thus been used in order to filter the objects which cannot be easily detected by means of 3D geometric analysis in a more efficient manner. Thus, Doneus *et al.* [DON 08] simply eliminate the 3D points characterized by excessively high echo width values and which in practice correspond to low vegetation. Only the genuinely high frequencies of the relief are preserved. Mücke [MUC 08] prefers to use this attribute in order to weight the 3D points in the progressive densification method of Kraus and Pfeifer [KRA 01], whereas Jalobeanu and Gonçalves [JAL 13] integrate it in a probabilistic Bayesian framework without requiring that thresholds be set in advance.

Figure 7.6. *Relevance of the echo width attribute for the purpose of filtering natural off-ground objects. Top: orthoimage of a peri-urban area (Lat: 48,0989° – Long: 9,7802°). Bottom left: 3D LiDAR point cloud colored according to altitude (Blue: 615 m → Red: 650 m). Bottom right: display according to the width attribute generated by the LiDAR waveform (Purple: 4 ns → White: 5 ns). LiDAR echoes widen with the altimetric dispersion of the targets in an environment of small dimensions. The width attribute is therefore a crucial item of information for the detection of vegetated areas. For a color version of this figure, see www.iste.co.uk/baghdadi/1.zip*

7.3. Interpolation/surface adjustment

A second step subsequent to the filtering operation consists of estimating the final surface on the basis of ground points. DEM is often limited to a 2.5 D representation in which a unique z value is assigned to the pixel $p(x,y)$ in a regular grid. Two distinct cases may be identified:

– the surface is adjusted as a function of the *ground* point sample obtained beforehand. This represents a classic data interpolation (resampling) or adjustment problem in two dimensions. A large number of solutions are provided. These solutions are not specific to LiDAR data (elastic grid, thin plate spline, kriging, etc.). The only characteristic of 3D point clouds is represented by their capacity to detect breaklines which can occur at the level of the ground surface;

– some of the methods described above are implemented in order to carry out the filtering and adjustment operation in an iterative manner and a ground surface is thus generated following the convergence process;

– in the case of irregular triangular meshes, it is possible to conduct a remodeling operation in a regular square mesh;

– in the case of other square mesh methods, the existing surfaces are preserved or deformed following the filtering operation in order to better fit to the slope breaks detected, or to take account of global morphological constraints [BRE 10, CHE 12].

The detection of structural lines at the level of raw 3D point clouds in order to facilitate the DTM generation and a better localization of discontinuities is possible due to the very high altimetric resolution of the LiDAR data, in both absolute and relative terms. The raster methods based on regular grids are not specific to LiDAR data. In the case of a pure 3D analysis, the most commonly used method consists of detecting the lines in 3D by means of an intersection of the 3D planes: the planes are adjusted at the level of the overlapping areas of interest which are located in the proximity of the breaklines, then the analysis is conducted locally with a view to increase the levels of speed and precision [BRI 04]. An improvement to this method is proposed by Brzank *et al.* [BRZ 05] so as to derive the changes of direction in an effective manner, and to directly detect the networks (in this case, represented by the channels in the coastal regions) represented in this case by sets of parallel lines.

A second solution consists of extracting the 3D lines directly within the raw point cloud. This method is increasingly used in robotics and anthropogenic environments in order to detect roads, roofs or the edges of buildings [LAF 11]: the techniques adopted in this specific case are based on a local analysis of the geometric distribution of 3D points, or seek to adjust the lines to the contours of the 3D points of a given label.

Finally, the adjustment of surfaces in three dimensions is required in the case in which the important elements of the relief are to be preserved (overhangs, cliffs, tunnels, etc.) and in this case, an irregular mesh of the surface is calculated, which excludes the standard definition of DTM. Although this subject has been widely addressed in the publications released by the computer graphics and machine vision communities over the past 20 years [HOR 06, LAL 13], only [PFE 05] discusses the case of LiDAR data by adopting a subdivision method: a triangulation is densified by assessing the adequacy between the raw data and the local surfaces of the order of 2: if the deviation between the two is low, the points are included, the density increases and the angles between the triangles decrease, thus leading to an increasingly smooth surface while at the same time taking account of the existing breaklines.

7.4. Evaluation of digital terrain models

There are three main error sources related to the generation of DEM: the generation of 3D point clouds, the filtering method and the surface interpolation/adjustment method.

The first error source is not addressed here, as this is considered as an occurrence prior to the adjustment and the calibration of the acquired data. The precision of a 3D point is conditioned by multiple factors such as the flight altitude, the precision in the determination of the aircraft trajectory and attitude, etc. The slope of the terrain also exerts a non-negligible influence and becomes preponderant, while the point density decreases, or at least when the latter is no longer sufficient in order to sample the ground surface in an effective manner (Figure 7.7). The vertical precision is estimated to be 10–15 cm in the case of a flight altitude of 1,000 m once the systematic errors have been eliminated or in the case of terrains which are uneven only to a lesser extent.

In terms of filtering, the main error sources have been indicated in the previous section and the quality varies to a large extent as a function of the type of surfaces and the 3D point density at ground level. The characterization is carried out in an extrinsic manner by using a manual ground/offground classification or a classification obtained by means of a different process. [SIT 06] have proposed a benchmark in order to compare the existing methods according to their management capacity in terms of individual errors, large off-ground areas associated with small objects, as well as discontinuities. The evaluation is conducted according to two types of errors, referred to as I and II, which correspond to the false negatives and false positives in a classification in each case [FOO 02]. The study has shown that the management of urban environments, including objects of different dimensions and the regions characterized by sharp discontinuities in natural environments, are the most problematic areas, and that in general, surface filters provide the best guarantee for quality. In the case of commercial applications or the generation of DTM over large scales, including strict specifications in terms of vertical precision, human operators are still required in order to correct the errors arising from automatic algorithms.

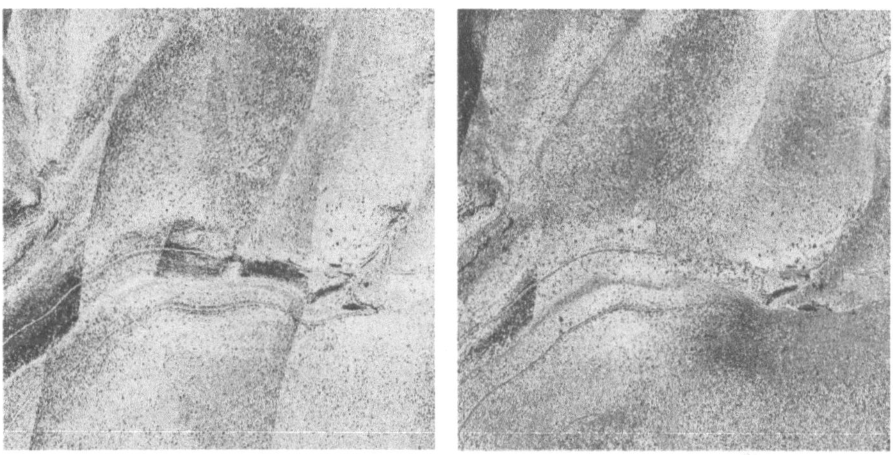

Figure 7.7. *Comparison of LiDAR DTM with different flight specifications in order to highlight random systematic bias. Left: DTM difference between a flight at 2,250 m and 1pt/m² and a flight at 1,530 m with 2 pts/m². Right: DTM difference between two flights at 1,530 m but with 0.5 and 2 pts/m². The altitude discrepancies between surfaces range from 0 (dark blue) to 50 cm (red). It is to be noted that the differences are derived from the calibration process including differences between LiDAR bands, slope differences (accentuated in the middle of the area) and marginally vegetated areas. This latter aspect is not implemented in the case of this area, given the fact that the distribution of LiDAR points at ground level is in fact sufficiently homogeneous. For a color version of this figure, see www.iste.co.uk/baghdadi/1.zip*

Finally, numerous studies have been conducted on the height quality of the DTM obtained following the surface adjustment procedure. These consist of measuring the distance using the points obtained by means of GNSS/GPS and topometry techniques. The complexity of this task is represented by the fact that the terrain measurement itself is subject to uncertainty and that the imprecision increases in the areas in which the LiDAR DTM are potentially of the lowest quality: below a dense forest canopy or in the case of slope areas. It is difficult to quantify the vertical precision of a DTM as this varies across several studies; however, it is generally accepted that in the absence of off-ground objects in the case of flat opaque surfaces characterized by high-quality georeferenced data, the precision value falls below 10 cm. The precision falls below 20–30 cm under a sparse forest canopy and increases by 10 cm in the case of low-quality georeferencing procedure. It is also necessary to add an additional loss related to the slope effect, which is conditioned by the incidence angle and altitude, and which can measure another 50 cm [HYY 05]. These values increase with the flight altitude and with the decrease in the "ground" point density [EST 11].

For the sake of completeness, it is to be noted that the altimetric quality of the DTM derived from airborne LiDAR data is higher than the value which can be obtained using the GLAS instrument carried on board the ICESat satellite. Deviations of the order of the meters may be identified. These can amount to approximately 2 m and can increase with the slope of the terrain given the fact that the footprint of such a sensor is estimated at 70 m [BAG 11, HIL 12]. A comparison with optical imagery and photogrammetric techniques remains a difficult task because LiDAR technology is mainly used in the cases in which the stereo optical imaging is not adapted: forest, coastal and river environments. In contrast, comparative studies with the interferometric RADAR and in particular the P-band radar have indicated deviations of 1 m in favor of the LiDAR, and have shown that in forest areas, the RADAR measurement was subject to tree height and biomass [HEU 05].

Independently from any quantitative evaluation of the altimetric bias between reality and models, a visual qualitative analysis is also possible in order to assess whether the protruding elements of the landscape (slopes, ditches and roads) are correctly restored. A point density at ground level of 0.8 pt/m^2 facilitates an implementation on minimally damaged grounds. The high quality of a DEM will thus depend both on pulse density and on the flying season and the type of vegetation present in the area: in the case of deciduous trees, flights at 1 and 4 pts/m^2 in winter, and respectively, in summer are sufficient (Figure 7.2).

7.5. Key points

– the generation of Digital Terrain Models on the basis of 3D LiDAR point clouds is now operational and it is carried out on a large scale in numerous countries. This represents the fundamental processing operation to be implemented prior to the extraction of high-level products, in particular in forest areas to which the LiDAR technology is best adapted;

– a vast number of methods are provided, along with the advantages and disadvantages thereof, and the techniques currently developed seek to address the identified problems: dense forest canopies, sharp slope breaklines, etc. The integration of additional information derived from the LiDAR waveforms, the DTM itself or derived from other remote sensing data sources, represents the most adequate solution to these problems in a research context;

– from a commercial perspective, the *a posteriori* human intervention in order to filter the LiDAR point clouds in a more effective manner, or in order to integrate the breaking lines or water surfaces remains a necessity. Under optimal conditions and in the case of well-defined surfaces, an altimetric precision lower than 10 cm may be envisaged in the forthcoming years.

7.6. Bibliography

[ASN 13] ASNER G.P., KELLNER J.R., KENNEDY-BOWDOIN T. *et al.*, "Forest canopy gap distributions in the Southern Peruvian Amazon", *PLoS One*, vol. 8, no. 4, p. e60875, 2013.

[AXE 00] AXELSSON P., "DEM generation from laser scanner data using adaptative TIN models", *International Archives of Photogrammetry and Remote Sensing*, vol. 33, no. B4/1, pp. 110–117, 2000.

[BAG 11] BAGHDADI N., LEMARQUAND N., ABDALLAH H. *et al.*, "The relevance of GLAS/ICES at elevation data for the monitoring of river networks", *Remote Sensing*, vol. 3, pp. 708–720, 2011.

[BAR 10] BARTELS M., WEI H., "Threshold-free object and ground point separation in LiDAR data", *Pattern Recognition Letters*, vol. 31, no. 10, pp. 1089–1099, 2010.

[BEN 03] BENEDIKTSSON J.A., PESARESI M., AMASON K., "Classification and feature extraction for remote sensing images from urban areas based on morphological transformations", *IEEE Transactions on Geoscience and Remote Sensing,* vol. 41, no. 9, pp. 1940–1949, 2003.

[BRE 10] BRETAR F., CHEHATA N., "Terrain modelling from LiDAR range data in natural landscapes: a predictive and Bayesian framework", *IEEE Transactions on Geoscience and Remote Sensing*, vol. 48, no. 3, pp. 1568–1578, 2010.

[BRI 04] BRIESE C., "Three-dimensional modelling of breaklines from airborne laser scanner data", *International Archives of Photogrammetry and Remote Sensing,* vol. 35, no. B3, pp. 1097–1102, 2004.

[BRZ 05] BRZANK A., LOHMANN P., HEIPKE C., "Automated extraction of pair wise structure lines using airborne laser scanner data in coastal areas", *International Archives of Photogrammetry and Remote Sensing,* vol. 36, no. 3/W19, pp. 36–41, 2005.

[BRZ 08] BRZANK A., HEIPKE C., GOEPFERT J. *et al.*, "Aspects of generating precise digital terrain models in the Wadden Sea from LiDAR – water classification and structure line extraction", *ISPRS Journal of Photogrammetry and Remote Sensing,* vol. 63, no. 5, pp. 510–528, 2008.

[CHE 07] CHEN Q., GONG P., BALDOCCHI D.D. *et al.*, "Filtering airborne laser scanning data with morphological methods", *Photogrammetric Engineering & Remote Sensing,* vol. 73, no. 2, pp. 175–185, 2007.

[CHE 09] CHENG W., GLENN N.F., "Integrating LiDAR intensity and elevation data for terrain characterization in a forested area", *IEEE Geoscience and Remote Sensing Letters*, vol. 6, no. 3, pp. 463–466, 2009.

[CHE 12] CHEN H., Digital terrain models generation from airborne LiDAR point clouds using a multi-scale terrain filtering method, PhD Thesis, University of Waterloo, Canada, 2012.

[CHE 13] CHEN C., LI Y., LI W. *et al.*, "A multiresolution hierarchical classification algorithm for filtering airborne LiDAR data", *ISPRS Journal of Photogrammetry and Remote Sensing*, vol. 82, pp. 1–9, 2013.

[DON 08] DONEUS M., BRIESE C., FERA M. *et al.*, "Archaeological prospection of forested areas using full-waveform airborne laser scanning", *Journal of Archaeological Science*, vol. 35, no. 4, pp. 882–893, 2008.

[EST 11] ESTORNELL J., RUIZ L.A., VELÁZQUEZ-MARTÍ B. *et al.*, "Analysis of the factors affecting LiDAR DTM accuracy in a steep shrub area", *International Journal of Digital Earth*, vol. 4, no. 6, pp. 521–538, 2011.

[EVA 07] EVANS J.S., HUDAK A.T., "A multiscale curvature algorithm for classifying discrete return LiDAR in forested environments", *IEEE Transactions on Geoscience and Remote Sensing,* vol. 45, no. 4, pp. 1029–1038, 2007.

[FAU 07] FAUVEL M., Spectral and spatial methods for the classification of urban remote sensing data, PhD Thesis, INP Grenoble, France, University of Iceland, Iceland, 2007.

[GOE 08] GOEPFERT J., SOERGEL U., HEIPKE C. *et al.*, "An approach for filtering LiDAR data in coastal vegetated areas using intensity information and multiple echoes", *International Archives of Photogrammetry and Remote Sensing*, vol. 37, no. B3b-1, pp. 219–225, 2008.

[HEU 05] HEURICH M., METTE T., KRZYSTEK P. *et al.*, "Comparison of DTM, DSM and DCM derived from LIDAR, InSAR and digital camera over a mixed mountain forest in central Europe", *Silvilscan*, Blacksburg, VA, 2005.

[HIL 12] HILBERT C., SCHMULLIUS C., "Influence of surface topography on ICESat/GLAS forest height estimation and waveform shape", *Remote Sensing*, vol. 4, pp. 2210–2235, 2012.

[HOR 06] HORNUNG A., KOBBELT L, "Robust reconstruction of watertight 3D models from non-uniformly sampled point clouds without normal information", *Symposium on Geometry Processing*, Cagliari, Italy, pp. 41–50, 2006.

[HYY 05] HYYPPÄ H., YU X., HYYPPÄ J. *et al.*, "Factors affecting the quality of DTM generation in forested areas", *International Archives of Photogrammetry, Remote Sensing and Spatial Information Sciences*, vol. 36, no. 3/W19, pp. 85–90, 2005.

[JAL 13] JALOBEANU A., GONÇALVES G., "Robabilistic terrain models from waveform airborne LiDAR", *American Society of Photogrammetry and Remote Sensing Annual Conference* (ASPRS'13), Baltimore, MA, 2013.

[KOB 07] KOBLER A., PFEIFER N., OGRINC P. *et al.*, "Repetitive interpolation: a robust algorithm for DTM generation from aerial laser scanner data in forested terrain", *Remote Sensing of Environment*, vol. 108, no. 1, pp. 9–23, 2007.

[KRA 01] KRAUS K., PFEIFER N., "Advanced DTM generation from LiDAR data", *International Archives of Photogrammetry and Remote Sensing*, vol. 36, no. 3/W4, pp. 23–30, 2001.

[LAF 12] LAFARGE F., MALLET C., "Creating large-scale city models from 3D-point clouds: a robust approach with hybrid representation", *International Journal of Computer Vision,* vol. 99, no. 1, pp. 69–85, 2012.

[LAF 13] LAFARGE F., ALLIEZ P., "Surface reconstruction through point set structuring", *Computer Graphics Forum*, vol. 32, no. 1, pp. 225–234, 2013.

[LEI 15] LEITOLD V., KELLER M., MORTON D.C. *et al.*, "Airborne lidar-based estimates of tropical forest structure in complex terrain: opportunities and trade-offs for REDD+", *Carbon Balance and Management*, vol. 10, no. 3, 2015.

[MAG 14] MAGUYA A.S., JUNTTILA V., KAURANNE T., "Algorithm for extracting digital terrain models under forest canopy from airborne LiDAR data", *Remote Sensing*, vol. 6, pp. 6524–6548, 2014.

[MAL 09] MALLET C., BRETAR F., "Full-waveform topographic lidar: state-of-the-art", *ISPRS Journal of Photogrammetry and Remote Sensing,* vol. 64, no. 1, pp. 1–16, 2009.

[MON 12] MONGUS D., ŽALIK B., "Parameter-free ground filtering of LiDAR data for automatic DTM generation", *ISPRS Journal of Photogrammetry and Remote Sensing*, vol. 67, pp. 1–12, 2012.

[MON 14] MONGUS D., LUKAČ N., ŽALIK B., "Ground and building extraction from LiDAR data based on differential morphological profiles and locally fitted surfaces", *ISPRS Journal of Photogrammetry and Remote Sensing*, vol. 93, pp. 145–156, 2014.

[MUC 12] MÜCKE W., Analysis of full-waveform airborne laser scanning data for the improvement of DTM generation, Master Thesis, Institute of Photogrammetry and Remote Sensing, TU Vienna, Austria, 2008.

[PFE 05] PFEIFER N., "A subdivision algorithm for smooth 3D terrain models", *ISPRS Journal of Photogrammetry and Remote Sensing*, vol. 59, pp. 115–127, 2005.

[PIN 13] PINGEL T., CLARKE K., MCBRIDE W., "An improved simple morphological filter for the terrain classification of airborne LiDAR data", *ISPRS Journal of Photogrammetry and Remote Sensing,* vol. 77, pp. 21–30, 2013.

[REE 15] REESE H., NORDKVIST K., NYSTRÖM M. *et al.*, "Combining point clouds from image matching with SPOT 5 multispectral data for mountain vegetation classification", *International Journal of Remote Sensing,* vol. 36, no. 2, pp. 403–416, 2015.

[SCH 15] SCHMITT M., SHAHZAD M., XIANG ZHU X., "Reconstruction of individual trees from multi-aspect TomoSAR data", *Remote Sensing of Environment,* vol. 165, pp. 175–185, 2015.

[SIT 05] SITHOLE G., Segmentation and classification of aiborne laser scanner data, Phd Thesis, ITC, Enschede, The Netherlands, 2005.

[SIT 06] SITHOLE G., VOSSELMAN G., "Experimental comparison of filter algorithms for bare-Earth extraction from airborne laser scanning point clouds", *ISPRS Journal of Photogrammetry and Remote Sensing* vol. 59, no. 1, pp. 85–101, 2006.

[SOH 08] SOHN G., DOWMAN I., "A model-based approach for reconstructing a terrain surface from airborne LiDAR data", *Photogrammetric Record*, vol. 23, no. 122, pp. 170–193, 2008.

[SOI 15] SOININEN A., Website of Terrasolid software, available at: http://www.terrasolid.fi, accessed 30 May 2015.

[TOV 05] TOVARI D., PFEIFER N., "Segmentation based robust interpolation – a new approach to laser data filtering", *International Archives of Photogrammetry and Remote Sensing,* vol. cXXXVI-3/W19, pp. 79–84, 2005.

[VOS 12] VOSSELMAN G., "Automated planimetric quality control in high accuracy airborne laser scanning surveys", *ISPRS Journal of Photogrammetry and Remote Sensing*, vol. 74, pp. 90–100, 2012.

[WAN 14] WANG Q., WU L., XU Z. *et al.*, "A progressive morphological filter for point cloud extracted from UAV images", *IEEE International Geoscience and Remote Sensing Symposium (IGARSS),* Quebec, Canada, pp. 2023–2026, 2014.

[ZHA 03] ZHANG K., CHEN S., WHITMAN D. *et al.*, "A progressive morphological filter for removing nonground measurements from airborne LiDAR data", *IEEE Transactions on Geoscience and Remote Sensing*, vol. 41, pp. 872– 882, 2003.

[ZHA 13] ZHANG J., LIN X., "Filtering airborne LiDAR data by embedding smoothness-constrained segmentation in progressive TIN densification", *ISPRS Journal of Photogrammetry and Remote Sensing*, vol. 81, pp. 44–59, 2013.

Glossary

2.5D: A representation of a geometric set in the form of a function $z = f(x, y)$. Very useful for returning the 3D geometry analysis to functional analysis, 2.5D is limited in the geometries that it can represent.

Above ground: Set of perennial objects or not, human or not (vegetation, buildings), located on or above the surface "ground" which are not taken into account when calculating a DTM.

Absolute calibration coefficient: Parameter that allows for converting digital count or digital number of an image into energy (luminance).

Absorption: Transformation of bright radiative energy into heat.

Absorption band: Relatively narrow spectral interval with low reflectance values due to a strong absorption.

Active system: The remote sensing active sensors are those that produce their own energy to illuminate the target.

Adherence: Adhesion is a current default in the restituted or obtained Digital Surface Model (DSM) by photogrammetry, which results in a slight elongation of the elements of the above ground. It is caused by the estimation methods of the disparity based on the matching blocks.

Aerodynamic temperature: Theoretical temperature of the average source of sensitive heat flux for a complex surface, corresponding to the extrapolation within the canopy of the air temperature profile observed above the canopy.

Aerosols: Particles suspended in the atmosphere capable of interacting with solar radiation through the processes of absorption and backscattering. They can be solid (dust) or liquid (mist) of organic nature (black carbon) or inorganic nature (rock eroded) and more or less large (ranging from a few tenths of a nanometer to a hundred micrometers).

Artifact: Image distortion arising from a defect of the acquisition medium.

Atmospheric correction: Method developed to eliminate the disruptive effects of the atmosphere (absorption and diffusion of solar radiation) that contribute to the signal received by a satellite.

Attribute: Descriptor of a pixel or a 3D point.

Azimuth angle: The angle in the horizontal plane between the direction of an object and a reference direction (geographical or magnetic north), according to Wikipedia.

Backscattering cross section: The ratio between the energy scattered in the direction of the source and the incident energy (irradiance). It is expressed in m^2/steradian.

Bathymetric LiDAR: Active remote sensing system dedicated to the analysis of fluvial and coastal environments. It is dedicated to the measurement of water depth by measuring the travel time between two laser pulses at two different wavelengths: infrared to describe the surface of the water and the green for the background.

Bidirectional Reflectance Distribution Function (BRDF): Ratio of outgoing radiance and incident irradiance.

Bistatic: Characteristic of an optronic system where the transmitter and receiver are separated are separate.

Brightness temperature: Temperature obtained by converting the radiance measured according to Planck's Law at the level of the sensor, weighted by its transfer function; it can be defined for a measurement in a given direction (called directional brightness temperature).

CALIOP: Cloud-Aerosol LiDAR Pathfinder Satellite Observation is the LiDAR system of the CALIPSO spatial mission.

CALIPSO: Cloud-Aerosol LiDAR Pathfinder Satellite Observation is a French-American spatial mission which was put into orbit on April 28, 2006, by the NASA.

Canopy Height Model (CHM): For a given portion of territory, the mapping (in raster format) of the maximum height of the forest area compared to the ground within each pixel. It is a surface resulting from the calculation of the difference between a Digital Surface Model (DSM) and a Digital Terrain Model (DTM). Excluding forest environments, it is commonly called Digital Elevation Model (DEM). Subtracting the DTM makes it possible to eliminate the effect of the terrain, and any area not composed of objects above-ground has zero height.

Catadioptric: Referring to a system that includes both mirrors and lenses.

Channel: An image file which often consists of several channels, each one transcribing different information. For example, a color digital photograph features three channels: blue, green and red. See *spectral band.*

Classification: Process oriented towards assigning a class to the elements of a given set, according to criteria usually defined in advance. It can be supervised or unsupervised.

Clustering (or data partitioning): Technique that allows separating data sets into homogeneous groups (as defined by previously established criteria).

Confusion matrix: Double entry table used to analyze the quality of a classification; the rows correspond to the class of evaluated objects and the columns correspond to the assigned class; the boxes in the table indicate the number of objects.

Deconvolution: Mathematical operation aimed to reverse the effects of convolution. It is used in the treatment of LiDAR waveforms to eliminate the influence of the LiDAR wave emitted in the backscattered one and used for the extraction of 3D echoes.

DIAL: Differential Absorption LiDAR, LiDAR system operating with at least two wavelengths, one wavelength is located at or near the absorption feature for the target, and the other is located away from the same absorption feature.

Differential Global Positioning System (DGPS): Instrument that allows measuring geographic positions in a differential manner using a constellation of satellites and a network of fixed ground-based reference stations.

Diffuse reflection: Diffuse reflection is a reflection of the wave in different directions and with a distribution of energy along these directions. The distribution between specular and diffuse reflection processes is determined by the roughness and the dielectric properties of the surface.

Digital count: Pixel coding in bits reflecting the measurement of radiation recorded and converted into an electrical signal by the sensor.

Digital Elevation Model (DEM): 3D representation of a terrain's surface. It represents the Earth's surface and all objects on it.

Digital Surface Model (DSM): A topographic surface model describing the buildings, vegetation, roads (above-ground) and/or the characteristics of the terrain (ground). For simplicity, the DSM is the generic term for models that describe the ground and/or above-ground.

Digital Terrain Model (DTM): Representation of the topography of a land area in a matrix form (image), where each cell (pixel) contains the height information generally in relation to a terrestrial reference system (usually expressed in meters above sea level). For simplicity, the DTM is the only ground without vegetation or building.

Diopter: Referring to an optical system containing only lenses.

Dynamics (of an image): Defines the interval in which the radiometry of an image lies. A strong dynamic indicates a significant ratio between the lowest values and the strongest ones. An important dynamic can lead to large errors in sampling (saturation, rounding errors).

Earth coordinate system: System of coordinates that allow locating a point in an absolute manner on the surface of the Earth.

Electromagnetic radiation: An energy wave that is propagated without any hardware support (in vacuum) and is characterized by a period (cycle length or wavelength – in seconds) and a frequency (number of cycles per second – in hertz). The electromagnetic spectrum is the decomposition of electromagnetic radiation according to its wavelength or frequency (Hz).

The long wavelengths (low frequencies) are the radio and microwaves while smaller wavelengths (high frequencies) are nuclear or gamma rays.

Electromagnetic spectrum: Continuous sequence of electromagnetic energy conceived in function of wavelengths or frequencies (for the same pixel).

Emissivity: The emissivity of a material is a dimensionless number (without units) that allows characterizing the radiative properties of a material. Thus, an ideal black body has an emissivity of 1 while any real material has an emissivity of less than 1.

Endmembers: Source of information of the image in the form of a spectrum of reference.

Equalization: Action of modifying the overall image radiometry so that the values of identical elements of the imaged scene becomes the same.

Evapotranspiration (ET): Amount of water transferred to the atmosphere by vegetation covers resulting from plant transpiration and soil evaporation. ET represents the actual water consumption of the crop; it depends on the needs of the crop (themselves dependent on its state of development and climatic demand) and water available in the soil. Its knowledge is essential for monitoring crop water balance.

Extinction: Optical measure of opacity by distance unit.

Field diaphragm: A thin opaque structure allowing the angular limitation of the field of view of an optical system. It is positioned in the focal plane.

Fluorescence: Inelastic scattering process. The incident wavelength is spontaneously re-emitted to a different wavelength.

Focal plane: Plane defined by the focal point and the perpendicular to the optical axis.

Focal point: The point at which initially collimated rays of light meet after passing through a convex lens, or reflecting from a concave mirror according to Wikipedia.

Form recognition: Set of methods oriented to identifying known elements in a dataset. For 3D point clouds, they are 2D or 3D parametric simple shapes such as lines, planes, cubes or cylinders.

Full waveform: For every LiDAR pulse, a signal reflected by Earth's surface and the objects that compose it, received and digitized by the LiDAR system.

Fusion (Data fusion): Method oriented to aggregating data from often heterogeneous sources. In remote sensing, it can be done when talking about observations, descriptors or decisions (classification).

Geodetic system: Reference system for expressing the positions in the vicinity of the Earth (source Wikipedia).

Geographic Information System (GIS): A geographic information system or geographical information system (GIS) is a system designed to capture, store, manipulate, analyze, manage, and present all types of spatial or geographical data (source Wikipedia).

Geoid: A geoid is the equipotential surface of the Earth's gravity field defining zero altitude. At sea, it coincides with the mean of sea level. In continental domains, it cannot be measured directly; it is calculated from other measures of gravity fields (gravity anomalies, deflections of the vertical, satellite orbit perturbations). It has a volume equal to that of the GRS80 ellipsoid. In other words, the sum of the geoid heights relative to the ellipsoid GRS80 is zero.

Geometric model: Analytical or sampled model giving out a relationship between column-line coordinates of an image and a position on the ground.

Geo-referencing: Operation which consists of properly positioning geographical data (like that from a satellite image) in a geographic coordinate system.

GeoTIFF: A metadata standard which allows georeferencing information to be embedded within a TIFF file, according to Wikipedia.

Global Positioning System (GPS): Geolocation system operating at a global scale and based on a measurement of receipt by remote electromagnetic signals transmitted by satellites in orbits around the Earth.

Ground truth: In image processing, information that is available on a set of pixels.

Hyperspectral remote sensing: Remote sensing from a sensor allowing the observation in many narrow spectral bands (a few nanometers) in the wavelength range from the visible, near infrared, mid-infrared to the infrared thermal.

Ice, Cloud and land Elevation satellite (ICEsat): NASA's spatial mission launched in 2003. The mission embarked a backscatter LiDAR system for the study of surfaces, primarily polar caps (GLAS).

Image anomaly: Set of pixels which are not in compliance to the rest of the image.

Image characterization: Detection, discrimination and/or identification of objects that form the image.

Image correlation: Image correlation is a method that allows measuring the displacement between two images. This method is classically used in stereoscopy to measure disparities of a stereoscopic pair in order to determine the elevation of a scene.

Image unmixing: For a hyperspectral image, it is the determination of the pure poles and weights that allow rebuilding each of the spectra of the image.

Imaging spectrometer (or hyperspectral camera): Optical instrument that allows making both spectroscopy and imaging at the same time.

Infrared: Electromagnetic radiation of a wavelength between 0.7 μm and 100 μm.

Interpolation: Mathematical operation allowing to construct a response from a finite number of input points. It is used in particular in 2D resampling on a regular grid size given a number of 3D measurements (here onwards, altitude) and generate models of Digital Surfaces, Terrains, or Elevations.

Intimate mix: Non-linear combination of several endmembers.

Ionosphere: Upper part of the atmosphere (of about 50 kilometers to a few hundred kilometers of altitude), conductive due to the ionization of the gas by ultraviolet sunlight. The ionosphere is a dispersive medium, in this sense, the speed of propagation of electromagnetic waves (and thus the error in the propagation time measurement) is function of on their wavelength.

JPEG2000: Image compression standard (with or without loss of information) using a multi-resolution approach based on wavelet decomposition of the signal contained in the image.

Kappa coefficient: A quality estimator. It measures the level of agreement between the classification and the actual data. It varies between 0 and 1. The closer it is to 1, the more the agreement is important.

Kernel trick: A method which consists of using a linear classifier to solve a nonlinear problem, transforming the representation space of the input data into a space of larger size, where the linear classifier is then used.

Lambertian: A material that reflects light in a uniform manner in all directions regardless of the direction of the incident light.

Laser intensity: Amount of radiation of the laser beam.

Laser spot: The area on a surface illuminated by a laser beam. Its size and shape, often elliptical, vary with the divergence of the transmitted beam, the distance between the source and the target and the orientation between the emitted beam and the target.

Leaf area index (LAI): A dimensionless quantity that is defined as the one-sided green leaf area per unit ground surface area (LAI = leaf area/ground area, m^2 / m^2) the "green" LAI is the part of the green vegetation LAI.

LiDAR amplitude: Numerical value of the waveform at the position of a given echo LiDAR.

LiDAR echo: Local energy maximum in a LiDAR waveform resulting from significant backscattering of a wave emitted from an object or a group of spatially close objects. They are called first, intermediate or last ones, depending on their position in the backscattered wave.

LiDAR equation: Fundamental *equation* to relate the received photon counts (or light power) with the transmitted laser photon counts (or laser power), light propagation in background atmosphere, physical interaction between light and objects, and LiDAR system efficiency and geometry. When discussing a topographic LiDAR, it is inspired by the radar equation.

LiDAR In-Space Technology Experiment (LITE): The first LiDAR spatial mission of the NASA, launched by the Discovery ship in August of 1994.

LiDAR intensity: Numerical value corresponding to the integral of the waveform for a given echo. It is also called energy.

LiDAR points filtering: For the creation of a Digital Terrain Model, it encompasses the operation of removing all LiDAR points outside of the ground or not being useful to calculate the DTM.

Light Amplification by Stimulated Emission of Radiation (Laser): Emits coherent light thanks to an amplified stimulated emission. The time between transmission and reception is converted into a distance by multiplying by the speed of light.

Light Detection And Ranging, Active Instrument (LiDAR): Measures the optical properties of a medium by scanning a laser beam returned to its transmitter.

Line of slope failure: Line segment in 3D marking an abrupt change in the slope of the terrain.

Litto3D: A precise altimeter digital model of land-sea continuum carried out jointly by SHOM (French Naval Hydrographic and Oceanographic Service) and IGN (French National Institute of Geographic and Forest Information).

Mie scattering: Mie scattering is a theory of the diffraction of light by spherical particles. Mie scattering refers to the scattering by particles whose radius is between 0.1 and 10 times the wavelength. Note that the scattered power is maximum when the particle is large in comparison to the wavelength.

Monostatic: Refers to an optronic system where the transmitter and receiver are identical.

Mosaic: For overlapping images, the action of only capturing a portion of each image so as to generate a single image whose support is the union of supports of the images used.

Multiple scattering (optical): A photon is released outside of the medium after it interaction with several particles.

Multispectral remote sensing: Remote sensing from a sensor allowing to observe several wide enough spectral bands (of the order of several tens of nanometers) in wavelengths ranging from the visible, near infrared, mid-infrared to the thermal infrared.

Net radiation: Algebraic sum of hemispherical radiations received and transmitted by a surface on the set of wavelengths; net radiation represents the radiative energy balance available (or lost) by the surface. SI Unit: Wm^{-2}.

Noise Equivalent Power (NEP): Is a measure of the sensitivity of a photodetector or detector system. It is defined as the signal power that gives a signal-to-noise ratio of one in a one hertz output bandwidth, according to Wikipedia.

Normalized Difference Vegetation Index (NDVI): This vegetation index is calculated from the reflectances in the Near Infrared (NIR) and in the red (R): NDVI = (NIR–R)/(NIR+R). It measures the intensity of the photosynthetic activity of the surface. It is efficient to distinguish the types of covers provided that they are not too dense as tropical rainforests.

Object-oriented classification: Image classification method aiming to segment an image into distinct polygonal objects. The objects are classified using the radiometric properties of the image as well as the geometric properties of the objects such as their size or form.

Occlusion: Refers to a visible element only in one of the images constituting a stereoscopic pair, this element being masked in the second view by another landscape component. Occluded pixels cannot be returned rendered to the Digital Surface Model (DSM) calculated from this pair, unless an additional image to its observation under a different view is brought.

Optical reflectance: Proportion of incident light (solar illumination) reflected by the Earth's surface. It is expressed as a percentage.

Optical thickness: The optical thickness of a layer (atmosphere, vegetation) measuring the degree of transparency of the medium. It characterizes the fraction of electromagnetic radiation (or light radiation) scattered or absorbed by the components of the layer traversed.

Orthoimage: Image corrected geometrically in a projection system given so that all of its pixels perfectly coincide with a geographic map. Deformations due to terrain, the roundness of the Earth and the conditions of viewing were corrected: each pixel seems to have been acquired by a sensor performing measurements perpendicular to the ground at any point in the observed area.

Panchromatic: A panchromatic image is a comprehensive record of the radiation reflected by the Earth's surface in the range of wavelengths of the visible spectrum without color separation. For example, a SPOT 5 panchromatic image is acquired between 0.48 and 0.71 μm. This data is more spatially resolved that the multispectral image acquired by the same satellite.

Paraxial: Refers to an optronic system, where transmitter and receiver are close and parallel.

Passive systems: Passive sensors are remote sensing devices that measure the available energy from the target, with no emissions from the sensor.

Photocurrent: Current generated by photoelectric effect.

Photodetector: A photo detector converts light signals that hit the junction into voltage or current.

Photodiode: Type of photodetector without internal amplification, based on a combination of semi-conductors.

Photo-electron: An electron ejected from the surface of a material by the photoelectric effect, according to Wiktionary.

Photointerpretation: Image analysis by a human operator, eventually assisted by image processing tools.

Photomultiplier: High gain photodetector by internal multiplication of photoelectrons.

Pixel: Individual element constituting an image (deriving of the English phrase "picture element"). Depending on the image type and the encoding used, the pixel "contains" one or more values corresponding to the digital pixel accounts in different image channels.

Polarization: The orientation of the electric field vector in the plane orthogonal to the wave propagation direction.

Potential evapotranspiration: Theoretical evapotranspiration of a well water-fed surface submitted to real climatic conditions.

Preprocessing: Necessary image enhancement technique prior to the analysis and extraction of information. The preprocessing generally includes radiometric and geometric corrections.

Principal Components Analysis (PCA): A statistical procedure that uses an orthogonal transformation to convert a set of observations of possibly correlated variables into a set of values of linearly uncorrelated variables called principal components. The number of principal components is less than or equal to the number of original variables (source Wikipedia).

Pupil: Element limiting the size of the light beam passing through an optical system.

Pupil plane: Plane of the pupil of an optical system.

Quantum efficiency: Conversion probability of an incident photon into a photoelectron in a photodetector.

Radiative surface temperature: Emission temperature of the surface considered as a gray body; it can be defined for a measurement in a given direction (referred to as directional radiative temperature).

Radiative transfer: Theory based on the conservation of energy flux through an elementary volume. It allows to describe the behavior of an electromagnetic wave passing through a scattering medium composed of dielectric particles.

Radio Detection and Ranging Instrument (Radar): Emits a microwave signal and is able to measure the characteristics of the reflected signal in its direction. It also makes it possible to determine the distance and the direction of an obstacle.

Radiometric resolution: It describes the ability of a sensor to distinguish small differences in the intensity of the received energy. The maximum number of levels of intensity depends on the number of bits used. For example, a RapidEye image is encoded in 16 bits or 65,536 numeric values available to transcribe the recorded intensity; a SPOT image is encoded in 8 bits or 256 digital values.

Raman effect: Physical phenomenon, corresponding to an energy exchange between the light beam and the medium, changing the frequency of light that circulates.

Rasterization: Process of performing raster graphics.

Rayleigh scattering: Scattering regime of electromagnetic waves on particles sized much smaller than the wavelength, also known as molecular scattering.

Real evapotranspiration: Vertical density of flux of water vapor corresponding to the sum of evaporation from the soil, the rain intercepted by the foliage and the transpiration.

Recovery function: The LiDAR overlap describes the overlap between the laser beam and the receiver field of view.

Reference ellipsoid: In geodesy, a reference ellipsoid is a mathematically defined surface that approximates the geoid, the truer figure of the Earth, or other planetary body. Because of their relative simplicity, reference ellipsoids are used as a preferred surface on which geodetic network computations are performed and point coordinates such as latitude, longitude, and elevation are defined (source Wikipedia).

Regression: Method that allows estimating the relationship between a variable (to explain) and a set of other variables (explanatory ones).

Relative radiometric normalization: Statistical method that allows reducing on satellite images of the effects of change due to atmospheric conditions in the absence of access to atmospheric parameters.

Remote sensing: The acquisition of information about an object or a phenomenon without making physical contact with the object.

Revisit time: The satellite revisit time is the time elapsed between observations of the same point on earth by a satellite. According to Wikipedia.

Sampling: In signal processing, sampling is the reduction of a continuous signal to a discrete signal.

Scale: The ratio of the distance on a map to the corresponding actual distance. According to Wikipedia. Often expressed as a fraction, a large scale is a ratio which is approximated to the integer 1.

Scattering (optical): Angular redistribution of solar radiation after its interaction with a component present in the atmosphere-to-surface system (i.e. air or water molecules, aerosols, hydrosols, etc.).

Segmentation: Process oriented to separating a given set in homogeneous subsets according to given criteria, in order to identify them.

Sensor (satellite): Device on board of the satellite that collects and measures the incident radiation coming from the observed surface.

Signal to Noise Ratio (SNR): Characterizes the level of useful signal in respect to the level of noise.

Simple scattering (optical): A photon is disseminated outside of the medium traversed as a result of its interaction with a single particle.

Solar zenith angle: The solar zenith angle is the angle between the zenith and the centre of the sun's disc. According Wikipedia.

Solid angle: In geometry, a solid angle (Ω) is the two-dimensional angle in three-dimensional space that an object subtends at a point. It is a measure of how large the object appears to an observer looking from that point. A solid angle is expressed in a dimensionless unit called a steradian (sr). According to Wikipedia

Spatial resolution: The spatial resolution of a geographical document is the actual size of the smallest element shown in that document. On a satellite image, the spatial resolution corresponds to the size of a side of a pixel.

Spatiotriangulation: The spatiotriangulation applied to satellite images enables the simultaneous geometric processing of a large number of images and strips together to reduce the control point requirement.

Spectral band: Channel of an image corresponding to a wavelength interval in the electromagnetic spectrum. For example, the RapidEye satellite has five spectral bands: blue, green, red, red edge and near infrared.

Spectral continuum: Convex hull of a spectrum.

Spectral filtering: Wavelength selection technique of electromagnetic radiation.

Spectral library: Set of labeled spectral signatures.

Spectral radiance of a surface dA: The energy in the elementary spectral range $[\lambda + d\lambda]$ confined in the elementary solid angle $d\Omega$ around the direction of propagation (θ, φ) relative to the normal to the surface dA. It is expressed in $W.m^{-2}.sr^{-1}$(watts per square meter per steradian).

Spectral resolution: Describes the finesse of a spectral band used by a sensor. If the spectral resolution is finer, the windows of the different channels of the sensor are narrower.

Spectral signature: Electromagnetic reflectance characteristic of an object in function of its wavelength.

Spectrometry: It is the experimental study of the spectrum of a physical phenomenon, which means, of its decomposition on an energy scale, or any other magnitude returning an energy (frequency, wavelength, etc.)

Specular reflection: The mirror-like reflection of light (or of other kinds of wave) from a surface, in which light from a single incoming direction (a ray) is reflected into a single outgoing direction (source Wikipedia).

Stereoscopic angle: The stereoscopic angle is also called the B/H, where B is the distance between the two viewing centers. It may encompass two different carriers (two satellites), or also two detectors in the same focal plane, two cameras on the same satellite, or two images successively acquired by an agile satellite. The height H represents the average altitude of the satellites.

Stereoscopy: The set of techniques used to reproduce a perception of a terrain from two-dimensional image planes of the same scene.

Supervised classification: Automatic learning technique that produces a classification map from a ground truth.

Supervised learning: Learning method where we look to automatically generate rules from a learning database containing examples (cases previously treated and validated).

Support Vector Machines (SVM): Name given to the techniques used to find a separator (linear or not) between several groups of variables. The separator depends only of a few points called "support vectors".

Telemetry: Technique of measurement of distances.

Temporal resolution: Frequency at which a remote sensing image acquisition system can acquire an image of the same point on the Earth's surface.

Tensor analysis: Process oriented to the definition of local geometry of a 3D point cloud through the extraction and combination of its 3 Eigenvalues, derived from a Principal Components Analysis.

Terrestrial laser scanner: Equipment operating on the basis of a laser allowing us to measure and digitize objects or large complex spaces in three dimensions.

Texture (of an image): A certain spatial arrangement of the image signal levels.

Texture index: Local statistic calculated in the neighborhood of a pixel of an image which takes into account spatial organization.

Thermal Infrared Emissivity: Average emission spectrum of bodies present on Earth (3–100 μm).

Topographic LiDAR: Active remote sensing system dedicated to the analysis of land surfaces. It mainly operates in the near infrared. The distances are obtained by measuring the time of return between the sensor and the target.

Triangular Irregular Network (TIN): Is a digital data structure used in a geographic information system (GIS) for the representation of a surface. A TIN is a vector-based representation of the physical land surface or sea bottom, made up of irregularly distributed nodes and lines with three-dimensional coordinates that are arranged in a network of nonoverlapping triangles. According to Wikipedia.

ULM: *Aéronef Ultra Léger Motorisé*, or Ultra-Light Motorized Aircraft.

Ultraviolet: Electromagnetic radiation of a wavelength between 10 and 380 nm, shorter than that of visible light, but longer than that of X-rays.

Unsupervised learning: Automatic learning method, sometimes referred to as "clustering", which divides a heterogeneous group of data in subsets of similar data.

Vegetation index: This is an index that is calculated, in general, on an optical image and combines, for example, the reflectance of several spectral bands. It allows identifying the special properties of vegetation (see NDVI, EVI, etc.).

Visible: Part of the electromagnetic spectrum visible to the human eye, between 380 and 780 nm.

Wave decomposition: Process to describe a waveform as a combination of echoes, each echo resulting from the backscattering of one or more objects.

Wavelength: Each electromagnetic wave is defined by its wavelength, which represents the spatial periodicity of the oscillations (maximum distance between two oscillations, for example). The wavelength is inversely proportional to the frequency and is expressed in meters.

Zenith and Nadir: Zenith is the highest point of the celestial sphere. The direction in which gravity pulls i.e. toward the opposite direction, is the nadir.

List of Authors

Nicolas BAGHDADI
TETIS - IRSTEA
Montpellier
France

Jean-Stéphane BAILLY
UMR LISAH
AgroParisTech
Montpellier
France

Xavier BRIOTTET
ONERA
Toulouse
France

Xavier CEAMANOS
ONERA
Toulouse
France

Patrick CHAZETTE
DSM/LSCE, CEA
Gif sur Yvette
France

Nesrine CHEHATA
ENSEGIG
University of Bordeaux
Pessac
France

Thomas CORPETTI
LETG-Rennes-COSTEL
CNRS
Rennes
France

Nicolas DAVID
MATIS, IGN
Saint-Mandé
France

Jean-Marc DELVIT
CNES
Toulouse
France

Laurent DEMAGISTRI
UMR Espace-Dev, IRD
Montpellier
France

Laurent HESPEL
DOTA/IODI, ONERA
Toulouse
France

Clément MALLET
MATIS, IGN
Saint-Mandé
France

Julien MICHEL
CNES
Toulouse
France

Kenji OSE
UMR TETIS
IRSTEA
Montpellier
France

Julien TOTEMS
DSM/LSCE, CEA
Gif sur Yvette
France

Silvia VALERO
CESBIO
University Paul Sabatier
Toulouse
France

Mehrez ZRIBI
CNRS – CESBIO
Toulouse
France

Index

A, B

C, D

E, F, G

H, I, K, L

M, N

O, P

Scientific Committee

Abdelaziz KALLEL, University of Sfax, Tunisia

Jean-Philippe GASTELLU, University of Toulouse, France

Gérard DEDIEU, CNES, France

Jean-Philippe CHEREL, University of Montpellier, France

Grégoire MERCIER, Telecom Bretagne, France

Danielle DUCROT, University of Toulouse, France

Stéphane DUPUY, CIRAD, France

Jean-Michel MOREL, École Normale Supérieure Cachan, France

Etienne BERTHIER, CNRS, France

Michel JORDAN, University of Cergy-Pontoise, France

Mathieu FAUVEL, Institut National Polytechnique de Toulouse, France

Jocelyn CHANUSSOT, Institut National Polytechnique de Grenoble, France

Josiane COSTERASTE, CNES, France

Jean-Christophe RAUT, CNRS, France

Gérard ANCELLET, CNRS, France

Michel ROUX, Telecom Paris-Tech, France

Yves PASTOL, SHOM, France

Jean-Baptiste FERET, Irstea, France

Nicolas BAGHDADI, Irstea, France

Mehrez ZRIBI, CNRS, France

FSC
www.fsc.org
MIX
Papier | Fördert gute Waldnutzung
FSC® C083411

Kolibri 360 GmbH
Ferdinand-Jühlke-Straße 7
99095 Erfurt, Deutschland
produktsicherheit@kolibri360.de